POLLUTION A to Z

POLLUTION | A to Z

pol • lu • tion (pə-lōo´shən) n. 1. The act or
process of polluting or the state of being polluted.
2. The contamination of soil, water, or the
atmosphere by the discharge of har...

Richard M. Stapleton,
Editor in Chief

volume **2**

Labor, Farm to Zero
Population Growth
Index

MACMILLAN
REFERENCE
USA™

THOMSON
★
GALE

New York • Detroit • San Diego • San Francisco • Cleveland • New Haven, Conn. • Waterville, Maine • London • Munich

THOMSON
™
GALE

Pollution A to Z
Richard M. Stapleton, Editor in Chief
Patricia Hemminger, Ph.D., Associate Editor
Susan L. Senecah, Ph.D., Associate Editor

LIBRARY OF CONGRESS CATALOGING-IN-PUBLICATION DATA

Pollution A to Z / Richard Stapleton, editor in chief.
   p. cm.
         Includes bibliographical references and index.
         ISBN 0-02-865700-4 (set : hardcover : alk. paper) — ISBN
         0-02-865701-2 (v. 1) — ISBN 0-02-865702-0 (v. 2)
         1. Pollution—Encyclopedias. I. Stapleton, Richard M.

         TD173.P65 2003
         363.73'03—dc21

                                          2003000078

This title is also available as an e-book.
ISBN 0-02-865905-8 (set)
Contact your Gale sales representative for ordering information.

Printed in Canada
10 9 8 7 6 5 4 3 2 1

Table of Contents

VOLUME 2

Preface

Can you see the Great Milky Way where you live? Most Americans cannot. The greatest vista known to humankind is obscured by the veil of light pollution that shrouds all but the least developed regions on Earth.

From the quality of life to life itself, there is not one person who is not affected in some way by pollution. Pollution affects our ability to swim in local waters or enjoy clear views in our national parks. More critically, pollution is responsible for waterborne diseases, birth defects, increased cancer incidence, and neurological problems ranging from loss of intelligence to madness itself. Pollution can kill instantly—over 8,000 died in just three days when methyl isocynate leaked from the Union Carbide facility in Bhopal, India—or it can take decades for the full impact to be known. Indeed, the number of lives cut short by the radiation released when the Chernobyl nuclear reactor exploded in the Ukraine in 1986 is still being counted.

The other fundamental truth about pollution is that we have no one to blame for it but ourselves. Yes, there are natural causes of pollution, and we include an article on *Natural Disasters*, but the preponderance of pollutant threats are anthropogenic—caused by man. From lead in paint to mercury in water, PCBs in rivers to VOCs in the atmosphere, from CFCs to greenhouse gases, the sources of pollution can be traced to the decisions of industry, government and, ultimately, the individual consumer/voter.

With that in mind, one entry deserves special mention. *Lifestyle* is less an article than an opinion essay. Its inclusion is meant to challenge the reader's social choices, to ask you to consider how your own personal lifestyle affects the environment. Do you use bottled ketchup or individual packets? Do you ride to school in an SUV or take a bus? The fact is that just as every person on the planet is affected by pollution, so each of us directly and indirectly creates pollution. Some of us just create more of it than others.

One caution: if you are looking to these volumes for the answers to all questions about pollution and its effects on human and environmental health, you will be disappointed. There are dozens—perhaps hundreds—of toxic substances, for example, for which we do not have health-based standards, meaning we do not know what is a "safe" level of exposure. And if we know little about these contaminants individually, we know virtually nothing about the cumulative (synergistic) impact of multi-contaminant exposure. Perhaps the most important thing we have learned in the last half-century is how little

A map of London, England, showing locations of pumps and deaths from cholera during the epidemic, 1854. See Health, Human; Snow, John; Water Treatment.

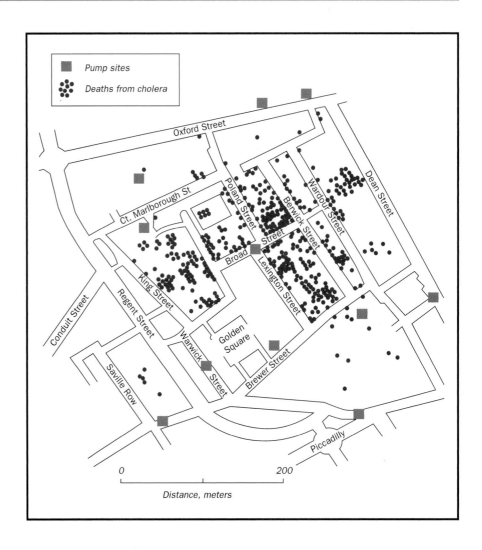

we know. There is no shortage of discovery left for the next generation to undertake.

Organization of the Material

As its title would suggest, *Pollution A to Z* is organized alphabetically with 267 articles presented in two volumes. Articles are cross-referenced. Authors were aware of (and sometimes wrote) related articles and, for the fullest understanding, the reader is encouraged to explore at least one level beyond the subject first selected. This is made easier with the inclusion of cross-references at the end of many articles. You will find that articles are balanced between hard science and social science. You can research the contaminants that pollute a river, learn the health impacts of the pollution, and then trace society's response, from activism through the political process required to enact legislation to the enforcement that ultimately slows or reverses the pollution.

Each entry has been commissioned especially for this work. Our contributors are drawn primarily from the ranks of academia and government, each chosen for his or her particular experience and expertise. Who better, for instance, to write about the first Earth Day than Denis Hayes, the man who organized it. Equally important, our authors were chosen for having the

uncommon ability to make their knowledge accessible to advanced high school students and university undergraduates. We also provide a glossary in the back matter of each volume, summarizing the definitions of teh terms in the margins throughout the set.

The two volumes are richly illustrated with charts, tables, maps, and line drawings. Each, along with the many photographs, was selected to amplify the text it accompanies. Historic photographs such as the one taken at noon during Donora, Pennsylvania's, killer smog are especially important; they convey far more about the state of our environment at its nadir than any words could. Finally, articles include selected lists of additional resources. The lists focus on materials that students can reasonably expect to locate, and each contains at least one Internet reference.

Clean-up efforts underway at Love Canal, May 22, 1980. (©Bettmann/Corbis. Reproduced by permission.) See Comprehensive Environmental Response, Compensation, and Liability Act (CERCLA); Environmental Movement; Gibbs, Lois; History; Laws and Regulations, United States; Mass Media; Politics;

Acknowledgements

There are so many people to thank for their commitment, encouragement, and patience along the way. First, the editorial team at Macmillan Reference

Boats approaching the oil-covered beach of Green Island, Alaska, following the 1989 *Exxon Valdez* oil spill. (©Natalie Fobes/Corbis. Reproduced by permission.) See Disasters: Chemical Accidents and Spills; Disasters: Oil Spills; History; Industry; Mass Media; Petroleum.

USA and the Gale Group. In particular, my thanks to Hélène Potter for her unflinching support, and to Marie-Claire Antoine, Michael J. McGandy, Shawn Corridor, Patti Brecht, and Frank Castronova. Their gracious patience, from the initial vision through searching for just the right authors to the endless tweaking of content, has been a much-appreciated constant. No one, of course, has been more patient than my wife, Andrea, and son, Matthew, who forgave me so many nights at the computer.

I trace my appreciation for the environment to growing up on a small New England dairy farm. To work the land is to connect with it; the intimate relationship between air, water, land, and life is seen every aspect of life. I have left the land behind now, both figuratively—I work in the city—and literally—for relaxation, we sail. It is the sailing that now seeds me with the environment, and it is a bittersweet connection. We sail by the grace of nature, propelled by balancing the forces of wind and water. But we sail in a nature disgraced by humans. To depart the harbor, we must first breach the

trash line, a floating windrow of plastic bottles, styrofoam cups, paper trash, old tires and worse. And the return means putting the clear ocean sky behind us to head instead for the orange-brown smudge that heralds yet another urban ozone-alert day.

My son, Matthew, is thirteen as I write this. He and his generation are making their own connections with the environment. My hope is that the information presented here will in some small way help them to be better stewards than their parents were.

Richard M. Stapleton

Topical Outline

Phytoremediation
Science
Scrubbers
Superfund

CULTURAL ISSUES

Consumer Pollution
Education
Environmental Movement
Green Marketing
Lifestyle
Mass Media
Popular Culture
Population
Poverty
Public Participation
Sprawl
Writers

ECONOMICS

Consumer Pollution
Cost-benefit Analysis
Economics
Emissions Trading
Energy
Enforcement
Green Chemistry
Green Marketing
Industrial Ecology
Industry
ISO 14001
Labor, Farm
Life Cycle Analysis
Limits to Growth
Pollution Shifting
Smart Growth
Sprawl
Sustainable Development
Tragedy of the Commons
World Trade Organization

EFFECTS OF POLLUTION

Acid Rain
Cryptosporidiosis
Endocrine Disruption
Fish Kills
Global Warming
Health, Human
Hypoxia
Smog

ENERGY

Antinuclear Movement
Arctic National Wildlife Refuge
Coal
Diesel
Disasters: Environmental Mining Accidents
Disasters: Oil Spills
Economics
Electric Power
Energy
Energy, Nuclear
Energy Efficiency
Fossil Fuels
Fuel Cell
Fuel Economy
Global Warming
Green Chemistry
Greenhouse Gases
Lifestyle
Light Pollution
Mining
Radioactive Waste
Renewable Energy
Vehicular Pollution
Waste to Energy

ENVIRONMENTAL HEALTH

Acid Rain
Air Pollution
Bioaccumulation
DDT (Dichlorodiphenyl trichloroethane)
Electric Power
Endocrine Disruption
Energy
Fish Kills
Hypoxia
Oxygen Demand, Biochemical
Pesticides
Phosphates
Sedimentation
Smart Growth
Sprawl
Water Pollution
Water Pollution: Freshwater
Water Pollution: Marine

GLOBAL ISSUES

CFCs (Chlorofluorocarbons)
Disasters: Nuclear Accidents
Earth Summit
Global Warming
Greenhouse Gases
Halon
ISO 14001

Consensus Building
Earth Day
Education
Environmental Justice
Environmental Movement
Environmental Racism
GIS (Geographic Information System)
Government
Green Party
Information, Access to
Legislative Process
Litigation
Mediation
National Environmental Policy Act (NEPA)
New Left
Nongovernmental Organizations (NGOs)
Politics
Progressive Movement
Property Rights Movement
Public Interest Research Groups
Public Participation
Public Policy Decision Making
Regulatory Negotiation
Right to Know
Unintended Consequences
Whistleblowing
Wise-Use Movement

POLLUTANTS

Adaptive Management
Arsenic
Asbestos
Carbon Dioxide
Carbon Monoxide
CFCs (Chlorofluorocarbons)
Coal
DDT (Dichlorodiphenyl trichloroethane)
Dioxin
Fossil Fuels
Greenhouse Gases
Halon
Heavy Metals
Household Pollutants
Infectious Waste
Lead
Mercury
Methane (CH_4)
NO_x (Nitrogen Oxides)
Nonaqueous Phase Liquids (NAPLs)
Particulates
PCBs (Polychlorinated Biphenyls)
Persistent Bioaccumulative and Toxic Chemicals
    (PBTs)
Persistent Organic Pollutants (POPs)
Phosphates

Sulfur Dioxide
VOCs (Volatile Organic Compounds)

POLLUTION PREVENTION

Beneficial Use
Bottle Deposit Laws
Catalytic Converter
Composting
Energy Efficiency
Enforcement
Environmental Impact Statement
Green Chemistry
Industrial Ecology
Integrated Pest Management
Life Cycle Analysis
Pollution Prevention
Pollution Shifting
Recycling
Renewable Energy
Reuse
Science
Systems Science
Technology, Pollution Prevention
Toxic Release Inventory
Waste
Waste Reduction
Waste to Energy

RADIATION

Disasters: Nuclear Accidents
Electromagnetic Fields
Energy, Nuclear
Radioactive Fallout
Radioactive Waste
Radon
Yucca Mountain

SCIENCE

Carson, Rachel
Carver, George Washington
Citizen Science
Colborn, Theo
Cousteau, Jacques
GIS (Geographic Information System)
Green Revolution
Politics
Risk
Science
Systems Science
Technology, Pollution Prevention
Toxicology
Union of Concerned Scientists

SOCIAL ACTION

Activism
Consensus Building
Earth Day
Earth First!
Ecoterrorism
Education
Environmental Impact Statement
Environmental Justice
Environmental Movements
Environmental Racism
Ethics
GIS (Geographic Information System)
Gauley Bridge, West Virginia
Green Party
Greenpeace
Information, Access to
Labor, Farm
Legislative Process
Lifestyle
Mass Media
National Toxics Campaign
New Left
Nongovernmental Organizations (NGOs)
Popular Culture
Poverty
Precautionary Principle
Progressive Movement
Property Rights Movement
Public Interest Research Groups (PIRGs)
Public Participation
Public Policy Decision Making
Settlement House Movement
Toxic Release Inventory
Union of Concerned Scientists
Warren County, North Carolina
Wise-Use Movement
Writers
Zero Population Growth

SOURCES OF POLLUTION

Agriculture
Consumer Pollution
Disasters: Chemical Accidents and Spills
Disasters: Environmental Mining Accidents
Disasters: Natural
Disasters: Nuclear Accidents
Disasters: Oil Spills
Dry Cleaning
Electric Power
Electromagnetic Fields
Energy, Nuclear
Incineration
Industry

Lifestyle
Mining
Nonpoint Source Pollution
Pesticides
Petroleum
Point Source
Smelting
Terrorism
Vehicular Pollution

TREATIES AND CONFERENCES

Agenda 21
CFCs (Chlorofluorocarbons)
Earth Summit
Environmental Crime
Ethics
Global Warming
Greenhouse Gases
Halon
Montréal Protocol
NAFTA (North American Free Trade
   Agreement)
Precautionary Principle
Treaties and Conferences

TYPES OF POLLUTION

Air Pollution
Light Pollution
Medical Waste
Mold Pollution
Noise Pollution
Plastic
Radioactive Waste
Soil Pollution
Space Pollution
Thermal Pollution
Vehicular Pollution
Visual Pollution
War
Water Pollution
Water Pollution: Freshwater
Water Pollution: Marine

VEHICULAR POLLUTION

Catalytic Converter
Diesel
Energy Efficiency
Fuel Cell
Fuel Economy
Ozone
Petroleum
Smog
Vehicular Pollution

WASTE

Beneficial Use
Biosolids
Burn Barrels
Hazardous Waste
Injection Well
Landfill
Medical Waste
Ocean Dumping
Plastic
Pollution Shifting
Recycling
Reuse
Solid Waste
Superfund
Waste
Waste Reduction
Waste to Energy
Waste, International Trade in
Waste, Transportation of
Yucca Mountain

WATER

Acid Rain
Agriculture
Biosolids
Clean Water Act
Cryptosporidiosis
Disasters: Oil Spills

Dredging
Dry Cleaning
Energy
Fish Kills
Groundwater
Hypoxia
Infectious Waste
Injection Well
Marine Protection, Research, and Sanctuaries
    Act
Mixing Zone
National Pollutant Discharge Elimination
    System (NPDES)
Nonpoint Source Pollution
Ocean Dumping
Ocean Dumping Ban Act
Oxygen Demand, Biochemical
PCBs (Polychlorinated Biphenyls)
Petroleum
Phosphates
Point Source
Sedimentation
Superfund
Thermal Pollution
Underground Storage Tank
Wastewater Treatment
Water Pollution
Water Pollution: Freshwater
Water Pollution: Marine
Water Treatment

For Your Reference

Below is a list of selected symbols, abbreviations, acronyms, and initialisms that are used regularly throughout the articles in this book.

ACh	acetylcholine
ACM	asbestos-containing materials
ACTION	Activists' Center for Training in Organizing and Networking
AEC	Atomic Energy Commission
AFL	Affiliated Federation of Labor
AFT	American Federation of Teachers
AHERA	Asbestos Hazard Emergency Response Act
AHERA	Asbestos Hazard Emergency Response Amendment
AMD	acid mine drainage
ANILCA	Alaska National Interest Lands Conservation Act
ANWR	Arctic National Wildlife Refuge
AOC	Area of Concern
APA	Administrative Procedures Act
APCA	Air Pollution Control Act
APHIS	Animal and Plant Health Inspection Service
As	arsenic
ATCA	Alien Torts Claims Act
ATSDR	Agency for Toxic Substances and Disease Registry
BCC	bioaccumulating chemical
BCC	bioaccumulative chemical
BCF	bioconcentration factor
BEAR	Business and Environmentalists Allied for Recycling
BHC	benzene hexachloride
BMP	best management practice

BOD	biochemical oxygen demand
BTNRC	Brookhaven Town Natural Resources Committee
BTU	British Thermal Unit
C	carbon
C	Celsius
C-BA	Cost-benefit analysis
C_2H_6	ethane
C_3H_8	propane
$(CH_3)_2Hg$	mercury—methylmercury compound
C_4H_{10}	butane
CAA	Clean Air Act
CAAA	Clean Air Act Amendments
CAFE	corporate average fuel economy
CAFO	concentrated animal feeding operation
CAP	Campaign Against Pollution
CCA	chromated copper arsenate
CCHW	Citizens Clearinghouse for Hazardous Wastes
Cd	cadmium
CDC	U.S. Centers for Disease Control
CEC	North American Commission for Environmental Cooperation
CEQ	[President's] Council on Environmental Quality
CERCLA	Comprehensive Environmental Response, Compensation, and Liability Act
CFC	chlorofluorocarbon
CFR	Code of Federal Regulations
CGIAR	Consultative Group on International Agricultural Research
CH_3Hg^+	mercury—methylmercury compound
CH_4	methane
ChE	cholinesterase
CHEJ	Center for Health, Environment and Justice
CHP	combined heat and power
CITES	Convention on International Trade in Endangered Species of Wild Fauna and Flora
CLEAR	Environmental Working Group Clearinghouse on Environmental Advocacy and Research
CO	carbon monoxide

Co	cobalt
CO_2	carbon dioxide
CPAST	Corporation for Public Access to Science and Technology
Cr	chromium
CRJ	United Church of Christ's Commission for Racial Justice
CSD	Commission on Sustainable Development
CSISSFRRA	Chemical Safety Information, Site Security and Fuels Regulatory Relief Act
CSO	Combined Sewer Overflow
CSO	Community Service Organization
Cu	copper
CWA	Clean Water Act
CWS	community water system
DBP	disinfection by-product
DDT	dichlorodiphenyl trichloroethane
DEA	Drug Enforcement Agency
DES	diethylstilbestrol
DHHS	U.S. Department of Health and Human Services
DNA	deoxyribonucleic acid
DNAPL	dense nonaqueous phase liquid
DO	dissolved oxygen
DOA	U.S. Department of Agriculture
DOE	U.S. Department of Energy
DOJ	U.S. Department of Justice
DOL	U.S. Department of Labor
DOT	U.S. Department of Transportation
E-coli	Escherichia coli
E-FOIA	Electronic Freedom of Information Act
EC	European Community
ECOSO	U.N. Economic and Social Council
ED	effective dose
EDA	Emergency Declaration Area
EDC	ethylene dichloride
EDF	Environmental Defense Fund
EEA	European Environment Agency
EF	ecological footprint

EF!	Earth First!
EFA	ecological footprint analysis
EIA	Energy Information Administration
EIS	environmental impact statement
ELF	Earth Liberation Front
ELF	extremely low frequency
EMF	electromagnetic field
EPA	U.S. Environmental Protection Agency
EPCRA	Emergency Planning and Community Right-to-Know Act
EPI	efflux pump inhibitor
EPR	extended producer responsibility
ERNS	Emergency Response Notification System
EU	European Union
F	Fahrenheit
FBI	Federal Bureau of Investigation
FDA	U.S. Food and Drug Administration
FeS_2	iron sulfide (incl. marcasite and pyrite)
FICAN	Federal Interagency Committee on Aviation Noise
FIFRA	Federal Insecticide, Fungicide, and Rodenticide Act
FOIA	Freedom of Information Act
FTC	Federal Trade Commission
FTIR	Fourier transform infrared spectroscopy
FWPCA	Federal Water Pollution Control Act
FWS	U.S. Fish and Wildlife Service
GAC	granular activated carbon
GASP	Group Against Smog and Pollution
GATT	General Agreement on Tariffs and Trade
GDP	gross domestic product
GEF	World Bank's Global Environmental Facility
GEO	geosynchronous Earth orbit
GHG	greenhouse gas
GIPME	Global Investigation of Pollution in the Marine Environment
GIS	Geographic Information System
GLOBE	Global Learning and Observations to Benefit the Environment

GMA	Grocery Manufacturers of America
GPS	Global Positioning System
H	hydrogen
H_2SO_4	sulfuric acid
HAA	hormonally active agent
HAP	hazardous air pollutant
HC	hydrocarbon
HFC	hydrofluorocarbon
Hg	mercury
HGP	Human Genome Project
HgS	cinnabar
HHS	U.S. Department of Health and Human Services
HHW	household hazardous waste
HPLC	high-performance liquid chromatography
HSWA	Hazardous and Solid Wastes Amendment
HYV	high-yielding variety
IADC	Inter-Agency Space Debris Coordination Committee
IAEA	International Atomic Energy Agency
IAP2	International Association for Public Participation
IARC	International Agency for Research on Cancer
IBI	Index of Biotic Integrity
ICC	International Chamber of Commerce
ICP-AES	inductively coupled plasma emission spectra
IDA	International Dark Sky Association
IEGMP	Independent Expert Group on Mobile Phones
IIED	International Institute for Environment and Development
IMF	International Monetary Fund
INPO	Institute of Nuclear Power Operations
IOS	International Organization for Standardization
IPCC	Intergovernmental Panel on Climate Change
IPM	integrated pest management
IR	infrared
ISO	International Organization for Standardization
IWI	index of water indicators
LC	lethal concentration

LC-72	London Convention 1972
LCA	life cycle analysis
LCA	life cycle assessment
LD	lethal dose
LEO	low Earth orbit
LEPC	Local Emergency Planning Committee
LNAPL	light nonaqueous phase liquid
LQG	large-quantity generator
LUST	leaking underground storage tank
MACT	Maximum Achievable Control Act
MACT	Maximum Achievable Control Technology
MARPOL	International Convention for the Prevention of Pollution from Ship 1973
MASSPIRG	Massachusetts Student Public Interest Research Group
MCL	maximum concentration load
MCL	maximum contaminant level
MEO	middle Earth orbit
MGD	million gallons per day
Mha	million hectare
MIT	Massachusetts Institute of Technology
MNA	Monitored Natural Attenuation
MPG	miles per gallon
MSDS	Material Safety Data Sheet
MSW	municipal solid waste
MSWLF	municipal solid waste landfill
MTBE	methyl tertiary-butyl ether
MTD	maximum tolerated dose
MW	megawatt
MWTA	Medical Waste Tracking Act
N	nitrogen
N_2	atmospheric nitrogen
N_2O	nitrous oxide
NAAEC	North American Agreement on Environmental Cooperation
NAAEE	North American Association for Environmental Education
NAAQS	National Ambient Air Quality Standards

NAFTA	North American Free Trade Agreement
NAPL	nonaqueous phase chemical
NAPL	nonaqueous phase liquid
NAS	U.S. National Academy of Sciences
NASA	U.S. National Aeronautics and Space Administration
NCA	Noise Control Act
NEA	National Education Association
NEMO	Nonpoint Education for Municipal Officials
NEPA	National Environmental Policy Act
NESHAP	National Emission Standards for Hazardous Air Pollutants
NFWA	National Farm Workers Association
NGO	nongovernmental organization
NH_3	methane
NHGRI	National Human Genome Research Institute
Ni	nickel
NIEHS	National Institute of Environmental Health Sciences
NIH	National Institutes of Health
NIMBY	not-in-my-backyard
NIOSH	National Institute for Occupational Safety and Health
NO	nitric oxide
NO_2	nitrogen dioxide
NO_3	nitrate
NO_x	nitrogen oxide
NOAA	National Oceanographic and Atmospheric Administration
NOEL	no observable effect level
NORM	Naturally Occurring Radioactive Material
NPDES	National Pollutant Discharge Elimination System
NPL	National Priorities List
NPL	National Priority List
NPPR	National Pollution Prevention Roundtable
NPRI	National Pollution Release Inventory
NPS	National Park Service
NRC	Nuclear Regulatory Commission
NRCS	Natural Resources Conservation Service
NRDA	Natural Resource Damage Assessment

NRDC	Natural Resources Defense Council
NSR	New Source Review
NSRB	Nuclear Safety Regulatory Board
NSTA	National Science Teachers Association
NTC	National Toxics Campaign
O	oxygen
O&M	operations and maintenance
O_2	molecular oxygen
O_3	ozone
ODA	Ocean Dumping Act
OECD	Organization for Economic Cooperation and Development
ONAC	Office of Noise Abatement and Control
OPA	Oil Pollution Act
OPEC	Organization of the Petroleum Exporting Countries
OPP	Oil Pollution Prevention Act
OSHA	Occupational Safety and Health Administration
OTEC	ocean thermal energy conversion
P2	pollution prevention
PAC	polycyclic aromatic compound
PACCE	People Against a Chemically Contaminated Environment
PAH	polycyclic aromatic hydrocarbon
Pb	lead
PBB	polybrominated biphenyl
PBT	persistent bioaccumulative and toxic chemical
PCB	polychlorinated biphenyl
PCC	primary combustion chamber
PCDD	polychlorinated dibenzo dioxin
PCDF	polychlorinated dibenzo furan
PCE	perchloroethylene
PCN	polychlorinated naphthalene
PCP	pentachlorophenol
PCP	Principia Cybernetica Project
PCSD	President's Council on Sustainable Development
PEM	proton exchange membrane
PERC	perchloroethylene

PET	polyethylene terephthalate
PIC	Prior Informed Consent
PIRG	Public Interest Research Group
PM	particulate matter
PO_4^{3-}	phosphate ions or groups
POP	persistent organic pollutant
POTW	publicly owned treatment works
PPA	Pollution Prevention Act
PPCP	pharmaceutical and personal care product
PPE	personnel protective equipment
PrepCom	preparatory committee
PS	polystyrene
PSAC	President's Science Advisory Committee
PV	photovoltaic
PVC	polyvinyl chloride
RCRA	Resource Conservation and Recovery Act
RDF	refuse-derived fuel
ReDo	Reuse Development Organization
RF	radio frequency
RMP	recommended agricultural practice
RRA	Resource Recovery Act
RTK	Right to Know
S	sulfur
SANE	Sane Nuclear Policy
SARA	Superfund Act
SARA	Superfund Amendments and Reauthorization Act
SB	styrene-butadiene
SCC	secondary combustion chamber
SDS	Students for a Democratic Society
Se	selenium
SEED	Schlumberger Excellence in Educational Development
SEJ	Society of Environmental Journalists
SEP	Supplemental Environmental Project
SERC	State Emergency Response Commission
SF_6	sulfur hexafluoride

SHAC	Stop Huntingdon Animal Cruelty
SIP	State Implementation Plan
SLAPP	Strategic Litigation against Public Participation
SMCRA	Surface Mining Control and Reclamation Act
SO_2	sulfur dioxide
SO_4	sulfate
SOC	soil organic carbon
SQG	small-quantity generator
Superfund	Comprehensive Environmental Response, Compensation and Liability Act
SUV	sport utility vehicle
SWDA	Safe Drinking Water Act
TBT	tributyltin
TCDD	tetrachloro dibenzo dioxin
TCE	trichloroethylene
TEF	Toxicity Equivalency Factor
TEPP	tetraethyl pyrophosphate
THM	trihalomethanes
TMDL	Total Maximum Daily Load
TMI	Three Mile Island
TOMS/EP	Total Ozone Mapping Spectrometer on the Earth Probe Satellite
TRI	Toxic Release Inventory
TSCA	Toxic Substances Control Act
TSP	total particulate matter
TT	Treatment Technique
U	uranium
UCC	United Church of Christ
UCS	Union of Concerned Scientists
UFW	United Farm Workers of America
UK	United Kingdom
UN	United Nations
UNCED	U.N. Conference on Environment and Development
UNCHE	U.N. Conference on the Human Environment
UNEP	U.N. Environmental Programme
US	United States of America

USDA	U.S. Department of Agriculture
USGAO	U.S. General Accounting Office
USGS	U.S. Geological Survey
UST	underground storage tank
UV	ultraviolet
VOC	volatile organic compound
WET	Whole Effluent Toxicity
WHB	Workers Health Bureau of America
WHO	World Health Organization
WPA	Works Progress Administration
WRI	World Resources Institute
WSP	Women Strike for Peace
WTC	World Trade Center
WTE	waste to energy
WTO	World Trade Organization
ZID	zone of initial dilution
Zn	zinc
ZPG	zero population growth

Contributors

David E. Alexander
University of Massachusetts
Amherst, Massachusetts

Paul T. Anastas
Executive Office of the President
Washington, D.C.

Sarah Anderson
Institute for Policy Studies
Washington, D.C.

Mary Jane Angelo
St. John's River Management District
Palatka, Florida

Phillip Anz-Meador
Viking Science and Technology, Inc.
Houston, Texas

Matthew Arno

William Arthur Atkins
Atkins Research and Consulting
Normal, Illinois

Jay Austin
Environmental Law Institute
Washington, D.C.

Pamela Baldwin
Owings, Maryland

Anne Becher
Boulder, Colorado

Elizabeth D. Blum
Troy State University
Troy, Alabama

Brigitte Bollag
Pennsylvania State University
University Park, Pennsylvania

Jean-Marc Bollag
Pennsylvania State University
University Park, Pennsylvania

Arline L. Bronzaft
Lehman College, City University of New York
New York, New York

Joanna Burger
Rutgers University
Piscataway, New Jersey

Dave Brian Butvill
Fontana, Wisconsin

Julie Hutchins Cairn
Seattle Public Utilities
Redmond, Washington

George Carlson
Canterbury, New Hampshire

Elizabeth L. Chalecki
California State University
Hayward, California

Ron Chepesiuk
Rock Hill, South Carolina

Christos Christoforou
Clemson University
Clemson, South Carolina

Allan B. Cobb
Kailua-Kona, Hawaii

Christopher H. Conaway
University of California
Santa Cruz, California

Stacie Craddock
U.S. Environmental Protection Agency
Washington, D.C.

James L. Creighton
Creighton and Creighton, Inc.
Los Gatos, California

José B. Cuellar
San Francisco, California

Raymond Cushman
New Hampshire Department of Environmental Services
Canterbury, New Hampshire

Kenneth A. Dahlberg
Western Michigan University
Kalamazoo, Michigan

Heinz H. Damberger
Illinois State Geological Survey
Boulder, Colorado

Lawrence C. Davis
Kansas State University
Manhattan, Kansas

Joseph E. de Steiguer
University of Arizona
Tucson, Arizona

Larry Deysher
Ocean Imaging
Solana Beach, California

Thomas D. DiStefano
Bucknell University
Lewisburg, Pennsylvania

Bruce K. Dixon
National Association of Science Writers
American Association for the Advancement of Science
Naperville, Illinois

Clive A. Edwards
Ohio State University
Columbus, Ohio

Robert M. Engler
U.S. Army Engineer Research and Development Center
Vicksburg, Mississippi

Christine A. Ennis
National Oceanic and Atmospheric Administration Aeronomy Laboratory
Cooperative Institute for Research in Environmental Sciences
Boulder, Colorado

Larry Eugene Erickson
Kansas State University
Manhattan, Kansas

Gary R. Evans
SE4 Consulting
Potomac Falls, Virginia

Jess Everett
Rowan University
Glassboro, New Jersey

John P. Felleman
State University of New York
Syracuse, New York

Adi R. Ferrara
Bellevue, Washington

Linda N. Finley-Miller
U.S. Army Corps of Engineers
Sacramento, California

A. Russell Flegal
University of California
Santa Cruz, California

David Frame
Oxford, United Kingdom

Ralph R. Frerichs
University of California
Los Angeles, California

David Friedman
Union of Concerned Scientists
Washington, D.C.

David Goldberg
Decatur, Georgia

Janice Gorin

Kevin Graham
Windom Publishing
Denver, Colorado

Robert F. Gruenig
Reynolds, Illinois

Janet Guthrie
National Institute of Environmental Health Services
Research Triangle Park, North Carolina

Charles Hall
State University of New York
Syracuse, New York

Dan Hamburg
Voice of the Environment
Ukiah, California

Ian Scott Hamilton
Texas A&M University
College Station, Texas

Burt Hamner
Seattle, Washington

Donald J. Hanley
Bechtel SAIC Company, LLC
Las Vegas, Nevada

Donald R. Hastie
York University
Toronto, Ontario, Canada

Richard A. Haugland
U.S. Environmental Protection Agency
Cincinnati, Ohio

Denis Hayes
Bullitt Foundation
Seattle, Washington

Patricia Hemminger
New York, New York

Paul Philip Hesse
NCI Information Systems, Inc.
Washington, D.C.

Annette Huddle
San Francisco, California

Craig R. Humphrey
Pennsylavania State University
University Park, Pennsylvania

Susan M. Jablonski
Texas Natural Resource Conservation Commission
Austin, Texas

Betsy T. Kagey
Frostburg State University
Cumberland, Maryland

Corliss Karasov
Madison, Wisconsin

James P. Karp
Syracuse University
Syracuse, New York

Sara E. Keith
State University of New York
Syracuse, New York

Suzi Kerr
Motu Economic and Public Policy Research
Wellington, New Zealand

Leeka Kheifets
World Health Organization
France

Stephen M. Kohn
Kohn, Kohn, & Calapinto
Washington, D.C.

Philip Koth

William Kovarik
Radford University
Radford, Virginia

Michael E. Kraft
University of Wisconsin
Green Bay, Wisconsin

Ashok Kumar
University of Toledo
Toledo, Ohio

Rishi Kumar
Global Educational and Consulting Services
Mississauga, Ontario, Canada

J. Michael Kuperberg
Florida State University
Tallahassee, Florida

Rattan Lal
Ohio State University
Columbus, Ohio

Deborah Lange
Carnegie Mellon University
Pittsburgh, Pennsylvania

Denise M. Leduc
West Bloomfield, Michigan

Terra Lenihan
Denver, Colorado

Peggy Leonard
King County Wastewater Treatment
Seattle, Washington

Lois Levitan
Cornell University
Ithaca, New York

Deena Lilya
Boise, Idaho

David Lochbaum
Union of Concerned Scientists
Washington, D.C.

Tim Lougheed
Ottawa, Ontario, Canada

Adrian MacDonald
Long Island City, New York

Peter S. Machno
Peter S. Machno LLC
Seattle, Washington

Daniel Barstow Magraw
Center for International Environmental Law
Washington, D.C.

Kenneth H. Mann
Bedford Institute of Oceanography
Halifax, Nova Scotia, Canada

Jack Manno
State University of New York
Syracuse, New York

Michael Mansur
The Kansas City Star
Kansas City, Missouri

Burkhard Mausberg
Environmental Defense Canada
Toronto, Ontario, Canada

Michael J. McKinley
U.S. Geological Survey
Reston, Virginia

Glenn McRae
CGH Environmental Strategies, Inc.
Burlington, Vermont

Martin V. Melosi
University of Houston
Houston, Texas

Peter Michaud
Gemini Observatory
Hilo, Hawaii

Bruce G. Miller
Pennsylvania State University
University Park, Pennsylvania

Joel A. Mintz
Nova Southwestern University
Davie, Florida

John Morelli
Rochester Institute of Technology
Rush, New York

Office of Solid Waste/U.S. Environmental Protection Agency
Washington, D.C.

Sunil Ojha
University of Toledo
Toledo, Ohio

Kenneth Olden
National Institute of Environmental Health Services
Research Triangle Park, North Carolina

Christine Oravec
University of Utah
Salt Lake City, Utah

Tim Palucka
Pittsburgh, Pennsylvania

Lee Ann Paradise
Lubbock, Texas

David Petechuk
Rifle, Colorado

P.A. Ramachandran
Washington University
St. Louis, Missouri

Stephen C. Redd
Centers for Disease Control and Prevention
Atlanta, Georgia

William E. Rees
University of British Columbia
Vancouver, British Columbia, Canada

Kevin Anthony Reilly
New York State Supreme Court
New York, New York

Joseph Richey
Boulder, Colorado

Heather V. Ritchie
Tallahassee, Florida

Marin Sands Robinson
Northern Arizona University
Flagstaff, Arizona

Mary Elliott Rollé
National Oceanic and Atmospheric Administration
Department of Commerce
Vermont Law School
Silver Spring, Maryland

Walter A. Rosenbaum
University of Florida
Gainesville, Florida

Joan Rothlein
Oregon Health and Science University
Portland, Oregon

Natalie Roy
The Environmental Council of the States
Washington, D.C.

Joseph N. Ryan
University of Colorado
Boulder, Colorado

Karen M. Salvage
State University of New York
Binghamton, New York

Joseph J. Santoleri
RMT-Four Nines
Plymouth Meeting, Pennsylvania

Michael G. Schechter
Michigan State University
East Lansing, Michigan

Susan L. Senecah
State University of New York
Syracuse, New York

Hollie Shaner
CGH Environmental Strategies, Inc.
Burlington, Vermont

William E. Sharpe
Pennsylvania State University
University Park, Pennsylvania

Lynne Page Snyder
National Academies of Science, Institute of Medicine
Washington, D.C.

Gina M. Solomon
University of California
San Francisco, California

James J. Stapleton
University of California
Parlier, California

Richard M. Stapleton
U.S. Environmental Protection Agency
Washington, D.C.

Donald Stedman
University of Denver
Denver, Colorado

Richard S. Stein
University of Massachusetts
Amherst, Massachusetts

Diana Strnisa
Five Rivers Environmental Education Center
Delmar, New York

Jacqueline Vaughn Switzer
Northern Arizona University
Flagstaff, Arizona

Dorceta E. Taylor
University of Michigan
Ann Arbor, Michigan

Kender Taylor
Seattle, Washington

Christopher M. Teaf
Tallahassee, Florida

Valerie M. Thomas
Princeton University
Princeton, New Jersey

Nathan Thrall

Iris Udasin
Environmental & Occupational Health Sciences Institute
Piscataway, New Jersey

Johan C. Varekamp
Wesleyan University
Middletown, Connecticut

Stephen J. Vesper
U.S. Environmental Protection Agency
Cincinnati, Ohio

Margrit von Braun
University of Idaho
Moscow, Idaho

Frank A. von Hippel
University of Alaska
Anchorage, Alaska

Ted von Hippel

Paul Wapner
American University
Washington D.C.

Linda Wasmer Andrews
Albuquerque, New Mexico

Richard J. Watts
Washington State University
Pullman, Washington

Stefan Weigel
University of Hamburg
Hamburg, Germany

Laura Westra
York University
Toronto, Ontario, Canada

Ross Whaley
State University of New York
Syracuse, New York

Christine M. Whitney
Watertown, Massachusetts

Labor, Farm

The rise of organized labor in agriculture is epitomized by the United Farm Workers of America (UFW), the largest and oldest union of agricultural laborers in the nation, and its influence on environmental public policy, operations, and worker conditions. Many salient actors, events, and campaigns have contributed to this influence.

The National Farm Workers Association (NFWA), precursor to the UFW, was cofounded in 1962 by César E. Chávez and Dolores Huerta. Their lifelong commitments were to win recognition and respect, better wages, and safer working conditions for agricultural laborers in California and elsewhere. Earlier efforts to organize agricultural labor in the United States, such as the National Agricultural Workers Union, which Chávez joined in 1947, were not successful. Moreover, no labor union in the United States had ever expressed much concern about the effects of pesticides on farm workers and their families.

La huelga en general (also known as the general strike) catapulted the Chávez-led UFW to international attention after September 16, 1965, when it joined a strike against grape growers started eight days earlier by a union of Filipino workers in Delano, California. From the beginning, Chávez expressed concern about the harmful effects of commonly used pesticides on farm workers. In 1969 he marched with several hundred other protesters to the national headquarters of the U.S. Food and Drug Administration (FDA) and demanded increased government surveillance of pesticide use on food crops.

By the early 1970s, following Chávez-led fasts, secondary boycotts, and protest marches, Huerta had negotiated UFW contracts with many central California grape growers that required protective clothing for workers laboring in fields sprayed with pesticides and effectively banned the use of **DDT** (dichlorodiphenyl trichloroethane) and other dangerous pesticides. These contracts also required longer periods before reentry into pesticide-sprayed fields—beyond state and federal standards—and also mandated the testing of farm workers on a regular basis to monitor for pesticide exposure, several years before comparable government rules were established. The UFW was also the first union to require joint union-management committees to enforce state safety regulations regarding the use of pesticides in vineyards.

From the beginning, Chávez urged farm workers like Pablo Romero and activists like Marion Moses to become physicians committed to addressing

DDT the first chlorinated hydrocarbon insecticide; it has a half-life of 15 years and can collect in fatty tissues of certain animals; for virtually all but emergency uses, DDT was banned in the U.S. in 1972

the many pesticide-related health threats to farm workers. After medical school, Romero worked as a physician at the UFW clinic in Salinas, California, where he also helped form a community task force that set new rules to minimize the risk of accidental pesticide exposure. Moses, a native West Virginian and former UFW nurse, became Chávez's personal physician and union researcher, after studying internal and occupational medicine. Moses later founded the Pesticide Education Center in San Francisco, California, with the mission of educating the public about the adverse health effects of exposure to pesticides in the home, within the community, and at work.

The UFW initiated another antipesticide boycott against grape growers in 1984 after research found hundreds of thousands of local residents suffering from pesticide-related illnesses and an unusually high incidence of cancer among children in central California. Chávez called on Americans to once again stop buying grapes until the industry stopped using pesticides known to cause, or suspected of causing, cancer in laboratory animals. The UFW used an innovative direct-mail campaign to carry Chávez's antipesticide plea to consumers all over North America.

Following two mid-1980s incidents near Salinas in which hundreds of farm workers received emergency hospital treatment after they were twice accidentally sprayed with pesticides, the UFW pushed for Monterey County's enactment of the toughest pesticide restriction laws in the nation, which prompted similar policy changes throughout the state of California. Thus, the UFW became the first labor union to demand government protection for farm workers and others from dangerous pesticides, including airplane-sprayed chemical drifts. After Chávez's unexpected death in 1993, the UFW's leadership maintained its strong antipesticide position by continuing to advocate for more protection for farm workers and other who work and live near and around the fields. SEE ALSO ACTIVISM; AGRICULTURE; CHÁVEZ, CÉSAR E.; PESTICIDES.

Bibliography

Griswold Del Castillo, Richard, and Garcia, Richard A. (1995). *César Chávez: A Triumph of Spirit.* University of Oklahoma Press.

Ferriss, Susan, and Sandoval, Ricardo. (1977). *The Fight in the Fields—César Chávez and the Farmworker Movement.* New York: Harcourt and Brace.

Ross, Fred. (1989). *Conquering Goliath—César Chávez at the Beginning.* Keene, CA: El Taller Grafico Books.

Internet Resources

Children's Environmental Health Network Web site. Available from http://www.cehn.org/cehn.

San Francisco State University Web site. "César E. Chávez Institute for Public Policy." Available from http://www.sfsu.edu/~cecipp.

United Farm Workers Web site. Available from http://www.ufw.org.

José B. Cuellar

Labor, Industrial   *See Labor Unions*

LaDuke, Winona

ENVIRONMENTAL ACTIVIST
(1959–)

Winona LaDuke, an Ojibwe Indian, is an internationally recognized, long-time environmentalist, feminist, and indigenous rights activist. She was vice presidential running mate for Ralph Nader's 1996 and 2000 U.S. presidential campaigns on the Green Party ticket. Through her speaking, writing, international conference participation, and activist activities, she has increased public awareness of the environmental degradation of Native American lands by nuclear and toxic dumping, water pollution, mining, and toxic exposure. She also builds support for self-determined solutions and protections that honor the cultural and spiritual values of Native Americans. Indian lands hold large supplies of uranium, coal, and timber, and the vast, isolated lands are attractive to industries searching for radioactive, hazardous, and other waste-disposal sites. LaDuke advocates for Native American environmental groups to wage a vigilant battle to protect their environment for future generations. She is the founding director of the White Earth Land Recovery Project and the program director of the annual Honor the Earth Foundation. In 1994, *Time* magazine named LaDuke one of its "50 for the Future." SEE ALSO ACTIVISM; ENVIRONMENTAL JUSTICE; ENVIRONMENTAL MOVEMENT; ENVIRONMENTAL RACISM.

Bibliography

White Earth Land Recovery Project. Available from http://www.welrp.org.

Susan L. Senecah

Winona LaDuke. (AP/Wide World Photos. Reproduced by permission.)

Landfill

A landfill is a large area of land or an excavated site that is designed and built to receive wastes. There were 3,536 active municipal landfills in the United States in 1995 according to the U.S. Environmental Protection Agency (EPA). Today, about 55 percent of America's trash (more than 220 million tons annually) is disposed of in landfills. Municipal solid-waste landfills (MSWLFs) accept only household, commercial, and nonhazardous industrial waste. Hazardous waste generated by industrial sources must be disposed of in special landfills that have even stricter controls than MSWLFs.

In the past, garbage was collected in open dumps. Most of these small and unsanitary dumps have been replaced by large, modern facilities that are designed, operated, and monitored according to strict federal and state regulations. These facilities may be distant from urban centers, requiring the large-scale transport of waste. About 2,300 municipal solid waste landfills were operating in the United States in 2000.

A typical modern landfill is lined with a layer of clay and protective plastic to prevent the waste and leachate (liquid from the wastes) from leaking to the ground or groundwater. The lined landfill is then divided into disposal cells. Only one cell is open at a time to receive waste. After a day's activity, the waste is compacted and covered with a layer of soil to minimize odor, pests, and wind disturbances. A network of drains at the bottom of the landfill collects the leachate that flows from the decomposing waste. The leachate is usually sent to a recovery facility to be treated. Methane gas, carbon dioxide, and other gases produced by the decomposing waste are monitored and collected to reduce their effect on air quality. EPA regulations require many larger landfills to collect and burn landfill gas. EPA's Landfill Methane Outreach Program was created in 1994 to educate communities and local government

DIAGRAM OF A PROPERLY CLOSED LANDFILL

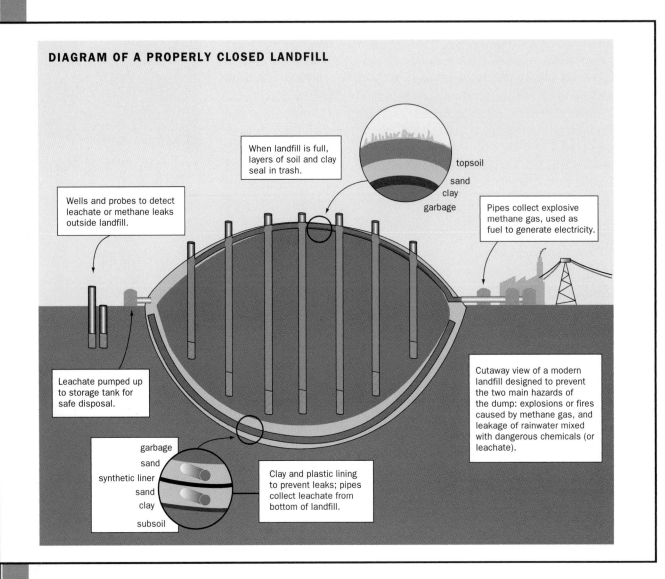

When landfill is full, layers of soil and clay seal in trash.

topsoil
sand
clay
garbage

Wells and probes to detect leachate or methane leaks outside landfill.

Pipes collect explosive methane gas, used as fuel to generate electricity.

Leachate pumped up to storage tank for safe disposal.

Cutaway view of a modern landfill designed to prevent the two main hazards of the dump: explosions or fires caused by methane gas, and leakage of rainwater mixed with dangerous chemicals (or leachate).

garbage
sand
synthetic liner
sand
clay
subsoil

Clay and plastic lining to prevent leaks; pipes collect leachate from bottom of landfill.

about the benefits of recovering and burning methane as an energy source. By 2002 the program had helped develop 220 projects that convert landfill gas to energy. Such projects, when analyzed in 2001, offset the release of carbon dioxide from conventional energy sources by an amount equivalent to removing 11.7 million cars from the road for one year.

Fresh Kills Landfill in Staten Island, the largest landfill in the United States, accepting approximately 27,000 tons of garbage a day in the late 1980s, closed in March 2001. Although landfills occupy only a small percentage of the total land in the United States, public concern over possible ground water contamination as well as odor from landfills makes finding new sites difficult. SEE ALSO SOLID WASTE; WASTE, TRANSPORTATION OF.

Internet Resources

Freudenrich, Craig C. "How Landfills Work." Available from http://www.howstuffworks.com/landfill.htm.

U.S. Environmental Protection Agency Office of Solid Waste Web site. Available from http://www.epa.gov/epaoswer.

Office of Solid Waste/U.S. Environmental Protection Agency

Laws and Regulations, International

The problems of pollution are not limited to the borders of any one country. Because the harmful effects of pollution often extend to areas beyond the country where the pollution originated, the international legal system is an important means of controlling pollution. (The text here refers to "countries," but the reader should be aware that countries are usually termed "states" in the parlance of international law.) International efforts to control pollution are numerous and complex. The following section identifies some of the main features of the system.

French police officer performing pollution test on a car in Paris, France. (©Le Segretain Pascal/Corbis Sygma. Reproduced by permission.)

International Legal System

The two primary sources of international law are custom and treaties, and both play a role in regulating international pollution. Customary international law emerges when countries engage in certain practices in the belief that those practices are required by international law. To become customary law, a practice must be generally followed, rather than just being the practice of a few countries. In contrast, treaties, which are often referred to as conventions or protocols, are legally binding agreements between countries or intergovernmental organizations. Treaties typically do not enter into force until a specified number of countries have expressed their consent to be bound by the treaty; even after the treaties enter into force, only the countries that expressed their consent are bound. A treaty is only effective to the extent it is implemented domestically by the parties to it. Each treaty raises its own questions of domestic implementation.

Customary International Law

Many environmental activists and other observers believe that countries have an obligation through customary international law to not cause transboundary environmental harm. Principle 21 of the Stockholm Declaration (1972) and Principle 2 of the Rio Declaration that emerged out of the 1992 Earth Summit both clearly state this principle. The Rio Declaration affirms that countries have "the sovereign right to exploit their own resources pursuant to their own environmental and developmental policies, and the responsibility to ensure that activities within their jurisdiction or control do not cause damage to the environment of other States or of areas beyond the limits of national jurisdiction." Under this principle, countries are prohibited from undertaking or allowing actions that will cause pollution in other nations.

Another important concept, known as the *precautionary principle* or *precautionary approach*, addresses circumstances where significant health, safety, or environmental risks may be involved although full scientific certainty is lacking. Many countries, especially those in Europe, consider the precautionary principle to be a part of customary international law, but this legal status is debated by other countries, such as the United States. Considerable controversy also exists over exactly what the precautionary principle means. Principle 15 of the 1992 Rio Declaration reads, "Where there are threats of serious or irreversible damage, lack of scientific certainty shall not be used as a reason for postponing cost-effective measures to prevent environmental degradation." Another formulation is that a country is not prohibited from taking measures to protect health or the environment because of the existence of scientific uncertainty. A more aggressive formulation is that countries should err on the side of caution when dealing with environmental problems rather than wait until a risk is certain to occur before acting, by which time it will often be too late to reverse the damage. For example, under this more aggressive interpretation of the precautionary principle, if there is evidence that a pollutant might be dangerous, even if the risk is not certain, a country should take action to prevent the risk involved despite the scientific uncertainty. Under any formulation, questions remain about what level of risk warrants precautionary action and what level of precaution may or should be taken.

Treaties and Regulations

There are hundreds of treaties and other international instruments relating to pollution. Some prominent examples include the following: The 2001 Stockholm Convention on Persistent Organic Pollutants (POPs) calls for an immediate ban on certain chemicals, severely restricts the use of others, and provides for POPs to be disposed of and managed using environmentally sound methods. To address the problem of climate change, which is caused by an increased concentration of carbon in the atmosphere, countries negotiated the United Nations Framework Convention on Climate Change, which entered into force in 1994, and finalized the Kyoto Protocol related to that convention in 1997 (not yet in force). A treaty that addresses other forms of air pollution is the Convention on Long-Range Transboundary Air Pollution formulated by the UN Economic Commission for Europe in 1979 and its protocols. The 1981 UN Convention on the Law of the Sea, several regional agreements on specific seas, and various other treaties address maritime pollution. The 1998 Convention for the Application of Prior Informed Consent (PIC) Procedure for Certain Hazardous Chemicals and Pesticides in International Trade (not yet in force) would ensure that countries have the opportunity to make informed decisions on whether to allow hazardous chemicals to enter their borders. There have also been important treaties regulating oil and nuclear pollution, such as the International Convention on Oil Pollution Preparedness, Response and Cooperation in 1990, and the International Atomic Energy Agency Convention on Nuclear Safety in 1994.

Since food is often imported and exported among countries, international regulations can be significant in reducing the amount of pollution contained in food that travels beyond national borders. The Codex Alimentarius Commission, created in 1963 by the United Nations, has as its highest priorities the protection of consumer health and guarantee of fair practices in trade. With those objectives in mind, it develops standards for, among other criteria, food labeling, food additives, contaminants, methods of analysis and sampling, food hygiene, nutrition and foods for special dietary uses, food import and export inspection and certification systems, residues of veterinary drugs in foods, pesticide-residue levels in food, and guidelines to protect consumer health. These standards are not automatically binding, either domestically or internationally. However, because most countries must at some point conform to international trade law—which requires that certain health-related standards be science-based and recognizes following codex standards as one way of meeting that requirement—some pressure exists for them to adopt codex-sponsored standards in their own regulations.

Enforcement

An important question is how international law regarding pollution can be enforced. There is no international police agency with the authority to enforce international law or any international court system with broad compulsory jurisdiction to make binding decisions on countries without their consent. Despite the lack of a central force, however, countries generally comply with their international legal obligations. Among other reasons, this is because countries will usually only assume obligations in the first place if they believe it is in their best interest to do so. In the event of noncompliance, economic sanctions may sometimes be imposed under the terms of certain agreements, and nonviolating countries may sometimes take other measures against countries that violate international law. The risk of negative publicity may also persuade countries to comply with their obligations.

Studies have shown that noncompliance, especially among developing countries, more often results from a lack of capacity than willful defiance. Compliance with international agreements regarding pollution usually requires a significant amount of scientific expertise that not all countries possess. In addition, some governments may not have the administrative capability necessary for monitoring actions, such as the emissions of pollutants, which take place within their countries, or a legal system capable of enforcing laws. Finally, countries, especially developing countries, also may be unable or unwilling to comply with their international legal obligations to restrict pollution because efforts to alleviate poverty in the immediate term take priority over environmental protection.

The United States, generally speaking, takes compliance with pollution-related treaties very seriously. For example, the United States has not become a party to the PIC and POPs Conventions, mentioned above, as well as the Basel Convention on the Transboundary Movement of Hazardous Waste (1989) because it does not have the domestic legal authority to implement those agreements fully. On the other hand, many environmentalists would argue that the United States has not fulfilled its obligations under the framework convention on climate change.

Voluntary Corporate Codes of Conduct

Transnational corporations exercise enormous economic power and engage in practices that result in the release of large amounts of pollution. However, the conduct of transnational corporations frequently is not effectively regulated by any environmental regime; since domestic law (especially in developing countries) often is not adequately enforced, it typically does not address the environmental activities of overseas corporations, and international law is not adequate to fill in the gaps. Given the lack of effective laws concerning pollution that govern transnational corporations, a recent trend has been the emergence of voluntary corporate codes of conduct. Although corporations have no legal obligation to follow these codes, the demands of the market may persuade international companies to adopt voluntary environmental codes in order to remain competitive. Compliance with these voluntary codes can result in reduced pollution.

The International Organization for Standardization (ISO), a nongovernmental body that develops worldwide standards to facilitate the international exchange of goods, has created a series, ISO 14000, of voluntary environmental management standards for corporations. ISO 14000 does not include specific environmental regulations for corporate compliance. Instead, the series contains general procedures for developing management systems that address the environmental impacts of corporate activities, including pollution, and thus can be adapted to different types of organizations. In order to become certified under ISO 14000, the top-level management of an organization must establish an environmental policy that takes into account all activities of the company which have environmental implications, and commits the organization, among other things, to the prevention of pollution. The environmental management system must have a planning process that creates specific environmental goals, methods of implementation and operation, and a system of monitoring and measuring environmental performance. Because ISO 14000 certification—like compliance with other voluntary codes of conduct—is sometimes contractually required by a company's customers to do business, ISO 14000 can encourage organizations to develop policies that reduce pollution.

Several other corporate codes of conduct relating to pollution prevention have been established. One example is the Ceres Principles, a moral code of environmental conduct that corporations can choose to adopt. It facilitates investment by shareholders in companies that have taken steps to improve their environmental performance. By 2000 approximately fifty-four major U.S. corporations, including General Motors, Ford Motor Company, Ben & Jerry's Ice Cream, and Domino's Pizza, had endorsed the Ceres Principles. The International Chamber of Commerce (ICC), a nongovernmental organization, has developed a set of environmental standards known as the Business Charter for Sustainable Development. The ICC also documents examples of successful environmental management practices for other companies to model. In addition, the United Nations has established the Global Compact, a set of voluntary corporate codes that incorporates principles from international environmental and human rights treaties. A final example is the Organization for Economic Cooperation and Development (OECD) Guidelines for multinational corporations, which include a chapter on the environment. SEE ALSO ENFORCEMENT; ENVIRONMENTAL CRIME; GOVERNMENT; ISO 14001; LAWS AND REGULATIONS, UNITED STATES; LEGISLATIVE PROCESS; PRECAUTIONARY PRINCIPLE; PUBLIC POLICY DECISION MAKING; RIGHT TO KNOW; TOXIC SUBSTANCES CONTROL ACT (TSCA).

Bibliography

Barber, Jeffrey. (1998). "Responsible Action or Public Relations? NGO Perspectives on Voluntary Initiatives," in *Industry and Environment*, 21 (United Nations Environment Programme, January-June).

Brown Weiss, Edith; Magraw, Daniel Barstow; and Szasz, Paul C., eds. (1992). *International Environmental Law: Basic Instruments and References*.

Brown Weiss, Edith; Magraw, Daniel Barstow; and Szasz, Paul C., eds. (1999). *International Environmental Law: Basic Instruments and References 1992–1999*.

Handl, Gunther, and Lutz, Robert E. (1989). *Transferring Hazardous Technologies and Substances: The International Legal Challenge*.

Magraw, Daniel Barstow, ed. (1991). *International Law and Pollution*.

Internet Resource

Center for International Environmental Law Web site. Available from http://www.ciel.org.

Daniel Barstow Magraw & Janice Gorin

Laws and Regulations, United States

Although pollution control laws have been in use in the United States for a century, it was not until the 1970s, the "Environmental Decade," that modern pollution-control laws began to take shape. The American public was awakened to the need for better pollution control through the 1967 publication of Rachel Carson's groundbreaking *Silent Spring* and environmental disasters such as Love Canal, New York; the Donora, Pennsylvania, inversion; and the Cuyahoga River fire in Ohio. In the late 1960s and early 1970s, citizens began to demand comprehensive environmental protection laws. In the thirty years since, those early environmental laws have been used as the broad framework on which national pollution control are based laws in the twenty-first century.

Overview of U.S. Pollution-Control Laws and Regulations

Pollution-control laws in the United States can take several different forms. Federal pollution-control statutes are enacted by Congress in response to domestic problems or needs, or to implement international treaties. They are complex laws that state a goal for lowering or eliminating the release of certain pollutants, generally within a specific medium. These laws assign a duty to an agency, typically the U.S. Environmental Protection Agency (EPA), to implement the law. The agency then creates rules and regulations to further establish and advance the statute's goals.

Virtually every federal pollution control law delegates authority to states, entrusting them to create their own programs for implementing the law. Usually states have some leeway in deciding how implementation of a federal statute is best achieved on the state level. However, as a general rule, state programs that are derived from a delegation of federal regulatory authority can

---

THE PERMITTEE EXPERIENCE

Owning a piece of land does not always mean having the freedom to do with it what you want. For example, if you want to build a boathouse on your lakefront property, you will have to follow a legal process before ensuring that such a project will be allowed under local, state, and/or federal law. Your property's proximity to the water may mean that it will be classified as wetlands. Under various wetlands protection laws, you will have to be granted permission from numerous government sources before going forward with your project.

Under the Clean Water Act, you will have to apply to the Army Corps of Engineers for a 404 permit. You will also need to provide an environmental assessment for NEPA purposes, certify that your project is consistent with your state's Coastal Zone Management Act program policies, and ensure that no other federal licenses will be required for your project. Then you will need to figure out which state and local regulations apply to your plan, and request any permits that may be required on the local level. While there will likely be some overlap in the state and federal requirements, it is often difficult to determine exactly what is needed before you can be assured that you are in compliance with all applicable laws. This complicated process can produce great frustration, and can lead applicants to avoid meeting legal requirements because complying with the rules is too difficult.

be more, but cannot be less stringent than the federal law. This leads to state and local laws and regulations that mirror their federal counterparts and allow for enforcement on the local level. Although this cooperative effort may ensure that federal environmental statutes reach a larger share of violators, it can also lead to confusion for individuals who try to comply with the law, and may make it difficult for agencies to apply the law with uniformity.

Major U.S. Pollution-Control Statutes

One of the first modern environmental protection laws enacted in the United States was the National Environmental Policy Act of 1969 (NEPA), which requires the government to consider the impact of its actions or policies on the environment. NEPA remains one of the most commonly used environmental laws in the nation. In addition to NEPA, there are numerous pollution-control statutes that apply to such specific environmental media as air and water. The best known of these laws are the Clean Air Act (CAA), Clean Water Act (CWA), and the Comprehensive Environmental Response, Compensation, and Liability Act (CERCLA) commonly referred to as Superfund. Among the many other important pollution control laws are the Resource Conservation and Recovery Act (RCRA), Toxic Substances Control Act (TSCA), Oil Pollution Prevention Act (OPP), Emergency Planning and Community Right-to-Know Act (EPCRA), and the Pollution Prevention Act (PPA).

Pollution-control laws focus on the regulation of activities that utilize materials that are potentially harmful to human health and the environment. These laws frequently vary in terms of their expectations and potential penalties for violators, depending on the risks associated with the materials involved. For example, CERCLA and RCRA are similar in terms of the activities they address. Both statutes focus on the storage, transport, and disposal of waste. However, the penalties for violating CERCLA are much more serious because that statute covers activities surrounding accidental or negligent releases of hazardous wastes, after the fact. RCRA's penalties are less severe, because the threat of harm is lower.

U.S. pollution-control statutes are numerous and diverse. Although many of the environmental statutes passed by Congress are useful tools in pollution prevention, they often need to be expanded before their impact is fully realized. Pollution-control laws are generally too broad to be managed by existing legal bodies, so Congress must find or create an agency for each that will be able to implement the mandated mission effectively. The statute then serves as a framework for the agency in organizing its agenda. At each level, the law becomes more specific and targeted.

Regulations: Role of the Agency in U.S. Pollution Control

Federal agencies in the United States are established through enabling legislation known as *organic acts*. These acts create and empower agencies, as well as define and limit their roles. Congress delegates a certain amount of authority to each agency, allowing its officials to develop regulations to ensure that the agency's duties will be achieved. Congress grants this authority to agencies because the legislature cannot always foresee all the elements that will be

MAJOR U.S. POLLUTION CONTROL LAWS

Medium	Statute	Goal/Description	Agency
Air	Clean Air Act (1970) 42 USC §§7401–7671q, 40 CFR Part 50	To prevent & control air pollution/Regulates air emissions through National Ambient Air Quality Standards (NAAQS)	EPA
Water	Clean Water Act (1977) 33 USC §121 et. seq. 40 CFR Parts 100–140; 400–470	To restore & maintain the integrity of U.S. waters/limits discharges to U.S. waters through National Pollutant Discharge System (NPDES)	EPA
Drinking Water	Safe Drinking Water Act (1974) 43 USC § 300f et. sec. 40 CFR Parts 140–149	To protect U.S. drinking water & supplies from contaminants/ Establishes safe standards for drinking water	EPA
Ocean	Oil Pollution Act of 1990 33 USC §6602 et. seq. 40 CFR Part	To prevent and clean up oil spills in U.S. waters/Establishes fund for response costs and requires vessels & facilities to make plans for responding to oil spills	EPA/Coast Guard
Ground/Toxics	Resource Conservation & Recovery Act (1976) 42 USC §321 et. seq. 40 CFR Parts 240–271	To promote protection of human health and the environment/ Oversees the handling of solid & hazardous wastes from "cradle to grave"	EPA
	Comprehensive Environmental Response, Compensation, & Liability Act (1980) 42 USC §§9601–9675 40 CFR Part 300	To oversee the clean up of the worst U.S. hazardous waste sites/ Establishes a "Superfund" to aid in the costs that arise in remediating CERCLA sites.	EPA
	Toxic Substances Control Act (1976) 15 USC §2601 et. seq. 40 CFR Parts 700–799	To understand the health risks of certain chemical substances/ Promotes the development of scientific health risk data	EPA
	Federal Insecticide, Fungicide, and Rodenticide Act (1972) 7 USC §§136–136y 40 CFR Parts 162–180	To prevent harm to human health and the environment from pesticide use/To register and classify all pesticides in use and analyze risks & benefits of use	EPA
	Food Quality Protection Act (1996) Public Law 104–170	To protect human health from the risks associated with exposure to pesticides/Uses a "risk cup" test for all pesticides & establishes maximum exposure levels for each	EPA/FDA
General	Pollution Prevention Act of 1990 42 USC §13101 et. seq.	To reduce or eliminate pollution/To improve technology & manufacturing & products in order to lower pollution levels	EPA
	Emergency Planning & Community Right-to-Know Act (1986) 42 USC §11011 et. seq.	To improve local solutions to pollution emergencies/Directs the creation of State Emergency Response Commissions (SERCs)	States
	Occupational Safety & Health Act (1970) 29 USC §61 et. seq.	To ensure that workers will be safe from harmful activities & hazardous exposures in the workplace/Establishes maximum exposure limits for workplace hazards	OSHA
	Noise Control Act (1972) 42 USC §4901 et. seq. 40 CFR Parts 204, 211	To prevent damage to human health from the effects of noise pollution/Establishes noise emissions standards and other noise-control measures; Congress has not funded the NCA since 1982, effectively gutting the law.	EPA
	National Environmental Policy Act (1969) 42USC 4321–4347	Established Council on Environmental Quality (CEQ); requires environmental impact statements (EIS) for all "legislation and major federal actions"	All Federal Agencies

necessary for pollution-control laws to be effective. Agencies can develop the expertise needed to execute their lawmaking and legally required oversight duties because they have a narrower focus than the legislature.

Agencies spend a great deal of time considering the effectiveness of their regulations. When an agency determines that its goals would be better achieved if its approach was changed or updated, the agency may propose that a new rule be created. The agency then must announce the proposed rule in the Federal Register, where the public is able to consider the change and return feedback on it to the agency. Federal law requires that agencies consider all public comments that are submitted regarding new rules before making their final decision. Any changes to the proposed rule must again be reported in the Federal Register, with new comments solicited from the public. When the final rule is complete, it is printed in the Federal Register as a new statute before it is *codified*, or entered into the Code of Federal Regulations (CFR).

Several federal agencies oversee pollution control in the United States. At the top of the regulatory pyramid of agencies focused on pollution control is the EPA, which is assigned the duty of coordinating and overseeing all environmental protection laws nationwide. EPA also monitors the implementation of a number of comprehensive pollution-control laws. In addition, there are numerous federal agencies that regulate more narrowly concentrated areas of pollution control law. These agencies include the U.S. Fish and Wildlife Service (FWS), U.S. Department of Agriculture (USDA), National Oceanographic and Atmospheric Administration (NOAA), Occupational Safety and Health Administration (OSHA), Food and Drug Administration (FDA), and Nuclear Safety Regulatory Board (NSRB).

Jurisdiction and Enforcement of U.S. Pollution-Control Laws

Agencies can achieve regulatory compliance through different approaches. One method is to enforce regulations through frequent inspections and stringent penalties. Another is to offer incentives to those who are out of compliance, in order to bring them in line with regulations. Several federal pollution-control statutes offer such alternatives to violators. For example, through the CAA, EPA offers emissions trading as an option to those whose emissions levels are above the agency's set limits. By making a deal with a neighboring industry whose emissions are similar in type, one plant can maintain its higher emissions levels in exchange for an agreement by the other to keep its emissions below the limit to a comparative degree. By allowing such agreements, EPA maintains acceptable emissions levels within corridors without drastically affecting the viability of individual industries.

EPA responds to all violations of pollution-control laws in one of four ways, depending on the severity of the violation. In the least extreme cases, EPA issues informal letters that advise violators to correct their behavior. The next level of violation leads to a formal agency response, a legal order that requires violators to come into compliance. For more severe violations, EPA initiates civil lawsuits, demands compliance, and imposes potential financial penalties. Finally, the agency may bring criminal charges against the most flagrant violators, leading to large fines and prison sentences. In all cases involving court actions, the U.S. Department of Justice takes over as attorney for the agency.

Although U.S. pollution control laws are very broad and complex, they are implemented in an organized system that focuses on the most effective strategies for approaching problems and bringing about compliance with the law's stated goals.

Because of the United States' comparatively long history of environmental regulation, it is ahead of many other nations of the world in certain aspects of pollution control. However, there are still many hurdles to overcome. Chemical corporations, pharmaceutical companies, the farm bureau, property-rights advocates, and other interested groups continually lobby Congress to weaken environmental laws. Such activities have had major impacts in some cases, including in 1982, when efforts by opponents of the Noise Control Act led to the effective gutting of that law. Organized lobbying groups also challenge existing laws when circumstances arise under which court cases can be won that will impact the application or effectiveness of a given law. Conversely, environmental and human health groups also lobby Congress in hopes of making pollution control laws even stricter. Such groups also bring a number of lawsuits each year to push for agency enforcement of existing laws. Ultimately, the effects of pollution control laws are usually visible, which suggests that they will stay in place for years to come.

SEE ALSO CARSON, RACHEL; COMPREHENSIVE ENVIRONMENTAL RESPONSE, COMPENSATION, AND LIABILITY ACT (CERCLA); DONORA, PENNSYLVANIA; ENFORCEMENT; ENVIRONMENTAL CRIME; GOVERNMENT; LAWS AND REGULATIONS, INTERNATIONAL; LEGISLATIVE PROCESS; NATIONAL ENVIRONMENTAL POLICY ACT (NEPA); POLITICS; PUBLIC POLICY DECISION MAKING; REGULATORY NEGOTIATION; RIGHT TO KNOW; TOXIC SUBSTANCES CONTROL ACT (TSCA).

Bibliography

Arbuckle, J.G., et al. (1983). *Environmental Law Handbook*, 7th edition. Rockville, MD: Government Institutes.

Percival, Robert V. (1996). *Environmental Regulation: Law, Science, and Policy*, 2nd edition. Boston: Little, Brown.

Internet Resource

U.S. Environmental Protection Agency. CFR Chapter 40, "Protection of the Environment." Available from http://www.epa.gov.

Mary Elliott Rollé

Lead

Lead (symbol Pb, atomic number 82) is a soft, dense, bluish-gray metal that melts at the relatively low temperature of 328°C (662°F). It has many beneficial uses in compounds as well as in its metallic form, but is toxic at almost any level in the body. Mentioned in the Bible, lead was one of the first known metals. Its toxicity was also recognized long ago; Greek physicians made the first clinical description of lead poisoning in the first century B.C.E., and lead is arguably the earliest known industrial pollutant.

Lead taken internally in any of its forms is highly toxic. At higher body levels, the symptoms of lead poisoning are anemia, weakness, constipation, colic, palsy, and often a paralysis of the wrists and ankles. At low levels, there may be no symptoms. Young children are especially at risk from lead, even at levels once thought safe. Low-level lead poisoning can reduce intelligence, delay motor development, impair memory, and cause hearing problems and troubles in balance.

Higher levels of lead poisoning are reduced with the use of **chelating agents** that help the body to excrete the lead in urine. Although this may

chelating agents chemicals that trap metal ions (*chele* = claw)

malleable able to be shaped and bent

alloy mixture of two or more metals

leach dissolve out

drier a compound that increases the drying rate

A little girl is standing and gazing out a window. Dust settling in windows may contain lead, a result of the use of lead-based paint in homes. (©2003, Robert J. Huffman, Field Mark Publications. Reproduced by permission.)

address the physical symptoms mentioned above, there is no cure for the loss of IQ and other neurological effects that lead poisoning has on young children.

Lead was used by the Romans to make water pipes and create elaborate urban water systems. The word plumbing comes from the Latin word for lead, *plumbum*. Lead was, and still remains, a natural choice for plumbing, Widely available, it is durable and easily **malleable**, and it does not rust. Water is still delivered to homes in many U.S. cities via lead supply pipes. **Alloys** of lead are also used in solder and in brass faucets and fixtures. Drinking water can **leach** lead out of a plumbing system, and this may be one source of lead exposure.

The most common uses of lead today are in lead-acid storage batteries and to shield against radiation. Computer screens are made of leaded glass to contain the electromagnetic radiation within, and as a consequence, two U.S. states have banned the disposal of CRT monitors in landfills and incinerators.

Lead is useful in many compounds. Lead carbonate, called white lead, has been used for over 2,000 years as a white pigment in paint and ceramic glazes, and other lead compounds have been used as pigments and **driers**. Lead-based paint was first identified as the source of deadly childhood poisoning in Australia in 1904.

Subsequently, lead-based paint was banned in Australia and much of Europe in the 1920s, but the United States did not prohibit its residential use for another fifty years. By 1971 it was determined that two hundred children a year died annually in this country as a result of lead poisoning. That year Congress passed the Lead-Based Poisoning Prevention Act, but delayed implementation of its official ban until 1977.

The lead-based paint applied to homes during the first two-thirds of the twentieth century continues to be the primary cause of childhood lead poisoning. Children who eat flakes of peeling and chipping paint in older, unmaintained housing are at serious risk. The National Survey of Lead and Allergens in Housing for 1998 to 2000 found that some 38 million housing units contain lead-based paint. Some 25 million of these units have "significant" lead-based paint hazards.

Even lead-based paint that is in good condition can pose a risk as the dust created by the friction of opening and closing windows may cause low-level lead poisoning. The renovation of an older home, when done improperly, can poison adults and children as well as pets living in that residence. Residential lead-based paint should never be sanded or burned off.

Lead poisoning is an important health problem, affecting an estimated 890,000 preschoolers, according to the U.S. Centers for Disease Control and Prevention. That means that about 4.4 percent of children aged one to five have unacceptably high levels of lead in their bodies. Although lead poisoning crosses all socioeconomic, geographic, and racial boundaries, the burden of this disease falls disproportionately on low-income families and those of color. In the United States, children from poor families are eight times more likely to be poisoned by lead than those from higher-income families.

Another compound, tetraethyl lead, was once routinely added to gasoline to prevent knocking or premature detonation in internal combustion engines. The lead survived the combustion process and became a significant contributor to air pollution. Leaded gasoline was phased out in the United States starting in 1976. All gasoline-powered cars and trucks now sold in this country must burn unleaded gasoline.

Leaded gasoline nevertheless remains a problem in many other countries. In 1995 fewer than thirty countries worldwide had banned leaded gasoline. In 1996 the World Bank called for the international phasing out of leaded gasoline, claiming that most of the 1.7 billion urban dwellers in developing countries were at risk from lead poisoning. The United Nations Commission on Human Settlements—known as Habitat—approved a resolution in 1999 that committed member nations to begin phasing out leaded gas. By 2001 forty-five nations worldwide had banned its use.

Because lead is an element; it does not **biodegrade**. Lead pollution from the dawn of civilization remains in the environment. Ice-core researchers in North Greenland have found layers of glacial ice contaminated with lead from ancient Rome's smelters. The lead pollution emitted by smelters can reach staggering levels. In Herculaneum, Missouri, where the nation's largest lead smelter has been in operation for more than one hundred years, health officials documented that almost 28 percent of children under seven have elevated levels of lead in their bloodstream; close to the facility that figure rose to 45 percent. Dust samples along the roads used by trucks serving the

Sign warning residents of high lead levels from Doe Run Smelting. (AP/Wide World Photos. Reproduced by permission.)

biodegrade to decompose under natural conditions

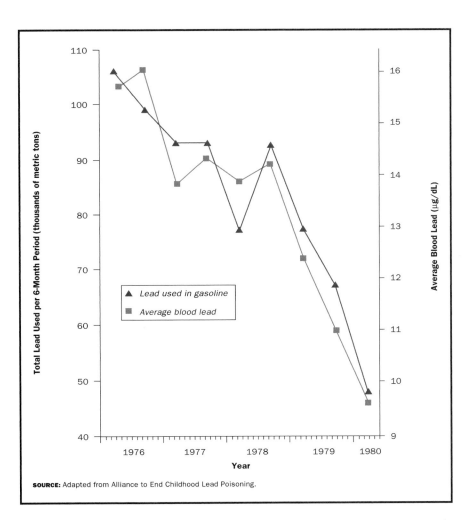

SOURCE: Adapted from Alliance to End Childhood Lead Poisoning.

smelter contained extremely high concentrations of lead (up to 300,000 parts per million), and the site has been declared an urgent public health hazard.

Enforcement actions by the U.S. Environmental Protection Agency (EPA) and the Missouri Department of Natural Resources led to an agreement by the Doe Run Company, the smelter's owner, to install new controls on air emissions, **remediate** lead contamination in residential yards, and stabilize a contaminated slag pile located in the Mississippi River flood plain.

Although lead is a persistent and widespread contaminant in both natural and man-made environments, lead poisoning is an entirely preventable disease. The key to prevention is the elimination of sources and **pathways**. The positive results of bans on leaded gasoline, lead in paints and glazes, lead solder, and lead plumbing can be seen in the reduction in the number of lead-poisoning cases as well as the decreased levels of lead found in the general population.

remediate reduce harmful effects; restore contaminated site

pathway the physical course a chemical or pollutant takes from its source to the exposed organism

Bibliography

Stapleton, Richard. (1994). *Lead Is a Silent Hazard.* New York: Walker and Company.

Warren, Christian. (2001). *Brush with Death; A Social History of Lead Poisoning.* Baltimore: Johns Hopkins University Press, 2001.

Internet Resources

Alliance to End Childhood Lead Poisoning. Available at http://www.aeclp.org.

CDC Childhood Lead Poisoning Prevention Program. Available at http://www.cdc.gov/nceh/lead/lead.htm.

Richard M. Stapleton

Legionnaires' Disease   *See Indoor Air Pollution*

Legislative Process

Simply, legislative process means the steps required for a proposed bill to become a law, but the whole process includes much more than what happens in Congress. At the federal level in the United States, this process has six major steps. First, a written draft of the proposed law, called a bill, is sponsored by a member in one of the two houses of Congress—House or Senate—and recommended for consideration. The presiding officer of the house puts the bill on the agenda and assigns it to a standing, or permanent, committee for consideration.

The standing committees consider all bills and oversee government actions on specialized issue areas. In the House of Representatives, committees that deal with environmental issues include the Agriculture, National Security, Resources, Science, and Appropriations Committees. In the Senate, standing committees relating to the environment include Agriculture, Nutrition and Forestry, Energy and Natural Resources, Environment and Public Works, and Judiciary Committees, although others in both houses may also consider related issues.

Within the standing committee, the bill goes to a subcommittee to study and modify; here, the bill is debated and edited in a line-by-line, and often word-by-word, manner, with the agreed upon changes literally written on the original draft bill.

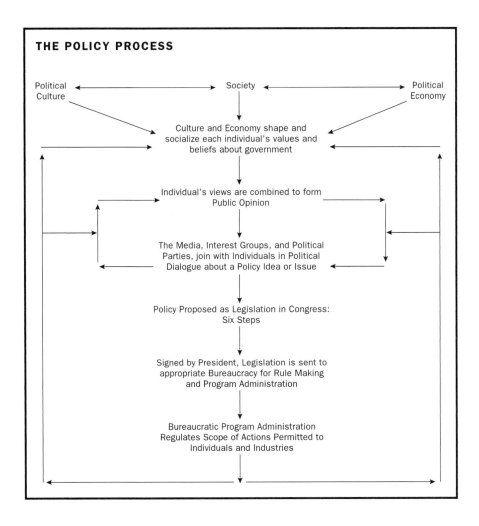

THE POLICY PROCESS

Political Culture ←————————→ Society ←————————→ Political Economy

Culture and Economy shape and socialize each individual's values and beliefs about government

Individual's views are combined to form Public Opinion

The Media, Interest Groups, and Political Parties, join with Individuals in Political Dialogue about a Policy Idea or Issue

Policy Proposed as Legislation in Congress: Six Steps

Signed by President, Legislation is sent to appropriate Bureaucracy for Rule Making and Program Administration

Bureaucratic Program Administration Regulates Scope of Actions Permitted to Individuals and Industries

Once modification is completed and the subcommittee and committee approve the bill, the committee sends the bill back to the full house for floor action, or more debate and a vote, this time will all the members of the house participating. After a bill passes in one house (gets a majority of "yea" votes), the bill is sent to the other house of Congress where the process begins all over again. Because of this process, it is unlikely that both houses will approve identical bills. However, when both houses have recommended a bill on the same issue, the two versions are sent to a conference committee of members of both houses, where the differences are discussed and argued over. If the conference committee arrives at a compromise bill, that bill is sent back to both houses for approval.

When this last legislative approval is obtained, Congress sends the bill to the president for approval or rejection. If the president agrees with the provisions of the bill, it is signed, and the bill becomes a law. However, if the president disagrees with the provisions in a bill, the bill will be vetoed, and a veto message will be sent with the bill back to Congress. Congress can override a presidential veto with a supermajority vote, or a vote in support of the bill by two-thirds of both houses.

The more complex legislative process begins before Congress drafts a bill and ends after legislation is signed. Other governmental institutions are involved, including various executive branch agencies, such as the Environmental Protection Agency (EPA). Outside government, other actors, such as the media and interest groups, are also involved in lawmaking. These external, unofficial actors help to mediate the political dialogue about what government should do.

Mediated politics occurs when there are institutions or individuals who carry the message between an individual and the representative, telling government what the public prefers. Legislative representatives receive messages in various forms from individuals (as letters, votes, and contributions), as well as from media reports and editorials, public opinion polls, political parties, lobbyists, and **interest groups**. Taken together, these convey what it is the public wants on an issue in addition to the level of interest. If an issue and opinion on it are compelling enough, then one or more legislators will introduce a bill in one of the houses of the legislature, and lawmaking begins.

Once a bill is passed by both houses of Congress and signed by the president, it goes to an executive agency where another part of the process begins. The agency creates a way to implement the policy, often by writing rules and regulations, stated in the law. Rule writing is based on implementation guidelines established in the Administrative Procedures Act, which include holding public hearings for citizen feedback. The legislative process ends here with an implemented and enforceable law.

Some members of society may be negatively affected by new legislation. These individuals may form interest groups, write their legislators, or go to court in order to get the law changed. This is where the policy process starts and where active citizens dissatisfied with what the government is, or is not, doing ask for a change in policy. SEE ALSO PUBLIC PARTICIPATION; PUBLIC POLICY DECISION MAKING.

interest groups corporate or citizen groups with a stake in influencing legislation

Bibliography

Douglas, Arnold R. (1990). *The Logic of Congressional Action.* New Haven, CT: Yale University Press.

Downs, Anthony. (1972). "The Issue-Attention Cycle." *The Public Interest* 28 (Summer):38–50.

Internet Resources

U.S. Congress. Legislative Process—How a Bill Becomes a Law. Available from http://www.house.gov/house/Tying_it_all.html.

Sara E. Keith

Life Cycle Analysis

impact a change to the environment resulting from a human activity or product

A typical product has a range of environmental **impact** arising from its manufacture, use, and disposal. A life cycle assessment (LCA) evaluates the entire environmental impact of a product through its life cycle. An LCA might, for example, compare the environmental impact of ordering an item online to going to a store to buy it. The analysis would include the environmental impact of having the item mailed to the purchaser's home directly from the distributor versus having it sent from the distributor to the store, and then having the customer drive to the store to buy it. In this example, an LCA has shown that it can be environmentally preferable to buy products online, but only if the item is sent by standard truck mail rather than by express airmail. Other LCAs have shown that lightweight plastic bumpers are superior to heavier steel bumpers for cars, and that the relative merits of cloth versus disposable diapers depend on how the cloth diapers are dried, because electric drying uses so much energy.

Life cycle analyses of products are typically coupled with efforts to reduce their environmental impact. Extended producer responsibility (EPR) is the concept that the producer of a product is also responsible for recycling the product. In Germany, producers are required to take back the packaging of their products, and in the Netherlands, the cost of cars incorporates a recycling tax. SEE ALSO RECYCLING; REUSE.

Internet Resources

Journal of Industrial Ecology. Available from http://www.yale.edu/jie.

U.S. Environmental Protection Agency. National Risk Management Research Laboratory, Life-Cycle Assessment Web site. "LC Access." Available from http://www.epa.gov/ORD.

Valerie M. Thomas

Lifestyle

It might be said that, whether conscious of it or not, everyone has a lifestyle. From this perspective, *lifestyle* refers simply to the defining characteristics or qualities of a particular way of life, be it of an individual, a nation, or an entire culture.

On the other hand, some argue that lifestyle is a Western concept, meaningful only to the citizens of affluent countries, not to those whose main concern is mere survival because of their absolute poverty. From this perspective, the concept of lifestyle applies only to variants of consumerism, a largely materialistic way of life that assumes: (1) that what one wants is entirely a matter of choice; (2) that almost all choices are within one's grasp; and (3) that consumer choices can and should be hierarchically ranked from the most to the least desirable, according to what the mass media and corporate enterprise decide is most worth having and doing. Underlying high-end consumerism is the belief that the most desirable lifestyle is dependent on having the most prestigious occupations, which are, in turn, associated with the highest incomes. The concept and its implications are closely connected to the values associated with extreme individualism, corporate capitalism, and an open market, preferably one that is global in its reach.

To critics, what is excluded from lifestyle is even more important than what is included. While most people would generally consider lifestyle to be a neutral or amoral concept, others, on looking more closely, see it as having an immoral side. Discussions of lifestyle generally exclude any thoughts of justice, respect for human rights, or fairness. In short, questions of "ways of being" are left out of the equation: We tend to forgo contemplation of what society has become and what it should be in pursuit of the favored lifestyle. This is innocent enough as long as people are truly ignorant of global circumstances, but it becomes increasingly inexcusable as the consequences of gross social inequity become better known and the income gap yawns ever wider.

Countries most closely identified with a consumer lifestyle include the United States (where shopping is the most popular leisure-time activity),

A consumer is selecting bulk foods. (M. Stone, U.S. EPA. Reproduced by permission.)

Canada, Western Europe, Japan, Australia, New Zealand, and a few others where post-Enlightenment "scientific materialism" has taken hold as the dominant way of seeing the world. In these generally democratic countries, the economy functions more or less according to the laws of supply and demand—if people buy a lot of some good or service, then private businesses organize to produce as much of that good as they can and still make a profit (keeping in mind that at least some of the demand may be stimulated by advertising in the first place). People spend their money as they see fit with little interference by governments. As a result, the economy produces what is *wanted* by citizens who have the money to pay rather than what might be *needed* by impoverished members of society who cannot "vote" in the marketplace. In the end, the citizens of free-market countries have access to the most prodigious outpouring of manufactured goods and consumer services, both necessary and trivial, ever made available to members of the human species. Little wonder that in most market democracies many citizens seek social status and define their self-worth in terms of the quality and quantity of their personal possessions, particularly automobiles, houses and furnishings (especially home entertainment products), and clothes. Indeed, the accumulation of private goods is a defining characteristic of a consumer lifestyle. It is often remarked that even the average person in the world's wealthy consumer societies enjoys greater personal comfort and convenience, if not outright wealth, than European monarchs of only a few centuries ago.

Given its pervasiveness in the West, many people will be surprised to learn that the consumer society was, in effect, deliberately constructed. In the years following World War II, North America was endowed with great industrial overcapacity (war-time factories) and large numbers of underemployed workers (returning soldiers). At the same time, the general population, having endured the material deprivation of the Depression and subsequent wartime rationing, was quite used to living modestly. To break people of their habit of "underconsuming," American industry purposefully organized to encourage North America to become a throw-away society and embrace a consuming way of life. In 1955, retail analyst Victor Lebow argued that Americans should make consumption their way of life. He suggested that if they succeeded in making the buying and use of goods into a kind of ritual, they would find spiritual satisfaction and ego gratification in consumption. His point was that to keep the economy going things had to be consumed, burned up, worn out, replaced and discarded at an ever-increasing rate. Today, a multibillion dollar advertising industry is still dedicated, in part, to creating needs that some new or improved product claims to meet.

Technology has also played a major role in helping industry to persuade people that material goods will help to fill the spiritual void that gnaws at the heart of techno-industrial society. For example, television has so successfully sold conspicuous consumption that the world consumed as many goods and services between 1950 (when commercial television was launched) and the mid-1990s as had all previous generations combined. For all that, a growing number of studies show that there is no correlation—indeed, there may even be a negative correlation—between growing incomes and subjective measures of "happiness" in the world's richest countries. It turns out that money really does not buy happiness.

The Pollution Connection

The promotion of consumerism, however it is portrayed in the media, leads to increasing pollution, resource scarcity, biotic impoverishment, and other forms of environmental degradation all over the world. Moreover, while a quarter of humanity enjoys the benefits of material plenty, the negative impacts of economic growth contribute to the loss of health and life among the poor in every country.

The economic production process often creates a vastly larger mass of waste than useful product. The packaging, distribution, use, and consumption of the product produce still more waste. Waste becomes pollution when the level of contamination impairs the aesthetic quality or productive capacity of the atmosphere, water, soil, or landscape; that is, when ecosystems are significantly damaged. Of course, as members of a human-centered, materialistic society, we tend to pay the most attention to pollution when it affects our own health or the health of other commercially valuable species. However they are perceived, waste and pollution are the inevitable and sometimes pernicious by-products of a consumer society, and the consumer lifestyle is spreading around the globe.

It is of little comfort that despite the best efforts of scientists, engineers, and technological optimists, progress in solving our waste and pollution problems has been decidedly erratic even in the world's most "advanced" economies. For example, studies show that although a greater number of people recycle, more waste than ever is also being hauled to landfills and incinerated. Often seeming improvements in one problem area are wiped out by worsening conditions in another—waste simply has to go somewhere. In 2002, the North American Commission for Environmental Cooperation

(CEC), an agency created under the North American Free Trade Agreement, reported in *Taking Stock* (its sixth annual study on pollution in Canada and the United States) that the total amount of toxic releases and transfers fell by three percent between 1995 and 1999. This slight decline was partially attributable to a 25-percent reduction in air emissions by the manufacturing sector. (It is probably also the result of economic restructuring, including the migration of some polluting industries or activities to developing countries.) However, reduced air pollution was offset by a 25-percent jump in on-site releases to land, a 35-percent surge in off-site releases—mainly to landfills, and a 26-percent rise in the waste dumped into lakes, rivers, and streams. Almost 3.4 million tons of toxic chemical waste were produced in 1999, roughly one million tons of that released on-site into the air. Almost 8 percent of total releases included chemicals known to cause cancer, birth defects, or other reproductive problems.

The following year, *Taking Stock* reported a continuing improvement overall—the reduction in industrial releases and transfers of chemicals in North America reached five percent in the six years from 1995 to 2000. However, there was a significant *increase* in toxic discharges among smaller manufacturing firms. A group of fifteen thousand industrial facilities across North America released and transferred 32 percent more toxic chemicals from 1998 to 2000. These facilities, with chemical releases and transfers up to 110 tons, represent the majority of polluters in Canada and the United States. Victor Shantora, acting executive director for the CEC, noted that "The small 'p' polluter might not grab the same headlines as a large power plant or chemical manufacturer, but their effect is being felt throughout the North American environment" (CEC 2003b). In Canada, these small 'p' polluters registered a 66-percent increase in chemical releases and transfers. In the United States, the same group recorded an increase of 29 percent.

Overall, wealthy industrial countries like the United States and Canada are responsible for more than 90 percent of the 350 million metric tons of hazardous waste produced globally each year. Approximately 65 percent of the world's economic production, consumption, and pollution is associated with cities in rich countries.

The World Resources Institute (WRI) describes the general problem of waste production in consumer economies in a particularly telling report, *The Weight of Nations*, released in 2000. This report documents the flow of certain materials through five of the world's most advanced, efficient, and wealthy industrial economies—Austria, Germany, Japan, the Netherlands, and the United States—over a twenty-year period up to 1996. The WRI analysis shows that, despite successful waste-reduction measures for some contaminants, significant improvements in the efficiency of material use and a slight reduction in resource throughput per unit of gross domestic product (GDP), both gross and per capita processed output (solid, liquid, and gaseous wastes) generally increased; the extraction and use of fossil energy resources dominated waste flows in all countries examined; and except in Germany, carbon dioxide emissions rose in both total and per capita terms in all countries studied (the data on Germany were distorted by unification and by that nation's one-time shift from coal to other more efficient hydrocarbon fuels).

These results may come as a surprise to those who believe that increased economic efficiency and resource productivity (technological efficiency),

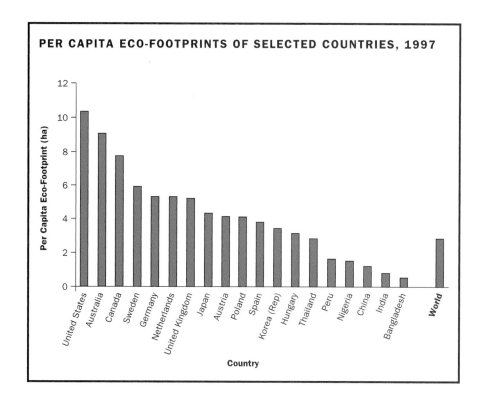

PER CAPITA ECO-FOOTPRINTS OF SELECTED COUNTRIES, 1997

combined with the shift to more "knowledge-based" sources of wealth creation, would significantly "decouple" the economy from nature. On the contrary, *The Weight of Nations* concludes unambiguously that the resource savings from efficiency gains and economic restructuring have been negated by population growth, growing consumption, and increasing waste output. Moreover, *The Weight of Nations* shows that despite the growing economic role of high-end services and other knowledge-based activities, modern industrial economies are carbon-based economies driven by fossil fuel; their predominant waste-generating activity is burning material.

The WRI study might actually be optimistic because it apparently examined only the energy and materials flows through the domestic economies of the countries studied. One may then ask how the data on resource consumption and waste generation would be affected if the calculations were corrected for trade flows. Does the embodied energy and material content of imported manufactured goods exceed that of exports? If so, the reduction in material consumption suggested by the modest decoupling of GDP growth from domestic energy and material use may be exaggerated.

Comparative Ecological Footprints

Ecological footprint analysis (EFA) provides another way to understand the problem of material throughputs in the modern world. The ecological footprint of a specified population may be defined as the area of productive land and water ecosystems required, on a continuous basis, to produce the resources that the population consumes and to assimilate its wastes, wherever on Earth the relevant land/water is located. Because of trade and natural flows, portions of any modern nation's eco-footprint are scattered all over the world.

Since eco-footprint estimates are based on the resource use and waste generation associated with final consumption by study populations, they provide a way to compare the ecological impacts of differing lifestyles. Recent national eco-footprint estimates underscore the fact that high-income countries—including the most technologically efficient economies examined in the WRI study—are the most material-intensive and polluting economies on Earth on a per capita basis. The bar graph shows the per capita ecological footprints (EFs) of a selection of countries across the income spectrum, from among the richest to the poorest on Earth. To facilitate comparison, the EFs are reported in hectares at world average productivity (data drawn from WWF 2002).

Note the enormous disparity between high-income "northern" countries and the poorer developing countries of the south. North Americans and Europeans typically consume ten to twenty or more times as much per capita of various resources as do the impoverished citizens of the poorest countries such as the people of Bangladesh and Sierra Leone; the wealthy therefore impose a correspondingly massive pollution load on the world's ecosystems.

Because of the finite volume of "ecological space" on Earth, it would not be possible to raise the entire world population to North American or western European material standards on a sustainable basis using prevailing technologies. The total eco-footprints of many densely populated high-income countries are already considerably larger than their domestic territories. Indeed, the world average eco-footprint is about 2.3 ha while there are fewer than 2 ha of productive land and water on Earth. Although the basic economic needs of a billion people have not yet been met, the world population has already overshot global carrying capacity. Humans are living and growing, in part, by depleting the biophysical resource base of the planet.

There can be little doubt that political factors help to maintain the disparity between high-income and developing countries. For example, the structural adjustment programs imposed by the International Monetary Fund (IMF) and the World Bank as a condition for development loans force borrowing countries to lower their standards of living and to export more minerals, timber, and food both to pay down their loans and to purchase imports from high-income countries. However, in the increasingly open global marketplace, developing countries must compete with each other and with first world subsidies for first world markets. This forces down the prices for developing countries' commodity exports in relation to the prices of the manufactured goods and services they must import. According to economist J.W. Smith, current terms of trade create a relative price difference that is even more effective than colonialism in appropriating the natural resources and in exploiting the cheap labor of less-developed countries. Remarkably, while developed countries claim to be financing the developing countries, the poor countries are actually financing the rich through low pay for equally productive labor, investment in commodity production for the wealthy world, and other dimensions of unequal trade.

Most significantly, many observe that the terms of trade and structural adjustments forced on third world countries are quite opposite to the policies under which the wealthy nations developed. This suggests that the power brokers of the developed countries know exactly what they are doing. Critics such as Smith claim that their grand strategy is to impose unequal trades on the world so as to lay claim to the natural wealth and labor of weaker nations. Intentional or not, the strategy is clearly effective: In the 1960s only $3 flowed north for every dollar flowing south; by the late 1990s the ratio was seven to one.

It is worth emphasizing here the extent to which wealthy industrialized countries are dependent on cheap commodities, particularly low-cost fossil fuel, to maintain their consumer economies. This reality is becoming an increasing strain on geopolitical stability. For example, both our highly productive intensive agriculture and almost all forms of transportation are directly or indirectly petroleum based. This dependence has, in turn, led to instances of aggression to control oil-producing countries thus assuring ready access to critical fuel supplies. (To some oil is certainly one of the motivating factors implicated in the 2003 war on Iraq.) It also encourages injustice, violations of human rights, and ecological degradation in order to extract oil as cheaply as possible. A clear example of this is the alleged genocide and ecocide committed by Royal Dutch Shell Oil in Ogoniland, Nigeria, a case that has been widely reported and is on trial in U.S. courts under the Alien Torts Claims Act (ATCA).

Although such gross human rights violations are particularly egregious, even normal day-to-day business activities that promote consumerism and the ever-expanding eco-footprints of wealthy consumers can be interpreted as a form of "institutionalized violence" if we continue these practices in full knowledge of the distant social and ecological consequences.

Eco-Apartheid

Worldwide, the urban poor tend to live in neglected neighborhoods, enduring pollution, waste dumping, and ill health, but lacking the political influence to effect improvements. Indeed, since the time of the Industrial Revolution in the late 1700s, the urban poor, particularly racial and ethnic minorities, have had neither the resources to avoid, nor the power to control, noxious hazards in the workplace or in their homes. These are the people who have borne the greatest ecological costs of two centuries of continuous material growth. Today, the consumer lifestyle of the world's wealthy elite imposes an unprecedented burden of pollution, ecological disintegrity, and global climate change on the world. The costs of this burden are paid most heavily by the most vulnerable members of the human family: the poor and people of color.

Indeed, some see an intensifying pattern of "eco-apartheid" throughout the world. Extreme examples of city-level environmental distress are found both in the industrial cities of the former socialist and communist economies and in middle- and low-income megacities in the developing world. Certainly, the urban environmental hazards causing the most ill health are those found in the impoverished homes, neighborhoods, and workplaces located principally in the poorer countries of the Southern Hemisphere.

The problem, however, is hardly confined to second and third world cities. Even in the United States, the geographic distribution of air pollution, contaminated waters and fish, toxic waste sites, and landfills, correlates strongly with the distribution of both racial minorities and poverty. People have therefore begun to speak passionately of the need to ensure environmental justice for environmentally beleaguered communities. Some analysts emphasize that the correlation between chronic exposure to ecological hazards and race is much stronger than that between exposure and income poverty. A National Wildlife Federation review of sixty-four studies of environmental inequity found sixty-three cases of disparity by race or income but race proved to be the more important factor. Similarly, the Argonne National Laboratory found that of U.S. population, 33 percent of whites, 50 percent of African-Americans, and 60 percent of Hispanics live in the 136 counties in which two or more air pollutants exceed standards.

To make matters worse, the evidence is clear that even in these enlightened modern times, rich neighborhoods are often better served by environmental law and regulatory agencies than are less advantaged ones. It seems that if a community is poor or inhabited largely by racial minorities, it will likely receive less protection than a community that is affluent or white. In his article "Decision Making," Robert Bullard has argued that:

> . . . the current environmental protection paradigm has institutionalized unequal enforcement, traded human health for profit, placed the burden of proof on the "victims" rather than on the pollution industry, legitimated human exposure to harmful substances, promoted "risky" technologies such as incinerators, exploited the vulnerability of economically and politically disenfranchised communities, subsidized ecological destruction, created an industry around risk assessment, delayed cleanup actions, and failed to develop pollution prevention as the overarching and dominant strategy. (p. 3)

It seems that in the United States economic privilege and power not only insulate the wealthy from the worst effects of ecological degradation, but also confer additional protection under the law.

Personal Responsibility

While overconsumption, particularly in northern rich countries, is a major contributor to accelerating human-induced global change, the situation is not totally hopeless. Human beings are consumers by nature—we have to consume to survive—but informed consumers can learn to consume responsibly. What, then, can the individual do to reduce his or her personal "load" on nature? The fact is that making careful consumer choices can greatly reduce the negative impacts of one's personal lifestyle. For example, the most ecologically harmful consumer activities are associated with fuel-guzzling private automobiles and light trucks, diets rich in industrially produced meat, poultry and other products of intensive agriculture, home heating and cooling (including water heating), modern appliances, home construction and household water/sewage. Personal transportation, food, and household operations alone account for between 59 and 80 percent of total household environmental impact in several categories of pollution and environmental damage (see table).

Deciding to take public transportation, walk, or bicycle (generally reducing automobile dependence) in the city, switching to a mostly organic low-meat diet, living in a modestly scaled house or apartment and ensuring that it is adequately insulated, and using only essential, certified high-efficiency appliances are some of the best ways for residents of high-income countries

ENVIRONMENTAL IMPACTS PER HOUSEHOLD

Activity	Climate Change Greenhouse gases	Air Pollution Common	Toxic	Water Pollution Common	Toxic	Habitat Alteration Water use	Land use
Transportation	32%	28%	51%	7%	23%	2%	15%
Food	12	17	9	38	22	73	45
Household operations	35	32	20	21	14	11	4
Subtotal	80%	77%	80%	67%	59%	86%	64%

SOURCE: Brower, M., and Leon, W. (1999). *The Consumer's Guide to Effective Environmental Choices.* New York: Three Rivers Press.

to shrink their personal ecological footprints. Consumers can also demand "fair trade" goods (such as coffee and other third world agricultural commodities) that ensure adequate returns to peasant producers in the developing world.

Unfortunately, shifting consumer preferences alone will not create a green and fair economy. For example, the unfettered market is unlikely to provide the financial incentives that are needed to stimulate the private sector to take advantage of technologies *that already exist* and that could be used to increase resource productivity (efficiency) and conservation. Citizens everywhere should therefore also support their governments to undertake the ecological tax reforms (e.g., pollution charges and resource depletion taxes) necessary to move their economies into a more efficient conservation mode. No country can go it entirely alone. International cooperation in this endeavor is necessary to create and maintain a level economic playing field.

There are of course, more radical solutions. Increasing numbers of people are taking an additional step to reduce their load on the earth in the movement toward "voluntary simplicity." These individuals adopt less hectic and materially simpler lifestyles in an effort both to reduce their ecological footprints and to provide the psychological space needed to enrich their lives spiritually. SEE ALSO ACTIVISM; ENVIRONMENTAL JUSTICE; INDUSTRY; MASS MEDIA; POPULAR CULTURE; POVERTY.

Bibliography

Brower, M., and Leon, W. (1999). *The Consumer's Guide to Effective Environmental Choices.* New York: Three Rivers Press (for The Union of Concerned Scientists).

Bullard, R. (1995). "Decision Making." Chapter 1 in *Faces of Environmental Racism*, edited by L. Westra and P. Wenz. Lanham, MD: Rowman and Littlefield.

CEC. (2002 and 2003a). *Taking Stock.* Montreal: North American Commission for Environmental Cooperation.

Colborn, T.; Dumanoski, D.; and Myers, J. P. (1994). *Our Stolen Future.* New York: Dutton.

Goldman, B. (1994). *Not Just Prosperity: Achieving Sustainability with Environmental Justice.* Washington, D.C.: National Wildlife Federation Corporate Conservation Council.

Hardoy, J.; Mitlin, D.; and Satterthwaite, D. (1992). *Environmental Problems in Third World Cities.* London: Earthscan.

Haughton, G. (1999). "Environmental Justice and the Sustainable City." Chapter 4 in *The Earthscan Reader in Sustainable Cities*, edited by D. Satterthwaite. London: Earthscan.

Lane, Robert. (2000). *The Loss of Happiness in Market Democracies.* New Haven, CT: Yale University Press.

McGranahan, G.; Songsore, J.; and Kjellén, M. (1999). "Sustainability, Poverty, and Urban Environmental Transitions." Chapter 6 in *The Earthscan Reader in Sustainable Cities*, edited by D. Satterthwaite. London: Earthscan.

Motavalli, J. (1966). "Enough!" *E Magazine* 7(2):28–35.

Rees, W. E. (1996). "Revisiting Carrying Capacity: Area-Based Indicators of Sustainability." *Population and Environment*, 17(3):195–215.

Rees, W.E. (2002). "Globalization and Sustainability: Conflict or Convergence?" *Bulletin of Science, Technology and Society*, 22(4):249–268.

Rees, W.E., and Westra, L. (2003). "When Consumption Does Violence: Can There Be Sustainability and Environmental Justice in a Resource-Limited World?" Chapter 5 in *Just Sustainabilities: Development in an Unequal World*, edited by Julian Agyeman, Robert Bullard and Bob Evans. London: Earthscan and Cambridge, MA: MIT Press.

Robins, N., and Kumar, R. (1999). "Producing, Providing, Trading: Manufacturing Industry and Sustainable Cities." *Environment and Urbanization*, 11(2):75-93.

Smith, J.W. (2000). *Economic Democracy: The Political Struggle of the 21st Century*. Armonk: NY: M.E. Sharpe.

Wackernagel, M., and Rees, W. E. (1996). *Our Ecological Footprint: Reducing Human Impact on Earth*. Philadelphia, PA: New Society Publishers.

Wackernagel, M.; Onisto, L.; Bello, P.; Linares, A. C.; Falfán, I. S. L.; Garcia, J. M.; Guerrero, A. I. S.; and Guerrero, M. G. S. (1999). "National Natural Capital Accounting with the Ecological Footprint Concept." *Ecological Economics*, 29:375–390.

Westra, L. (1998). *Living in Integrity*. Lanham, MD: Rowman Littlefield.

Westra, L. (2000). "Institutionalized Environmental Violence and Human Rights." Chapter 16 in *Ecological Integrity: Integrating Environment, Conservation and Health*, edited by D. Pimentel, L. Westra, and R. Noss. Washington, D.C.: Island Press.

World Resources Institute. (2000). *The Weight of Nations*. Washington, D.C.: World Resources Institute.

World Wide Fund for Nature. (2002). *Living Planet Report 2002*. Gland, Switzerland: World Wide Fund for Nature (and others).

Internet Resource

CEC. (2003b). "Latest News." Montreal: North American Commission for Environmental Cooperation. http://www.cec.org/news/details/index.cfm?varlan=english&ID=2529.

William E. Rees and Laura Westra

Light Nonaqueous Phase Liquid (LNAPLs) *See Nonaqueous Phase Liquids (NAPLs)*

Light Pollution

As humankind enters the twenty-first century, ours is the first generation where the majority of children cannot routinely see the night sky in all its splendor and glory. The problem is caused by light pollution, excess or misdirected artificial light that alters the natural night sky. In the night sky, light pollution causes an atmospheric phenomenon known as skyglow. You may have seen overhead clouds at night glowing with strange pink or orange colors; this is wasted light reflecting off the water particles that form clouds. Even without clouds, light shoots into the sky and reflects off of tiny airborne dust and moisture particles. The skyglow phenomenon directly affects the scientific research of amateur and professional astronomers. It also affects everyone else who simply enjoys a dark night sky abundant with stars overhead. Scientists say that nearly two-thirds of the U.S. population can no longer see the Milky Way.

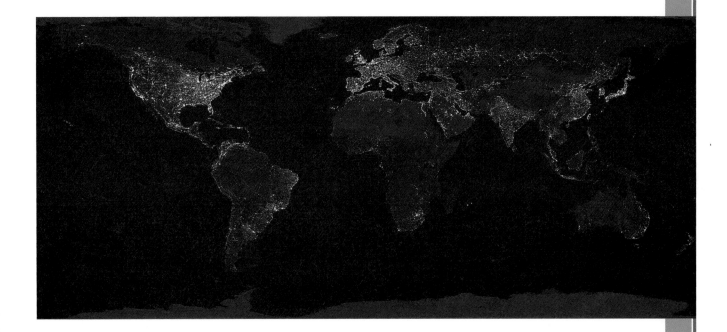

On Hawaii's big island, Mauna Kea towers almost fourteen thousand feet above sea level and is home to the greatest collection of large telescopes on the earth. In several communities around the base of the mountain, the types of allowed nighttime lights have been restricted to keep the skies dark at the mountaintop—assuring that this site will remain one of the best in the world for astronomical research. This approach has also been successful in communities such as Tucson, Arizona, whose nearby Kitt Peak National Observatory has been in operation since the late 1960s.

On a somewhat smaller scale, in Springfield, Vermont, an annual telescope makers convention named Stellafane was threatened in the late 1990s by the lights of a nearby, newly constructed planned prison called Vermont Southern State Correctional Facility. Since 1920 as many as 3,000 telescope makers and stargazers from around the world have converged at this site to scan the dark New England skies with their homemade telescopes. When stargazing was threatened by the corrections facility's bright lights, telescope makers worked closely with prison officials to install appropriate lighting that maintained security, while minimizing its impact on the dark night sky.

Unfortunately, when the growth of lighting has gone unchecked, as in parts of California, instruments of great historical value, such as the two-hundred-inch Hale Telescope on Mount Palomar, have had their usefulness severely limited.

In an effort to assess the magnitude of the light pollution problem, a comprehensive World Atlas of Artificial Night Sky Brightness was produced in 2001 by researchers at the University of Padua, Italy, and the National Oceanic and Atmospheric Administration (NOAA). Thanks to work like this, light pollution is rapidly gaining recognition as a global economic issue. Although the problem is most pronounced among developed industrialized nations, it is also responsible for squandering the limited resources of poor and developing nations that can least afford the waste.

In 1988 a nonprofit educational research organization known as the International Dark Sky Association (IDA) was founded to increase awareness of

Map of Earth at night, taken from NASA space satellites. Bright areas are those that are more developed. (Data courtesy Marc Imhoff of NASA GSFC and Christopher Elvidge of NOAA NGDC. Image by Craig Mayhew and Robert Simmon, NASA GSFC. Reproduced by permission.)

and offer solutions to the problems associated with light pollution. The IDA currently has almost ten thousand members around the world, and has created the definitive collection of resources for the study of light pollution and its impact on the planet. Just a glance through the archives of the IDA reveals that light pollution has an impact on everything from humans to moths.

For example, in Florida, sea turtle hatchlings become disoriented by inland artificial lighting that confuses them during their first trip to the sea. It appears that artificial lights can distract the young turtles from needed optical clues (e.g., the sky reflecting off the ocean) which would normally lead them to the sea shortly after hatching. In some cases, nesting females also go astray for the same reason. This problem has prompted sixteen counties and forty municipalities in Florida to adopt coastal lighting ordinances. However, even with these ordinances, in 1998 almost 20,000 hatchlings were reported to have become disoriented, and it is suspected that even this number might be underestimated.

Another example of light pollution's impact on wildlife may be found on the Hawaiian island of Kauai. Here, young birds called Newell's Shearwaters become disoriented by artificial lighting as they try out their wings for the first flight from mountainside nests to the ocean. The result is that many of the endangered seabirds die or collapse from exhaustion before making it to sea. In 1998 it was determined that 819 shearwaters had been disoriented by nighttime lighting on Kauai. Fortunately, a volunteer rescue effort saved 89 percent of these exhausted or injured birds.

These examples are representative of a much larger global problem that extends well beyond Florida and Hawaii. In 2002 a conference entitled the Ecological Consequences of Artificial Night Lighting was sponsored by the Urban Wildlands Group and the UCLA Institute of the Environment. Findings from a wide range of research focusing on the effects of light pollution on wildlife demonstrated that nighttime lighting is having a profound (and usually negative) impact on animals in both urban and rural areas. Although there is no question that wild animals are affected by light pollution, there is emerging evidence that humans might share some of the same light-induced (or dark-induced) chemical reactions that affect other animals. A key finding are the chemical bases for the circadian rhythm that regulates sleep/awake cycles in some insects. If humans share a similar photochemical basis for the sleep/awake cycle, then how does excess nighttime lighting influence this natural cycle in humans? There are currently more questions than answers regarding the impact of artificial nighttime lighting on humans, but it now appears possible that nighttime lighting could influence human lives well beyond the ability to play baseball at night!

It is estimated that each year the total value of wasted light in the United States alone is equal to about $1 billion. This is clearly a significant waste of resources. When the environmental impact of energy generation is considered, then light pollution is observed to have considerable secondary effects as well. A dramatic illustration of this problem occurs whenever one flies over a metropolitan area at night. While the thousands of tiny lights below might look impressive, all the light visible from an airplane window represents wasted illumination (and energy).

Obviously, light pollution is a problem with many negative ramifications. Fortunately, however, it is also a problem that has many positive solutions. One is to shield nighttime lighting and direct it appropriately so that all the light is directed down to the ground where it is needed rather than up into the sky. By doing this, lights with lower wattage can be used and a significant amount of energy and money saved.

Another solution is to simply reduce the number and brightness of lights. Often it is argued that this will reduce security, but research has shown that if done properly, less light can actually increase visibility (and thus security) by reducing glare and eliminating dark, high-contrast shadows.

Finally, replacing inefficient fixtures with modern energy-efficient (and shielded) models, as well as using motion sensors and timers, can all help to save energy and reduce wasted light.

These examples illustrate what can be done to reduce light pollution. Lawmakers are beginning to address these issues more and more, as many communities, parks, and even entire countries are enacting lighting controls, ordinances, and regulations.

As of 2002 many national parks throughout the United States enforce strict lighting plans to protect wildlife and to ensure that visitors will experience the outdoors and nighttime sky under natural (often only celestial) illumination. The same year the Czech Republic became the latest country to enact a national light-control policy, as Australia previously had. Many towns and cities worldwide have enacted local lighting-control laws to protect the night sky for reasons that include aesthetics, economics, security, and even astronomical research and amateur stargazing. Once light pollution is addressed, it leaves no residual pollutants behind and results in saved energy and better visibility. SEE ALSO ELECTRIC POWER.

In 2002 the Czech Republic became the first country to enact national light pollution legislation. The Czech law requires the use of fully shielded light fixtures—fixtures that "emit no light above the horizontal direction." Czech legislation was patterned after the "Lombardy Law," enacted after some 25,000 citizens of Italy's Lombardy region signed petitions demanding that action be taken against the glare caused by ineffective outdoor lighting.

Bibliography

Mizon, Bob. (2002). *Light Pollution–Responses and Remedies.* London and New York: Springer-Verlag.

Sky and Telescope. September 1998.

Internet Resources

Cinzano, P., Falchi, F., and Elvidge, C.D. *The First World Atlas of the Artificial Night Sky Brightness.* Available from http://xxx.lanl.gov/abs/astro-ph/0108052.

International Dark Sky Association Web site. Available from http://www.darksky.org/index.html.

Urban Wildlands Group Web site. Available from http://www.urbanwildlands.org/conference.html.

Peter Michaud

Limits to Growth, The

The Limits to Growth, written in 1972 by a team of researchers from the Massachusetts Institute of Technology (MIT), presented the results of a study in which a **computer model** attempted to predict the fate of society. The model studied the interrelationships between the world's population, agricultural production, natural resources, industrial production, and pollution. The results of the modeling effort were generally pessimistic, indicating a depletion of natural resources accompanied by a rapid decline in human population. The team argued that technological innovation could not halt the pending collapse. Instead, imposed limits to population growth and limits to investment in industrialization were the only solutions.

computer model a program that simulates a real event or situation

fatalistic of a person who believes that nothing one does can improve a situation

neo-Malthusians modern adherents to the ideas of Thomas Malthus

The Limits to Growth sold four million copies and brought notoriety to the research team. However, the study was criticized by other scholars and computer modelers who said that **fatalistic** assumptions had been programmed into the model, thus predetermining the pessimistic outcome. The team leaders stood behind their study, although they admitted a negative, Malthusian view of society's future. The book's findings were rejected by those who believed that technology would solve all problems, but they served to reinforce the views of **neo-Malthusians**. Its greater contribution was the innovative use of computers to model complex social, economic, and ecological systems for purposes of environmental policy analysis. SEE ALSO EHRLICH, PAUL; MALTHUS, THOMAS ROBERT; POPULATION; TRAGEDY OF THE COMMONS.

Bibliography

Steingraber, Sandra. (1997). *Living Downstream: An Ecologist Looks at Cancer.* Boston: Addison-Wesley.

Thomas, Janet. (2000). *The Battle in Seattle: The Story Behind and Beyond the WTO Demonstrations.* Golden, CO: Fulcrum.

Joseph E. de Steiguer

Litigation

Litigation, a case, controversy, or lawsuit, is a contest authorized by law, in a court of justice, for the purpose of enforcing a claimed right. Participants (plaintiffs and defendants) in lawsuits are called litigants. Litigation is often highly adversarial and can take a great deal of time, energy, and money, even when the case does not go to court (90 percent of all lawsuits are settled without trial). Many states and governments have enacted, or are considering, reforms directed at avoiding litigation, shortening the time a case takes to go to trial and minimizing the expense traditionally associated with litigation. Among these reforms are requiring that certain types of cases be arbitrated or directed to alternative dispute resolution procedures such as mediation and regulatory negotiation. SEE ALSO CITIZEN SUITS; CONSENSUS BUILDING; ENFORCEMENT; LAWS AND REGULATIONS, INTERNATIONAL; LAWS AND REGULATIONS, UNITED STATES; MEDIATION; PUBLIC POLICY DECISION MAKING; REGULATORY NEGOTIATION.

Internet Resource

U.S. Institute for Environmental Conflict Resolution Web site. Available from http://www.ecr.gov.

Susan L. Senecah

London Smog   *See Smog*

Los Angeles Smog   *See Smog*

Thomas Robert Malthus. (Corbis-Bettmann. Reproduced by permission.)

Malthus, Thomas Robert

ENGLISH CLASSICAL ECONOMIST AND CLERGYMAN (1766–1834)

Thomas Robert Malthus is best remembered for his 1798 treatise titled *An Essay on the Principle of Population as it Affects the Future Improvement of Society.*

In that work, he argued that because food production increased **arithmetically** and human population increased in a more rapid **geometric** manner, society would ultimately face dire consequences because of decreasing **per capita** food availability. Because of this thesis, economics has been coined "the dismal science." Detractors note that it is only a hypothesis and not a proven theory. Furthermore, they argue that human conditions since Malthus have improved in many ways due to technological innovation. The Malthusian hypothesis today remains influential in environmental thought because of its warning about unrestrained population growth. SEE ALSO EHRLICH, PAUL; LIMITS TO GROWTH; POPULATION; TRAGEDY OF THE COMMONS.

arithmetic increase by addition, e.g.., 2, 4, 6, 8 . . . as opposed to geometric, in which increase is by multiplication, e.g.., 2, 4, 8, 16 . . .

geometric by multiplication, e.g.., 2, 4, 8, 16 . . ., as opposed to arithmetic, in which increase is by addition, e.g.., 2, 4, 6, 8 . . .

per capita per individual person in the population

Internet Resource

The International Society of Malthus Web site, edited by Ronald Bleier. Available from http://www.igc.org/desip/malthus.

Joseph E. de Steiguer

Marine Protection, Research, and Sanctuaries Act

Although officially named the Marine Protection, Research, and Sanctuaries Act of 1972, this statute is better known by its common name, the Ocean Dumping Act. An amendment known as the "Ocean Dumping Ban Act of 1988" significantly superceded certain aspects of the original act. The Marine Protection, Research, and Sanctuaries Act arose from international treaty commitments, specifically negotiations resulting in the London Convention of 1975. Signing states agreed to take measures to prevent marine pollution and particularly to ban the dumping of identified toxins that could not be rendered harmless by natural processes. The statute's enactment also addressed the chronic and previously unfettered ocean dumping of municipal garbage, and industrial and commercial wastes, which by the 1970s was devastating marine ecosystems and fouling coastal beaches.

The Marine Protection, Research, and Sanctuaries Act regulates the disposal of any material in the U.S. territorial sea or contiguous zone, regardless of its point of origin; and the marine disposal anywhere of wastes and other material that originated in U.S. territory (expansively defined) or was transported on American vessels or aircraft. Although the U.S. Environmental Protection Agency (EPA) is the designated lead agency, the U.S. Army Corps of Engineers has statutory responsibilities, and enforcement often requires the services of the U.S. Coast Guard. Citizen plaintiffs may also sue to enforce the act. Unlike the case with many federal environmental statutes, states enjoy only a limited and generally advisory role. For marine disposals governed by the act, a permit is required. The disposal of high-level radioactive wastes, medical wastes, and radiological, chemical, or biological warfare agents is banned. Permits for various toxins, including mercury, cadmium, and halogens known to be carcinogens, mutagens, or **teratogens**, generally will be denied, unless present only in trace amounts or compounds known not to bioaccumulate.

teratogen something that causes birth defects, may be radiation, a chemical or a virus

For the most part, the act has been successful within U.S. waters, as evidenced by significantly cleaner coastal areas and more robust marine ecosystems since the 1990s. A series of public scares arising from medical wastes that washed up along the eastern seaboard in the late 1980s prompted greater scrutiny of coastal dumping and more exacting tracking mechanisms for the disposal of medical waste. The penalties for ocean dumping of medical wastes are much harsher than those for other violations. SEE ALSO OCEAN DUMPING; U.S. COAST GUARD; WATER POLLUTION: MARINE.

Internet Resource

U.S. Environmental Protection Agency Web site. "Ocean Dumping Ban Act of 1988." Available from http://www.epa.gov/history.

Kevin Anthony Reilly

Mass Media

Before the 1960s, the media reported sporadically on the environment—often then referred to as the 'ecology' issue.

But Rachel Carson's 1962 book, *Silent Spring*, which raised deep concerns about the nation's increasing reliance on synthetic pesticides, sparked the United States' modern environmental movement and, in turn, increased media scrutiny of its issues.

Before *Silent Spring*, some major pollution events, notably the "killer fog" of Donora, Pennsylvania, and the black afternoon smog of major industrial towns such as Pittsburgh and St. Louis, had largely been the limits of media coverage.

"Throughout most of the Sixties, unless a river was on fire or a major city was in the midst of a weeklong smog alert, pollution was commonly accepted by both the press and the general population as a fact of life," wrote David B. Sachsman in the *SEJournal*, the quarterly publication of the Society of Environmental Journalists (SEJ).

"Until the late Sixties, conservationists were thought of as eccentric woodsmen and environmentalists were considered unrealistic prophets of doom," continued Sachsman, a communications and public affairs professor at the University of Tennessee at Chattanooga.

With this new environmental interest, pioneers on the environmental beat began to distinguish themselves in the 1960s and 1970s. They included the *New York Times*' Gladwin Hill and the *Houston Post*'s Harold Scarlett. More reporters quickly followed.

"The year 1969 was pivotal for this growing media and public interest in the state of our environment," Sachsman concluded. That year, the *New York Times*, soon followed by other major newspapers, created an environment beat. *Time* and *Saturday Review* developed regular environment sections, *Look* devoted an entire issue to the "ecology crisis." *National Geographic* offered a nine-thousand-word article on humankind's environmental problems. As the 1970s dawned, Walter Cronkite presented the television feature "Can the World be Saved?" and Paul Ehrlich's book *The Population Bomb* had also become a best-seller.

About this time, television was coming into its own as a powerful new medium. Its coverage lent fuel to the growing environmental movement. Images of oil-soaked birds on the Santa Barbara beach, the result of the Channel-Union Oil spill in 1969; stories on the "death" of Lake Erie; giant fish kills in the Great Lakes; and the burning Cuyahoga River in Ohio cemented in the nation's mind that an important new political, business, and social issue had awakened.

"The mission of the Society of Environmental Journalists is to advance public understanding of environmental issues by improving the quality, accuracy, and visibility of environmental reporting."

—http://www.sej.org/

In turn, an estimated 20 million Americans gathered on April 22, 1970, for the first Earth Day. As a single event related to the environment, it would not be matched for two decades.

Such political action quickly prompted federal legislation, including the Clean Air Act in 1970 and the Clean Water Act in 1972. This legislative attention gave legitimacy to the issue, spawning more media coverage.

During the mid-1970s, the hot environment story was the threat of chemical pollution from the nation's industrial plants and the pollution such operations had left behind. The coverage of Love Canal, New York, in the late 1970s and, in 1983, the evacuation of tiny Times Beach, Missouri, put into headlines and daily conversation such insidious chemical names as "dioxin."

In 1989, the year of the *Exxon Valdez* oil spill in Alaska, television images again riveted the nation, showing oil-drenched birds struggling to survive on pollution-fouled beaches. Global warming, concern over endangered species, and air and water quality combined to increase coverage in all media. That year, 774 minutes of environmental coverage on the three major broadcast networks' nightly news set a new record, according to the *Tyndall Report*, an analysis of network news coverage.

In 1991, former *New York Times* environment reporter Phil Shabecoff, founded the nation's first environmental news service, known then as *Greenwire*. "The environment isn't a one-shot news story—it's something that needs to covered in-depth, day after day," Shabecoff later told the *Columbia Journalism Review*.

During the late 1980s, a group of daily reporters covering environmental issues began the SEJ, an organization formed by journalists to help other journalists do a better job on the difficult environment beat. Among the founders were some of the nation's distinguished environment reporters, including Jim Detjen of the *Philadelphia Inquirer*, Rae Tyson of *USA Today*, Noel Grove of *National Geographic*, Shabecoff, and Teya Ryan of Turner Broadcasting. Eighteen reporters attended the group's first organizational meeting.

"We doubt that we will ever become a slick operation," Detjen wrote in 1990. Today, the SEJ boasts more than 1,200 members—journalists, academics, and students, an annual budget of nearly $800,000, and a host of programs for journalists and students, including an annual conference, a quarterly journal, and website updated daily with the latest environmental reports.

In 1990, the twentieth anniversary of Earth Day marked the single largest global demonstration on the environment, winning coverage from Mt. Everest to Kansas. But a backlash against the issue and those who cover it soon developed.

"It is becoming trendy to ask whether environmental laws, not polluters, are the real public enemy," wrote Kevin Carmody, a founding SEJ board member, in the *Columbia Journalism Review* in 1995. "In newsrooms throughout the country, the hot story is the 'high cost of environmental regulation,' not the people or resources harmed when that regulation fails."

Indeed, journalists caught in the 1990 frenzy to celebrate Earth Day may have forgotten some basic journalistic principles—such as, question everything—opening the door for criticism. John Stossel, an ABC consumer and environment reporter, attracted sixteen million viewers in 1994 with a special report entitled "Are We Scaring Ourselves to Death?" The *Los Angeles Times*

> "The enthusiasms of Earth Day 1970 have been institutionalized in legislation, regulation, litigation, political dynamics and new personal values, and woven into the fabric of national life."
>
> —Gladwin Hill, *New York Times*, December 30, 1979

devoted seven full pages to a series by media critic David Shaw, called "Living Scared: Why Do the Media Make Life Seem So Risky?"

By 1993, minutes on the television networks devoted to environmental coverage had dwindled by 60 percent. Even so, environmental stories would reap ten Pulitzer Prizes in the 1990s, compared to just nine in the three previous decades.

When a new Republican president was elected to the White House in 2000—George W. Bush—environment coverage quickly picked up again. From January to May 2001, *New York Times* reporter Douglas Jehl wrote sixty stories on the environment, many of them displayed on page one.

"I didn't expect this," Jehl told the *Columbia Journalism Review*. "No matter how you measure it, in terms of volume of copy or prominence of play, there is a lot of environmental coverage today."

The *New York Times*, the *Washington Post*, and the *Los Angeles Times* added environment reporters, anticipating major conflicts thanks to the new Bush Administration. The *Tyndall Report* found evening news coverage of the issue back up, to nearly six hundred minutes.

"This renewed interest came at a time when the beat was in need of some new twists," said Bud Ward, then executive director of the Environmental Health Center in Washington, D.C. "There was a feeling on the part of some editors that we're talking about the same problems as twenty years earlier," Ward said. "Environmental problems today are more subtle than smog over Pittsburgh."

PULITZER PRIZES AWARDED FOR ENVIRONMENTAL REPORTING

- 1967—PUBLIC SERVICE

Milwaukee (WI) Journal: For its successful campaign to stiffen the law against water pollution in Wisconsin, a notable advance in the national effort for the conservation of natural resources.

- 1971—PUBLIC SERVICE

Winston-Salem (NC) Journal and Sentinel: For coverage of environmental problems, as exemplified by a successful campaign to block strip mining operation that would have caused irreparable damage to the hill country of northwest North Carolina.

- 1979—NATIONAL REPORTING

James Risser of the *Des Moines (IA) Register:* For a series on farming damage to the environment.

- 1992—PUBLIC SERVICE

Sacramento (CA) Bee: For "The Sierra in Peril,"

reporting by Tom Knudson that examined environmental threats and damage to the Sierra Nevada mountain range in California.

- 1996—PUBLIC SERVICE

News & Observer, Raleigh, NC: For the work of Melanie Sill, Pat Stith and Joby Warrick on the environmental and health risks of waste-disposal systems used in North Carolina's growing hog industry.

- 1996—EDITORIAL WRITING

Robert B. Semple, Jr. of the *New York Times:* For his editorials on environmental issues.

- 1998—INVESTIGATIVE REPORTING

Gary Cohn and Will Englund of the *Baltimore Sun:* For their compelling series on the international ship-breaking industry, that revealed the dangers posed to workers and the environment when discarded ships are dismantled.

The September 11, 2001, terrorist attacks, a war in Iraq, and the nation's sputtering economy (which might be used to rally support for decreased environmental protections) will present a new challenge to the coverage and interest in environmental issues. The environmental beat also faces an internal pressure. More newsroom staffs are being pared as the economy contracts and media competition increases.

But the last forty years have shown that each time interest in the topic wanes, enterprising reporters rekindle it. Their future attention or lack of it may play a pivotal role in how much larger the issue becomes in national politics.

"To report news about global warming in 10 inches of copy presents daunting challenges to even the most knowledgeable and skilled environmental reporter and editing team," Ward wrote in a recent issue of *Nieman Reports* that explored coverage of environmental issues.

Ward continued: "But the ways in which reporters and editors, correspondents and producers confront these challenges—the ones inside and outside the newsroom—will have a large effect in determining how Americans and their government anticipate and respond to continuing environmental pressures." SEE ALSO POPULAR CULTURE.

Bibliography

Hill, Gladwin. (1973) *Madman in a Lifeboat: Issues of the Environmental Crisis.* New York: John Day Co.

Keating, Michael. (1993). *Covering the Environment: A Handbook on Environmental Journalism.* Ottawa, Ontario, Canada: National Round Table on the Environment and the Economy.

Shabecoff, Philip. (2000). *Earth Rising: American Environmentalism in the 21st Century.* Washington, D.C.: Island Press.

Internet Resource

Society of Environmental Journalists Web site. Available from http://www.sej.org.

Michael Mansur

Mediation

Mediation is a facilitated negotiation in which a skilled, impartial third party seeks to improve relations between parties to resolve a conflict by improving communication, identifying interests, and exploring possibilities for a mutually agreeable resolution. The mediator has no power to impose any solution. Instead, the disputants remain responsible for negotiating a settlement. However, once signed, mediated agreements typically enter the regulatory process to become binding. The mediator's role is to assist the process in ways acceptable to the parties. Mediation most often is a voluntary process, but in some jurisdictions may be mandated by court order or statute. Many believe that mediation is more cost effective and produces better resolutions than settling a dispute out in court. SEE ALSO ARBITRATION; CONSENSUS BUILDING; ENFORCEMENT; LITIGATION; PUBLIC POLICY DECISION MAKING; REGULATORY NEGOTIATION.

Internet Resource

U.S. Institute for Environmental Conflict Resolution Web site. Available from http://www.ecr.gov.

Susan L. Senecah

Medical Waste

Medical wastes are generated as a result of patient diagnosis and/or treatment or the immunization of human beings or animals. The subset of medical waste that potentially could transmit an infectious disease is termed *infectious waste*. The Centers for Disease Control (CDC), the U.S. Environmental Protection Agency (EPA), and the World Health Organization (WHO) concur that the following wastes should be classified as infectious waste: sharps (needles, scalpels, etc.), laboratory cultures and stocks, blood and blood products, pathological wastes, and wastes generated from patients in isolation because they are known to have an infectious disease. Medical wastes can also include chemicals and other hazardous materials used in patient diagnosis and treatment. In some cases this subset of medical waste is classified as hazardous waste. Hospitals, clinics, research facilities, diagnostic labs, and other facilities produce medical waste. The bulk of the wastes generated by most health care facilities, however, is municipal solid waste (MSW), or trash. MSW includes large quantities of paper, cardboard and plastics, metals, glass, food waste, and wood. Medical waste, though a smaller portion of the total health care waste stream, is of special concern because of the potential hazards from pathogens that may be present, or from hazardous chemicals.

Risk and Health Care Waste

In the late 1980s there were a series of syringe wash ups on beaches along the East Coast of the United States, which were mistakenly attributed to health care facilities. The federal Medical Waste Tracking Act (MWTA) was passed and the EPA attempted to set standards for managing the infectious waste component of medical waste that they renamed *regulated medical waste*. Few states adopted its stringent guidelines. The MWTA expired in the early 1990s, making each state responsible for establishing its own classification and management guidelines for medical waste.

There are very few documented cases of disease transmission from contact with medical waste. The notable exception is needle stick, or "sharps" injuries. Paralleling the concern over beach wash ups of medical waste, was a growing awareness of the increase in HIV-AIDS and other cases of infectious diseases being diagnosed and treated in health care settings. This, along with a series of events, led to the Occupational Safety and Health Administration (OSHA), which established rules designed to protect health care workers (OSHA blood-borne pathogen standards and universal precautions) by stipulating the need for such personnel to wear protective clothing and equipment, and to take special precautions when handling or disposing of sharps. The interpretation of rules surrounding worker safety regulations led to some confusion over waste classification, thus causing a greater amount of wastes to be considered as potentially infectious. (For example, under the OSHA universal precautions guidelines, a worker handling a bandage with a single drop of blood on it should wear gloves, but the waste itself would most likely not be classified as infectious.)

Noting that there are multiple risks inherent in medical waste including toxic chemicals and radioactive materials, the WHO has chosen to use the term *health care risk waste* instead of *medical waste*.

Regulation

Many regulations govern the labeling, handling, treatment, transport, storage, and disposal of medical waste, including: Department of Transportation (DOT) rules for the packaging and transportation of wastes; OSHA guidelines for worker safety, waste labeling and handling; the Resource Conservation and Recovery Act (RCRA), which governs the management of hazardous materials and wastes, including hazardous pharmaceutical wastes; Nuclear Regulatory Commission (NRC) radioactive waste management practices, Drug Enforcement Agency (DEA) regulations for handling and disposing of controlled substances such as narcotics; the Clean Air Act, which regulates emissions from incinerators; the Clean Water Act, which defines what may be disposed of down the drain; state environmental and health rules that define certain types of waste and determine the specifics of waste treatment, as well as requirements for storage, labeling, handling, and segregation. Most other countries have similar multitiered regulatory regimes, such as Australia, where a national standard defines *clinical waste* (what is termed *medical waste* in the United States). However, the particulars of regulation are left to the discretion of individual Australian states.

Proper Management, Treatment, and Disposal

There is general consensus among professional health care organizations, the waste management industry, and regulators that proper management starts with the identification of wastes requiring special handling and treatment

Three syringes found on a beach in New London, Connecticut. (©Todd Gipstein/ Corbis. Reproduced by permission.)

MEDICAL WASTE

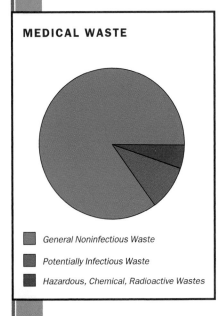

■ *General Noninfectious Waste*

■ *Potentially Infectious Waste*

■ *Hazardous, Chemical, Radioactive Wastes*

because of their hazardous nature (biological, chemical, or radioactive). Waste identification is necessary for proper segregation, so that only those wastes needing special treatment and handling are treated. Proper management of all waste streams enhances worker safety, protects the environment, and can reduce costs.

Wastes that are deemed potentially infectious may be treated prior to disposal by a number of different technologies that either disinfect or sterilize them. These technologies include incineration, steam sterilization, dry heat thermal treatment, chemical disinfection, irradiation, and enzymatic (biological) processes among others. In 2002 there were more than one hundred specific technologies in use. In order for treatment systems to work properly, distinctive protocols for the classification and segregation of wastes must be in place. Most treatment technologies for infectious wastes cannot process chemical or radioactive waste. Misclassification and inappropriate treatment of infectious wastes can result in significant harm to the environment and human health; for example, residual chemotherapeutic agents are should not be treated in autoclaves, but rather should be set aside and treated by either incineration (hazardous waste incinerators) or chemically neutralized where feasible.

The EPA has cited medical waste incinerators as among the top sources of mercury and dioxin pollution. New regulations governing the operation of, and emissions from, medical waste incinerators in the late 1990s have resulted in the closure of most such incinerators in the United States. Other countries such as the Philippines have completely banned incineration because of its adverse environmental impacts.

The health care industry is rapidly changing in ways that continue to have significant impact on the volume and characteristics of wastes produced.

- New (e.g., laproscopic and laser) surgical techniques result in procedures that produce very little blood-contaminated waste.

- Advances in cancer treatment have produced many drugs used in chemotherapy that are highly toxic in small quantities, producing more hazardous chemical wastes.

- Patient residence time in hospitals has declined. Procedures that previously required an extended stay now commonly occur on an outpatient basis without necessitating an overnight stay.

- Home care continues to grow, shifting the location of service delivery. Dialysis, chemotherapy, and hospice care are but a few examples of health care that often take place in a home setting, the result being that many wastes regulated as infectious or hazardous waste in a hospital are being disposed of as ordinary trash at curbside. (Household waste is exempt from many regulations.)

- As hospitals close their incinerators, biohazardous and sometimes (inadvertently) hazardous wastes are being hauled significant distances to centralized facilities for treatment and disposal.

All of these changes represent new challenges in continuing efforts to properly define, classify, regulate and manage medical wastes. SEE ALSO DIOXIN; ENDOCRINE DISRUPTION; HAZARDOUS WASTE; INCINERATION; INFECTIOUS WASTE; MERCURY; OCCUPATIONAL SAFETY AND HEALTH ADMINISTRATION (OSHA); RESOURCE CONSERVATION AND RECOVERY ACT.

PHARMACEUTICAL WASTE

Pharmaceutical wastes are diverse and in some cases trace amounts can be discarded as medical waste. Certain pharmaceuticals are hazardous wastes when disposed, and some common ones are "acute" hazardous wastes under RCRA regulations (e.g., Epinephrine, Nitroglycerin, Warfarin (>0.3%)).

THE HAZARDOUS WASTE STREAM

Hazardous Material	Point of Generation	Point of Use and Disposal	Common Disposal
Chemomtherapy and antineoplastic chemicals	Prepared in central clinic or pharmacy	Patient care areas Pharmacy Special clinics	Incineration as RMW Disposal as HW
Formaldehyde	Pathology Autopsy Dialysis Nursing units	Pathology Autopsy Dialysis Nursing units	Diluted and flushed down sanitary sewer
Photographic chemicals	Radiology Satellite clinics offering radiology services	Radiology Clinics offering radiology services	Developer and fixer is often flushed down sanitary sewer X-ray film is disposed of as solid waste
Solvents	Pathology Histology Engineering Laboratories	Pathology Histology Engineering Laboratories	Evaporation Discharged to sanitary sewer
Mercury	Throughout all clinical areas in thermometers, blood pressure cuffs, cantor tubes, etc. Labs	Clinical areas Labs	Broken thermometers are often disposed in sharps containers If no spill kits are available, mercury is often disposed of as RMW or SW Often incinerated
Anesthetic gases	Operating theater	Operating theater	Waste gases are often direct vented by vacuum lines to the outside
Ethylene oxide	Central Sterile Reprocessing Respiratory Therapy	Central Sterile Reprocessing Respiratory Therapy	Vent exhaust gas to the outside
Radio nuclides	Radiation Oncology	Radiation Oncology	Storage in secure area–disposal by national Atomic Energy Commission
Disinfecting cleaning solutions	Hospital-wide environmental services Facilities management Operating theater	Diagnostic areas Operating theater Facilities management	Dilution, disposal in sewer

Maintenance:
  Waste oil
  Cleaning solvents
  Leftover paints
  Spent florescent lamps
  Degreasers
  Paint thinner
  Gasoline
	Maintenance	Maintenance	Solid waste
			Sewer

Bibliography

Bisson, Connie Leach; McRae, Glenn; and Gusky Shaner, Hollie. (1993). *An Ounce of Prevention: Waste Reduction Strategies for Health Care Facilities.* Chicago: American Hospital Association.

Health Care without Harm. (2001). "Non-Incineration Medical Waste Treatment Technologies." Washington, D.C.

McRae, Glenn, and Gusky Shaner, Hollie. (1996). *Guidebook for Hospital Waste Reduction Planning and Program Implementation.* Chicago: American Hospital Association.

Pruss, A.; Giroult, E.; and Rushbrook, P. (1999). *Safe Management of Wastes from Health-care Activities.* Geneva: World Health Organization.

Rutala, William A., and Mayhall, C. Glen. (1992). "Medical Waste: The Society for Hospital Epidemiology of America Position Paper." *Infection Control and Hospital Epidemiology* 13:38–48.

Internet Resources

Centers for Disease Control Web site. Available from http://www.cdc.org.

U.S. Environmental Protection Agency Web site. Available from http://www.epa.gov.

Hollie Shaner and Glenn McRae

Mercury

Mercury is a metal with chemical similarities to zinc and cadmium. The metal is liquid at room temperature, with a freezing point at $-31°C$, and it is one of the most volatile metals. It occurs as the element Hg^0 and as the mercuric ion Hg^{++}, which has a great affinity for reduced sulfur (sulfide, $S^=$). Most mercury ore deposits consist of the very insoluble mineral cinnabar (HgS), with little droplets of elemental Hg. Mercury also occurs as impurities in many other ore minerals, creating mercury contamination when these minerals are mined or processed. Most common rocks have very low Hg contents, about ten to one hundred parts per billion (ppb) Hg . Elemental mercury is barely soluble in pure water, with only twenty-five ppb Hg dissolving at room temperature, but it is more soluble at higher temperatures. The mercuric ion is very soluble in most ambient waters, but very insoluble in the presence of sulfide. Natural enrichments of mercury occur in and around ore deposits and in geothermal hot spring areas and volcanoes. Bacteria in coastal waters convert inorganic Hg ions back into the elemental state, which then evaporate from the water back into the atmosphere. The physical transport of mercury from ore regions and the vapor transport from geothermal areas and the oceans provide the natural background contamination of mercury.

Mercury is a toxic element that damages the human nervous system and brain. Elemental mercury is less dangerous when it is ingested than when it is inhaled. The use of mercury in felt-making led to widespread elemental mercury poisoning of hatmakers ("mad as a hatter"), which was expressed by tremor, loss of hair and teeth, depression, and occasional death. The organic forms of mercury—methylmercury compounds, CH_3Hg^+ and $(CH_3)_2Hg$—are very bioavailable or are easily taken up by living organisms and rapidly enter cells, and are therefore the most hazardous. Minamata disease was an episode of mercury poisoning of a small coastal community in Japan (1954) through the direct industrial release of methylmercury in the bay. Another infamous episode of mercury contamination occurred in Iraq, where people ate wheat that was treated with a mercury-containing fungicide. The continuous flux of mercury from the atmosphere results in the low level of mercury pollution nationwide. A small fraction of the Hg^{++} from atmospheric deposition is converted by bacteria into the very dangerous methylmercury form. The methylmercury is then taken up by the lowest life forms and makes its way up the food chain and bioaccumulates in the larger fish. As a result, large predator fish such as bass, tuna, shark, and swordfish have the highest levels of Hg in the methylmercury form. Most states in the United States have advisories for eating only limited amounts of freshwater fish. Limiting intake of mercury-contaminated fish is especially important for pregnant women and young children. The current U.S. legal limit for Hg in fish for consumption is 1 ppm. Limits for Hg in soils vary from state to state but generally range from 10 to 20 ppm, whereas the Environmental Protection Agency's limit for drinking water is 2 ppb Hg. The Occupational Safety and

The most common exposure to mercury in the home comes when a mercury thermometer is dropped and broken. Children should be removed from the room immediately. DO NOT VACUUM SPILLED MERCURY. Vacuuming will disperse the mercury into the air; inhaling mercury poses high risk. Mercury naturally beads and if it is on a hard surface, it can be scooped up with index cards or a file folder. Seal in a ziplock bag and call the health department or a hospital to arrange safe disposal. Call the health department if mercury has spilled on a carpet or other fabric.

Health Administration limits for Hg in the air in the workplace (for an eight-hour average) are 0.01 mg organic Hg/m^3 air.

Modern sources of mercury contamination from human activities are subdivided into the following groups:

1. High-temperature combustion processes such as coal-fired power plants, incineration of solid household waste, medical waste, sewage sludge, and ore smelting.

2. Industrial waste effluents, such as from chlor-alkali plants that use liquid mercury as electrodes.

3. Effluents of wastewater treatment plants.

4. Point sources of specific industries, many of them no longer active today (such as hat making, explosives, mercury lights, herbicides, and plastics).

An overview of modern **anthropogenic** Hg fluxes into the environment shows that more than 80 percent of mercury is injected into the atmosphere through such combustion processes as coal-fired power plants. The combustion releases mercury as elemental vapor into the atmosphere, where it has an average residence time of about one year before it is oxidized to the mercuric form. The oxidized mercury attaches itself to small dust particles and is removed by wet and dry atmospheric deposition. As a result of this massive injection of Hg into the atmosphere—more than 100 tons of Hg per year in the United States in the late 1990s—the contaminant is distributed all over the globe. Even the polar ice caps show evidence of mercury contamination over the last 150 years, from atmospheric dispersal and deposition from anthropogenic sources. There are almost no places on earth that are not contaminated by anthropogenic mercury.

anthropogenic human-made; related to or produced by the influence of humans on nature

Mercury contamination is a matter of ongoing concern, and an extensive study was done for the U.S. Congress to summarize the sources, pathways, and sinks of mercury in the outdoor environment. There are several initiatives to limit the anthropogenic flux of Hg from coal-fired power plants, such as switching to mercury-poor coals and scrubbing the stack gases. Limiting or banning the production of mercury-containing materials, including switches, thermometers, thermostats, and manometers, both in the household as well as in the medical profession, would also reduce the mercury recycled back into the atmosphere from garbage incineration. SEE ALSO BIOACCUMULATION; HEALTH, HUMAN; INCINERATION; ISHIMURE, MICHIKO; MEDICAL WASTE; PERSISTENT BIOACCUMULATIVE TOXIC CHEMICALS (PBTs); SUPERFUND.

Internet Resource

U.S. Environmental Protection Agency. "Mercury Study Report to Congress." Available from http://www.epa.gov/oar/mercury.html.

Johan C. Varekamp

Methane (CH$_4$)

Methane is an invisible, odorless, and combustible gas present in trace concentrations in the atmosphere. It is the major component of natural gas, a

Chemical structure of methane (CH_4).

greenhouse gas a gas, such as carbon dioxide or methane, which contributes to potential climate change

anthropogenic human-made; related to or produced by the influence of humans on nature

fossil fuel commonly used for heating and cooking. The molecule consists of one carbon atom bonded to four hydrogen atoms (CH_4), making it the simplest member of a chemical family known as hydrocarbons. Other hydrocarbons include ethane (C_2H_6), propane (C_3H_8), and butane (C_4H_{10}).

As a **greenhouse gas**, methane ranks second to carbon dioxide. Methane levels, based on ice core samples, have more than doubled since 1750 (from 0.7 to 1.7 parts per million), largely due to human activity. On a molecule-for-molecule basis, methane is twenty-three times more potent as a greenhouse gas than carbon dioxide. Both gases are targeted for emissions reduction in the Kyoto Protocol.

Methane enters the atmosphere from both natural (30 percent) and **anthropogenic** (70 percent) sources. Methanogens (methane-producing bacteria in swamps and wetlands) are the largest natural source.

Anthropogenic sources of methane include leaks during fossil fuel mining, rice agriculture, raising livestock (cattle and sheep), and municipal landfills. Methanogens thrive in the oxygen-free (anaerobic) environment of landfills, releasing the gas in significant quantities. The gas is purposefully ignited to prevent explosion or captured for its commercial value as a fuel.

Livestock such as sheep, goats, camel, cattle, and buffalo currently account for 15 percent of the annual anthropogenic methane emissions. These grass-eating animals have a unique, four-chambered stomach. In the chamber called the rumen, bacteria break down food and generate methane as a by-product. Better grazing management and dietary supplementation have been identified as the most effective ways to reduce livestock methane emissions because they improve animal nutrition and reproductive efficiency. This general approach has been demonstrated by the U.S. dairy industry over the past several decades as milk production increased and methane emissions decreased. SEE ALSO Fossil Fuels; Global Warming; Greenhouse Gases; Landfill; Petroleum.

Bibliography

DeLong, Eward F. (2000). "Resolving a Methane Mystery." *Nature* 407:577–579.

Simpson, Sarah. (2000). "Methane Fever." *Scientific American* 282(2):24–27.

Turco, Richard P. (1997). *Earth under Siege: From Air Pollution to Global Change.* New York: Oxford University Press.

Internet Resource

Intergovernmental Panel on Climate Change, Working Group I. "Atmospheric Chemistry and Greenhouse Gases." *Climate Change 2001: The Scientific Basis.* Available from http://www.ipcc.ch.

Marin Sands Robinson

Mexican Secretariat for Natural Resources

The Mexican Secretariat for Natural Resources (La Secretaría del Medio Ambiente y Recursos Naturales or SEMARNAT) is the government office in Mexico responsible for creating sound national environmental policy, reversing existing damage to the environment, and establishing programs for sustainable development. SEMARNAT oversees the management of natural resources and coordinates development with other agencies. It works to restore ecosystems while taking into account the social and economic needs for natural resources.

Environmental policy in Mexico began in the 1940s, but was often over-shadowed by the push to industrialize the country. Little attention was given to protecting natural resources until the 1980s. At that point, the Mexican government created a series of agencies charged with protecting various natural resources. Finally, in November 2000, SEMARNAT was created to oversee the different agencies addressing environmental issues and establish national environmental policies.

In 2002 SEMARNAT and the U.S. Environmental Protection Agency (EPA) established a program called Border 2012. It is designed to strengthen the management of the environment and resources along the 2,000-mile border between the United States and Mexico. The program calls for the open exchange of information relating to natural resource issues and pollution prevention along the border. Border 2012 also involves regional workgroups so that ancillary programs may be tailored to individual needs and problems at a regional and local level. SEE ALSO ENVIRONMENT CANADA; U.S. ENVIRONMENTAL PROTECTION AGENCY

Internet Resources

"New U.S.-Mexico Border Environmental Program: Border 2012." Available from http://www.epa.gov/usmexicoborder.

SEMARNAT Web site. Available from http://www.semarnat.gob.mx./web_ingles.

Allan B. Cobb

Mining

Modern mining is an industry that involves the exploration for and removal of minerals from the earth, economically and with minimum damage to the environment. Mining is important because minerals are major sources of energy as well as materials such as fertilizers and steel. Mining is necessary for nations to have adequate and dependable supplies of minerals and materials to meet their economic and defense needs at acceptable environmental, energy, and economic costs. Some of the nonfuel minerals mined, such as stone, which is a nonmetallic or industrial mineral, can be used directly from the earth. Metallic minerals, which are also nonfuel minerals, conversely, are usually combined in nature with other materials as ores. These ores must be treated, generally with chemicals or heat to produce the metal of interest. Most bauxite ore, for example, is converted to aluminum oxide, which is used to make aluminum metal via heat and additives. Fuel minerals, such as coal and uranium, must also be processed using chemicals and other treatments to produce the quality of fuel desired.

There are significant differences in the mining techniques and environmental effects of mining metallic, industrial, and fuel minerals. The discussion here will mostly concentrate on metallic minerals. Mining is a global industry, and not every country has high-grade, large, exceptionally profitable mineral deposits, and the transportation infrastructure to get the mined products to market economically. Some of the factors affecting global mining are environmental regulations, fuel costs, labor costs, access to land believed to contain valuable ore, diminishing ore grades requiring the mining of more

Aerial view of shoreline, showing a stream polluted with waste water runoff from strip mining flowing into the Ohio River. (© Charles E. Rotkin/Corbis. Reproduced by permission.)

raw materials to obtain the target mineral, technology, the length of time to obtain a permit to mine, and proximity to markets, among others. The U.S. mining industry is facing increasing challenges to compete with nations that have lower labor costs—for example, less stringent environmental regulations and lower fuel costs.

Mining Life Cycle

Minerals are a nonrenewable resource, and because of this, the life of mines is finite, and mining represents a temporary use of the land. The mining life cycle during this temporary use of the land can be divided into the following stages: exploration, development, extraction and processing, and mine closure.

Exploration is the work involved in determining the location, size, shape, position, and value of an ore body using prospecting methods, geologic mapping and field investigations, remote sensing (aerial and satellite-borne sensor systems that detect ore-bearing rocks), drilling, and other methods. Building access roads to a drilling site is one example of an exploration activity that can cause environmental damage.

The development of a mine consists of several principal activities: conducting a feasibility study, including a financial analysis to decide whether to abandon or develop the property; designing the mine; acquiring mining rights; filing an Environmental Impact Statement (EIS); and preparing the site for production. Preparation could cause environmental damage by excavation of the deposit to remove overburden (surface material above the ore deposit that is devoid of ore minerals) prior to mining.

Extraction is the removal of ore from the ground on a large scale by one or more of three principal methods: surface mining, underground mining, and *in situ* mining (extraction of ore from a deposit using chemical solutions). After the ore is removed from the ground, it is crushed so that the valuable mineral in the ore can be separated from the waste material and concentrated by flotation (a process that separates finely ground minerals from one another by causing some to float in a froth and others to sink), gravity, magnetism, or other methods, usually at the mine site, to prepare it for further stages of processing. The production of large amounts of waste material (often very acidic) and particulate emission have led to major environmental and health concerns with ore extraction and concentration. Additional processing separates the desired metal from the mineral concentrate.

The closure of a mine refers to cessation of mining at that site. It involves completing a reclamation plan and ensures the safety of areas affected by the operation, for instance, by sealing the entrance to an abandoned mine. Planning for closure is often required to be ongoing throughout the life cycle of the mine and not left to be addressed at the end of operations. The Surface Mining and Control Act of 1977 states that reclamation must "restore the land affected to a condition capable of supporting the uses which it was capable of supporting prior to any mining, or higher or better uses." Abandoned mines can cause a variety of health-related hazards and threats to the environment, such as the accumulation of hazardous and explosive gases when air no longer circulates in deserted mines and the use of these mines for residential or industrial dumping, posing a danger from unsanitary conditions. Many closed or abandoned mines have been identified by federal and state governments and are being reclaimed by both industry and government.

Environmental Impacts

The environmental responsibility of mining operations is protection of the air, land, and water. Mineral resources were developed in the United States for nearly two centuries with few environmental controls. This is largely attributed to the fact that environmental impact was not understood or appreciated as it is today. In addition, the technology available during this period was not always able to prevent or control environmental damage.

Air. All methods of mining affect air quality. **Particulate** matter is released in surface mining when overburden is stripped from the site and stored or returned to the pit. When the soil is removed, vegetation is also removed, exposing the soil to the weather, causing particulates to become airborne through wind erosion and road traffic. Particulate matter can be composed of such noxious materials as arsenic, cadmium, and lead. In general, particulates affect human health adversely by contributing to illnesses relating to the respiratory tract, such as emphysema, but they also can be ingested or absorbed into the skin.

particulate fine liquid or solid particles such as dust, smoke, mist, fumes, or smog, found in air or emissions; they can also be very small solids suspended in water, gathered together by coagulation and flocculation

tailings residue of raw material or waste separated out during the processing of mineral ores

Land. Mining can cause physical disturbances to the landscape, creating eyesores such as waste-rock piles and open pits. Such disturbances may contribute to the decline of wildlife and plant species in an area. In addition, it is possible that many of the premining surface features cannot be replaced after mining ceases. Mine subsidence (ground movements of the earth's surface due to the collapse of overlying strata into voids created by underground mining) can cause damage to buildings and roads. Between 1980 and 1985, nearly five hundred subsidence collapse features attributed to abandoned underground metal mines were identified in the vicinity of Galena, Kansas, where the mining of lead ores took place from 1850 to 1970. The entire area was reclaimed in 1994 and 1995.

Water. Water-pollution problems caused by mining include acid mine drainage, metal contamination, and increased sediment levels in streams. Sources can include active or abandoned surface and underground mines, processing plants, waste-disposal areas, haulage roads, or **tailings** ponds. Sediments, typically from increased soil erosion, cause siltation or the smothering of streambeds. This siltation affects fisheries, swimming, domestic water supply, irrigation, and other uses of streams.

Acid mine drainage (AMD) is a potentially severe pollution hazard that can contaminate surrounding soil, groundwater, and surface water. The formation of acid mine drainage is a function of the geology, hydrology, and mining technology employed at a mine site. The primary sources for acid generation are sulfide minerals, such as pyrite (iron sulfide), which decompose in air and water. Many of these sulfide minerals originate from waste rock removed from the mine or from tailings. If water infiltrates pyrite-laden rock in the presence of air, it can become acidified, often at a pH level of two or three. This increased acidity in the water can destroy living organisms, and corrode culverts, piers, boat hulls, pumps, and other metal equipment in contact with the acid waters and render the water unacceptable for drinking or recreational use. A summary chemical reaction that represents the chemistry of pyrite weathering to form AMD is as follows:

$$\text{Pyrite} + \text{Oxygen} + \text{Water} \rightarrow \text{"Yellowboy"} + \text{Sulfuric Acid}$$

"Yellowboy" is the name for iron and aluminum compounds that stain streambeds. AMD can enter the environment in a number of ways, such as free-draining piles of waste rock that are exposed to intense rainstorms, transporting large amounts of acid into nearby rivers; groundwaters that enter underground workings which become acidic and exit via surface openings or are pumped to the surface; and acidic tailings containment ponds that may leach into surrounding land.

Major U.S. Mining Laws and Regulations

Some major federal laws and regulations affecting the mineral industry include the Comprehensive Environmental Response, Compensation and Liability Act (CERCLA), commonly known as Superfund, enacted in 1980. This law requires operations to report releases of hazardous substances to the environment and requires cleanup of sites where hazardous substances are found. The Superfund program was established to locate, investigate, and clean up the worst abandoned hazardous waste sites nationwide and is currently being used by the U.S. Environmental Protection Agency (EPA) to clean up mineral-related contamination at numerous locations. The Federal Water Pollution Control Act, commonly referred to as the Clean Water Act, came into effect in 1977. The act requires mining operations to meet standards for surface water quality and for controlling discharges to surface water. The Resource Conservation and Recovery Act (RCRA), enacted in 1976, regulates the generation, storage, and disposal of solid waste and hazardous waste, using a "cradle-to-grave" system, meaning that these wastes are governed from the point of generation to disposal. The National Environmental Policy Act (NEPA), enacted in 1970, requires federal agencies to prepare EIS for major federal actions that may significantly affect the environment. These procedures exist to ensure that environmental information is available to public officials and citizens before actions are taken. NEPA applies to mining operations requiring federal approval.

Comparison of U.S. and International Mining Laws and Regulations

The European Union (EU) has developed a set of environmental directives that have had a significant effect on the mining industries of member nations. Each country's environmental laws derive from these directives. Among the key directives are the Environmental Impact Assessment Directive (similar to the EIS requirements of the United States), the Water Framework Directive (addresses concerns similar to those of the U.S. Clean Water Act), and the Waste Framework, Hazardous Waste, and Landfill Directives (all address concerns similar to those of the U.S. RCRA).

Examples of Mining Pollution and Reclamation

The Bunker Hill Mine complex is located in northwest Idaho in the Coeur d'Alene River Valley, and has a legacy of nearly a hundred years of mining-related contamination since 1889. Operations ceased in 1982, and the EPA declared much of the area a Superfund site in 1983. The complex produced lead, zinc, cadmium, silver, and gold, as well as arsenic and other minerals and materials. Much of the mining pollution was caused by the dispersal of mining wastes containing such contaminants as arsenic, cadmium, and lead into the floodplain of the Coeur d'Alene River, acid mine drainage, and a leaking tailings pond. The metals contaminated soils, surface water, groundwater, and air, leading to health and environmental effects. Lead, in particular, was noted for its health effects on children in the area. EPA reports concerning lead poisoning state that experts believe blood levels as low as 10 micrograms per deciliter (μg/dl) are associated with children's learning and behavioral problems. High blood lead levels cause devastating health effects, such as seizures, coma, and death. Blood levels of children in areas near the complex ranged from about 35 to 65 μg/dl in the early 1970s to less than 5 percent in 1999, as remediation efforts progressed. EPA reports also state that children are at a greater risk from exposure to lead than adults because, among other reasons, children absorb and retain a larger percentage of ingested lead per unit of body weight than adults, which increases the toxic effects of the lead. Efforts by the federal government, the state of Idaho, and industry to remediate contaminated areas associated with the site are ongoing.

COAL-BED METHANE

Methane, a potent greenhouse gas trapped inside coal, can be released into the atmosphere when coal is mined. The 1993 President's Climate Change Action Plan encouraged the recovery of a possible 100 trillion cubic feet of this coal-bed methane for energy. This would reduce methane and carbon dioxide emissions overall, because burning methane produces less carbon dioxide than burning fossil fuels. Scientists from the United States Geological Survey are studying how to extract coal-bed methane without harming the environment. Current difficulties include how to dispose of the water that permeates coal beds and must be pumped off before methane can be released, and how to prevent methane migration. Methane, possibly from coal-bed methane mining, has been discovered in groundwater in residential neighborhoods.

There are also many mines with successful reclamation plans. For example, the Ruby Hill Mine, which is an open pit gold mine in Eureka, Nevada, won a state award in 1999 for concurrent reclamation practices, such as using revegetation and employing mitigation measures to offset potential impacts to local wildlife.

The mining of asbestos, either as the primary mineral or included as an unwanted material while mining for the "target" mineral, is one of the more controversial issues facing the mining industry in the United States. Asbestos is the name given to a group of six naturally occurring fibrous minerals. Asbestos minerals have long, strong, flexible fibers that can be spun and woven and are heat-resistant. Because of these characteristics, asbestos materials became the most cost effective ones for use in such items as building materials (roof coatings and shingles, ceiling and floor tiles, paper products, and asbestos cement products) and friction products (automobile clutch, brake, and transmission parts).

Unfortunately, it has been found that long-term, high-level exposure to asbestos can cause asbestosis and lung cancer. It was also determined that exposure to asbestos may cause mesothelioma, a rare form of cancer. Workers can be exposed to asbestos during mining, milling, and handling of ores containing asbestos or during the manufacture, installation, repair, and removal of commercial products that contain asbestos. One of the more recent controversies involving asbestos is the exposure of workers and the local residents to asbestos found in vermiculite ore mined in Libby, Montana. The vermiculite ore was shipped nationwide for processing and was used for insulation, as a lightweight aggregate, in potting soils, and for agricultural applications. Mining of the Libby deposit ended around 1991 but elevated levels of asbestos-related disease have been found in the miners, millers, and the local population. Another major area of concern is naturally occurring asbestos found in rock outcrops in parks and residential areas. SEE ALSO CLEAN WATER ACT; DISASTERS: ENVIRONMENTAL MINING ACCIDENTS; MINING LAW OF 1872; NATIONAL ENVIRONMENTAL POLICY ACT; RESOURCE CONSERVATION AND RECOVERY ACT; SMELTING; SUPERFUND.

Bibliography

Kesler, Stephen E. (1994). *Mineral Resources, Economics and the Environment.* New York: Macmillan.

Marcus, Jerrold J. (1997). *Mining Environmental Handbook: Effects of Mining on the Environment and American Environmental Controls on Mining.* London: Imperial College Press.

Ripley, Earle A.; Redman, Robert E.; and Crowder, Adele A. (1996). *Environmental Effects of Mining.* Delray Beach, FL: St. Lucie Press.

Sengupta, Mritunjoy. (1993). *Environmental Impacts of Mining: Monitoring, Restoration, and Control.* Boca Raton, FL: CRC Press.

Internet Resources

Brosius, Liz, and Swain, Robert S. (2001). "Lead and Zinc Mining in Kansas." Public Information Circular 17, Kansas Geological Survey. Available from http://www.kgs.ukans.edu.

Bureau of Land Management. (2001). "Abandoned Mine Lands Cleanup Program." Available from http://www.blm.gov/aml.

National Institute for Occupational Safety and Health. (1995)."Report to Congress on Worker's Home Contamination Study." NIOSH Report No. 95-123. Available from http://www.cdc.gov/niosh.

Michael J. McKinley

Mining Law of 1872

The General Mining Law of 1872 was enacted to promote the exploration and development of domestic mineral resources, primarily in the West. The law permits U.S. citizens to freely prospect for hard rock minerals, such as copper and gold, on federal lands not closed to or withdrawn from mining. Once a **deposit** is discovered, the prospector can stake a claim for ownership of the deposit, develop it, and obtain a **patent** for the land and mineral rights to the **claim**. Once the patent has been granted, the claim becomes private property for a small fee to the government.

The law, and whether or not it should be reformed, is hotly debated in both the public and private sectors. The lack of environmental controls under the Mining Law is a major issue that has spurred a host of reform proposals. Supporters of the law make the point that existing federal and state antipollution requirements are sufficient without creating new and possibly redundant laws. Also, much of the contention is centered on the patenting and claim system, and whether the government should assess a **royalty** for the extracted minerals. Because of the absence of royalties, critics view the existing system as a giveaway of federal lands. Proponents of maintaining the existing system argue that an incentive is still necessary for those who take the substantial financial risk to develop a mineral deposit, because mining the entire process is lengthy and involves high costs. They cite that to find and develop a new mineral deposit in the United States can take from four to eight years. The long duration is primarily owing to the lengthy permitting process that must be completed prior to establishing whether the site can be profitably developed.

Law-reform efforts address such issues as the institution of royalty fees, reserving federal land for a specific use that may preclude mineral development, and forcing public lands miners to bear the entire cost for the cleanup of past practices. SEE ALSO DISASTERS, MINING; LAWS AND REGULATIONS, UNITED STATES; MINING.

deposit concentration of a substance, i.e., mineral ore

patent legal document guaranteeing the right to profit from an invention or discovery

claim legal statement of intent

royalty money paid by a user to an owner

Internet Resource

Humphries, Marc. "Mining on Federal Lands." Congressional Research Service Issue Brief IB89130. Available from http://www.house.gov/price.

Michael J. McKinley

Mixing Zone

A mixing zone is an area of a lake or river where pollutants from a point source discharge are mixed, usually by natural means, with cleaner water. In the mixing zone, the level of toxic pollutants is allowed to be higher than the acceptable concentration for the general water body. The mixing zone is an area where the higher concentration is diluted to legal limits for water quality. Outside the mixing zone, the pollutant levels must meet water quality standards. A typical mixing zone consists of two parts: the zone of initial dilution (ZID), near the outfall, and the chronic mixing zone from the ZID out to where water quality criteria are met. The discharge into the mixing zone may be effluent from water treatment plants, chemicals, or hot water from cooling towers.

The U.S. Environmental Protection Agency (EPA) is taking steps to ban the use of mixing zones for toxic chemicals. The Great Lakes Initiative (2000) also bans the discharge of twenty-two chemicals considered to be bioaccumulative. Bioaccumulative chemicals (BCCs) are those that become more concentrated as they move up through the food chain, for instance, from aquatic insects to fish to humans. As the release of BCCs into water bodies is phased out, industries will need to treat the discharge at the source. SEE ALSO: Bioaccumulation; Dilution; Point Source; Water Pollution.

Bibliography

"Identification of Approved and Disapproved Elements of the Great Lakes Guidance Submissions From the States of Michigan, Ohio, Indiana, and Illinois, and Final Rule." (2000). In Federal Register 65:151.

Internet Resource

Great Lakes Initiative Fact Sheet. Available from http://www.epa.gov/ost/GLI/mixingzones/finalfact.html.

Diana Strnisa

Mold Pollution

Mold pollution is the growth of molds in a building resulting in damage to or the destruction of the structure itself (or its contents) and adverse health effects on the building's occupants. It is estimated that about 10 percent of U.S. buildings may suffer from mold pollution.

microorganism bacteria, archaea, and many protists; single-celled organisms too small to see with the naked eye

ovoid shaped like an oval or egg

Molds, also known as fungi, are **microorganisms** that generally have threadlike bodies called mycelium and reproduce by producing spores. Spores are generally round or **ovoid** single cells (but in some cases are multicellular). Spores can be colorless or pigmented and vary in size. While a human hair is approximately one hundred microns in diameter, spore size ranges from one to five microns.

substrate surface on which an organism, i.e. mold, grows

There are about fifty to one hundred different molds typically found growing indoors in water-damaged buildings. Water problems in buildings are generally the result of leaks from roofs or plumbing, condensation, and flooding. When building materials or furnishings such as wood, drywall, ceiling tiles, or carpets become wet, causing molds to grow on them. The types of **substrates** and the amount of moisture will often determine the kinds of molds that grow. For example, some molds like *Stachybotrys* require a highly water-saturated substrate. For other molds such as *Aspergillus*, only small amounts of excess moisture are necessary for growth. Thus, moisture control is key to controlling mold growth and eliminating their effects on the building or its occupants.

sick building syndrome shared health and/or comfort effects apparently related to occupation of a particular building

Mold growth can cause structural integrity problems in buildings constructed of wood. This generally goes under the misnomer of dry rot. The dry rot molds, like *Merulis lacrymans*, are the natural decomposers of leaves, stems, and trees in nature. If structural wood in buildings becomes wet, these molds may grow. The name dry rot comes from the powdery residue that is left after the wood is destroyed. Wood can be protected by the use of chemicals like creosote or by the use of sealants. Mold pollution in buildings may result in adverse health effects including infections, allergies, and asthma. Bleeding, memory loss, and a condition known as **sick building syndrome**

may also result from mold pollution, but such health effects remain controversial. **Epidemiological** studies have linked molds to these conditions; however, a direct causal relationship has not been established.

When health effects from molds occur, it is generally as a result of inhaling mold spores. For example, aspergillosis is an infection of the lungs caused by some species of *Aspergillus*, which can result in difficulty breathing. If left untreated, it can spread through the bloodstream to other organs, resulting in death. It is probably the most common type of building-acquired infection. Individuals with impaired immune systems are most susceptible to this infection. Mold infections can be acquired in health care facilities (nosocomial infections). Careful attention to removing spores from the air and water may be the best method to protect the public from these kinds of infections.

Occasionally, mold infections result from animals and birds inhabiting buildings. For example, bats or pigeons may deposit **guano** containing such molds as *Histoplasma capsulatum* and *Cryptococcus neoformans*. Disturbing this guano without respiratory protection can result in infection. The best defense against this kind of mold pollution is to keep these creatures out of the building.

In addition to infections, allergic diseases are associated with mold pollution. Asthma is the most common chronic disease of childhood and is the leading causes of school absenteeism, accounting for over ten million missed

Black mold at the bottom of a wall in a home on the Turtle Mountain Indian Reservation. (AP/Wide World Photos. Reproduced by permission.)

epidemiological epidemiology: study of the incidence and spread of disease in a population

guano solid or semisolid waste from birds and bats, rich in nutrients

allergen a substance that causes an allergic reaction in individuals sensitive to it

school days per year. For most elementary school children with asthma, **allergens** are the primary trigger for asthma, and their disease is thought to result from early exposure and sensitization to common allergens in their environment (e.g., dust mites, cockroaches, and molds). To prevent allergic disease, excessive mold growth must be controlled or eliminated.

The elimination of molds from structures requires first that water problems be corrected. Then, the mold-infested material must be removed using proper protection. In some cases, heavily mold-infested structures have had to be demolished or burned. In order to make the best decision on how to treat a mold-polluted structure, it is important to understand what molds are present and in what amount. A mycologist (scientist who studies molds) can often identify and count mold spores collected from indoor air, dust, or surfaces either by culturing them or by observing them under a microscope. However, these are slow and difficult processes.

In order for mycologists to improve their knowledge about molds in the indoor environment, mold DNA (i.e., moldgenomes) are being sequenced. Sequencing of DNA is the process of deciphering the spelling of the DNA alphabet that makes each organism unique. Like the sequencing of the human genome, this knowledge of mold genomes allows molecular biologists to develop easier and faster methods for the detection and quantification of molds. This is important because all molds in the indoor environment cannot be eliminated. If molds can be monitored, experts can find out when mold concentrations are at dangerous levels. Measures can then be taken to reduce the mold pollution in the environment. SEE ALSO ASTHMA; INDOOR AIR POLLUTION.

Bibliography

Heid, Christian A.; Stevens, Junko; Livak, Kenneth J.; and Williams, P. Mickey. (1996). "Real Time Quantitative PCR." *Genome Research* 6:986–994.

Persing, David H.; Smith, Thomas F.; Tenover, Fred C.; and White, Thomas J. (1993). *Diagnostic Molecular Microbiology: Principles and Applications.* Washington, D.C.: American Society for Microbiology.

U.S. Environmental Protection Agency. (2001). *Mold Remediation in Schools and Large Buildings.* Washington, D.C.: Author.

Internet Resources

U.S. Environmental Protection Agency. Indoor Air Quality Web site. Available from http://www.epa.gov/iaq.

Stephen J. Vesper and Richard A. Haugland

Montréal Protocol

protocol in government: agreement establishing rules or code of conduct; science: a series of formal steps for conducting a test

Following the discovery of the Antarctic ozone hole in late 1985, various governments recognized the need for stronger measures to reduce the production and consumption of a number of chlorofluorocarbons (CFCs). CFCs, which are human-made chemicals widely used in manufacturing, have been found to deplete the ozone layer that shields the surface of Earth from harmful forms of solar radiation. During the mid-1980s negotiations began on the Vienna Convention for the Protection of the Ozone Layer—a framework treaty focused on cooperation in research, information exchange, and scientific assessment of the atmospheric ozone (O_3) problem—government representatives discussed drafting a **protocol** controlling the use of CFCs, human-made chemicals widely used in manufacturing that deplete the ozone layer. However, no consensus could be reached. The Executive Director of the United Nations Environmental Programme (UNEP) established a working group to begin drafting such a protocol. The final agreement, which was concluded on September 16, 1987, reflects the contentious nature of the negotiations. For example, by Article V, developing countries with low consumption rates (e.g., Brazil, India, and Vietnam) that feared the protocol would hinder their economic development are allowed a ten-year delay in required **compliance** with targets and timetables for reducing ozone emissions.

compliance in law: meeting the terms of a law or regulation

However, countries have generally been aggressive and effective in implementing the protocol. By the time it came into effect on January 1, 1989, countries were already contemplating the protocol's modification and strengthening. Amendments and adjustments were agreed to in London (1990), Copenhagen (1992), Vienna (1995), Montréal (1997), and Beijing (1999). These modifications shortened the timetables for phasing out consumption of listed chemicals, added and funded the Montréal Protocol Fund, established the Implementation Committee, developed noncompliance procedures, and expanded the Technology and Economic Assessment Panels. These panels have addressed new issues as they have arisen, such as recycling and international smuggling of CFCs. SEE ALSO CFCs (CHLOROFLUOROCARBONS); OZONE; TREATIES AND CONFERENCES.

Bibliography

Benedick, Richard Elliott. (1998). *Ozone Diplomacy: New Directions in Safeguarding the Planet.* Cambridge, MA: Harvard University Press.

Weiss, Edith Brown. (2000). "The Five International Treaties: A Living History." In *Engaging Countries: Strengthening Compliance with International Environmental Accords,* edited by Edith Brown Weiss and Harold K. Jacobson. Cambridge, MA: The MIT Press.

Internet Resource

Ozone Secretariat of the United Nations Environment Programme. "The Montréal Protocol on Substances That Deplete the Ozone Layer." Available from http://www.unep.ch/ozone.

Michael G. Schechter

Nader, Ralph

AMERICAN CONSUMER ADVOCATE AND ENVIRONMENTALIST (1934–)

When a young Ralph Nader wrote a book about automobile safety, it made him a household name across America. The experience sparked a lifetime of service to numerous safety, political, and environmental causes.

One of the consumer activist's first major accomplishments involved the formation of the Public Interest Research Groups (PIRGs) in the 1970s. These student-led groups, funded by college activity fees and supported by paid professional staffs, serve as law offices working in the public's interest. PIRGs operate today in twenty-four U.S. states, tackling issues such as recycling, pollution, and public health and safety.

Nader's work also played a major role in the creation of the Coal Mine Health and Safety Act and the Occupational Safety and Health Act, both of

Ralph Nader. (Alex Wong/Getty Images. Reproduced by permission.)

which continue to save lives. His other early work focused on food safety, nursing homes, and water and air pollution.

Arguably, the most effective group that Nader founded is Public Citizen. This organization, with the support of 150,000 members, serves as a lobbying group—working to present ideas and critical information to members of Congress, all in an effort to persuade them to vote in favor of public-interest issues and, many times, against the wishes of major U.S. corporations.

Nader also led the fight against nuclear power in the 1970s and 1980s. He, Public Citizen, and other groups that he helped form played a major role in stopping the spread of nuclear power.

In 1996 and 2000, Nader ran for president on the Green Party ticket and brought his views on environmental issues and social justice to a larger audience. Nader's campaign played a role in the close 2000 election as he pushed a progressive agenda and brought plenty of new people—many of them young—into the political process. He continues his work today from an office in Washington, D.C. SEE ALSO PUBLIC INTEREST RESEARCH GROUPS (PIRGs).

Bibliography

Graham, Kevin. (2000). *Ralph Nader: Battling for Democracy*. Denver: Windom Publishing.

Internet Resource

Essential Information. Available from http://www.essential.org.

Kevin Graham

NAFTA (North American Free Trade Agreement)

On December 17, 1992, Canada, Mexico, and the United States entered into a historical trade pact called the North American Free Trade Agreement (NAFTA). It aims to increase trade by expanding market access and reducing investment barriers across North American borders. Of the many aspects of the debate in the United States over the **ratification** of NAFTA, none received as much attention as the potential impact of the agreement on the environment. A number of issues including **labor market** disruptions fueled intense debate over NAFTA, especially in the United States. But no issue received as much attention as the impact of NAFTA on the environment. Debate focused on (1) possible threats posed to previously signed U.S. domestic environmental laws and international environmental agreements; (2) concern that harmonization of environmental standards would result in acceptance of the least common denominator; and (3) fear that U.S. industries would establish pollution havens in Mexico, where labor is cheaper and enforcement of regulations is weaker than in the United States.

In order to allay such concerns, several provisions were added to the NAFTA text. For example, the preamble commits governments to undertake increased trade in "a manner consistent with environmental protection and conservation," and the agreement's dispute-settlement provisions can place the burden on the country challenging an environmental regulation. In addition, prior to NAFTA entering into force on January 1, 1994, the

ratification formal approval

labor market the area or pool of workers from which an employer draws employees

participating governments agreed to the North American Agreement on Environmental Cooperation (NAAEC), which obliges each country to "ensure that its laws and regulations provide for high levels of environmental protection and to strive to continue to improve those laws and regulations." It also ensures access by private persons to fair and equitable administrative and judicial proceedings on matters pertaining to the environment. The NAAEC established the Commission for Environmental Cooperation (CEC), which has three institutional components: a Council, a Secretariat, and a Joint Public Advisory Committee. The Council, assisted by the Secretariat, is charged with monitoring NAFTA's environmental impacts. When they uncover adverse environmental impacts, they publicize them in various ways, including posting notices on their web site. The aim of the council is that, by means of this public shaming, countries will take action to remedy these situations. SEE ALSO ECONOMICS; LAWS AND REGULATIONS, INTERNATIONAL; TREATIES AND CONFERENCES.

Bibliography

Audley, John N. (1997). *Green Politics and Global Trade: NAFTA and the Future of Environmental Politics.* Washington, D.C.: Georgetown University Press.

Magraw, Daniel. (1995). *NAFTA and the Environment: Substance and Process.* Washington, D.C.: American Bar Association.

Internet Resource

NAFTA Secretariat Web site. Available from http//:www.nafta-sec-alena.org.

Michael G. Schechter

NAPLs   *See Nonaqueous Phase Liquids*

National Environmental Policy Act (NEPA)

When signed into law in 1970, the National Environmental Policy Act (NEPA) was a visionary and wide-reaching statute that required U.S. agencies to fully identify, analyze, and weigh the environmental impacts of their decisions. Insofar as most modern land-use planning requires agency approvals, and industrial and commercial activity that results in pollution typically requires agency-issued permits, the NEPA-mandated environmental review process has dramatically affected modern lifestyles, the American economy and, obviously, the environment.

NEPA is essentially procedural, in that it simply requires agencies to proceed through certain steps of environmental review. It does not create substantive legal rights. Although NEPA is a federal statute, many states and even municipalities have enacted their own environmental review statutes. While statutes may differ from state to state (New York's, for instance, provides for substantive obligations and enforcement mechanisms), they incorporate many of NEPA's basic elements. NEPA, in fact, is the model for numerous similar laws in other countries.

The core of NEPA is the Environmental Impact Statement (EIS), a document that has significantly affected modern American business, and even political, practices. NEPA requires that an EIS be prepared and disseminated before any "major federal action significantly affecting the quality of the human environment" may proceed. "Major federal actions" often include the granting of a permit.

mitigation measures taken to reduce adverse impacts

Mitigation measures, if feasible, are also identified, although NEPA does not mandate that any particular form of mitigation be employed. This information is intended not only to aid the agency in its decision making, but also to put the public on notice as to the environmental consequences of various potential government responses.

NEPA has been dramatically effective as an informational device, especially to the extent that the public is included in the process and thereby given the tools necessary to shape political action. However, NEPA has been criticized in many quarters because it lacks significant enforcement capability. Nevertheless, the information-driven process it generates has proven to be an indispensable resource for not only the public, but also agencies presented with proposals that invariably have social, economic, and also environmental importance. SEE ALSO ACTIVISM; CITIZEN SUITS; ENVIRONMENTAL IMPACT STATEMENT; ENVIRONMENTAL MOVEMENT; PUBLIC PARTICIPATION; LAWS AND REGULATIONS, UNITED STATES.

Bibliography

Weinberg, Philip, and Reilly, Kevin A. (1998). *Understanding Environmental Law.* New York: Matthew Bender & Co.

Internet Resource

"Recent NEPA Cases." Available from http://www.naep.org/NEPAWG.

Kevin Anthony Reilly

National Oceanic and Atmospheric Administration (NOAA)

Established in 1970 under the Department of Commerce, the National Oceanographic and Atmospheric Administration (NOAA) guides the United States' use and protection of its air and water resources. With respect to air resources, the agency conducts research and gathers data about the earth's air, and engages in subsequent technical analyses. Specific agency concerns are air pollution, acid rain, and global warming, all greatly influenced by human activity. With respect to water resources, the agency conducts research and gathers data about marine environments, and provides technical analyses of the human activities affecting such environments. Specific agency concerns are ocean dredging and dumping, which can have an adverse effect on marine environments.

For both air and water issues, the agency has adopted policies to address the adverse effects of human activities and provide recommendations to limit or eliminate them. For example, the agency's policy of requiring trawl fishermen to use turtle excluder devices has served to protect sea turtles. Aside from its policy initiatives, the agency enforces a number of laws and treaties (e.g., Coastal Zone Management Act, Endangered Species Act, Magnuson Fishery Conservation and Management Act, Marine Mammal Protection Act, and Ocean Dumping Act), all of which promote the environmental protection of both the atmosphere and the earth's marine environments. SEE ALSO ACID RAIN; AIR POLLUTION; GLOBAL WARMING; OCEAN DUMPING; WATER POLLUTION.

Bibliography

Natural Research Council, Committee on Global Change Research. (1999). *Global Environmental Change: Research Pathways for the Next Decade.* Washington, D.C.: National Academy Press.

Internet Resource

National Oceanographic and Atmospheric Administration Website. Available from http: www.noaa.gov/fisheries.html.

Robert F. Gruenig

National Park Service

Established in 1916 under the National Park Service Organic Act, the National Park Service (NPS) manages over 83.6 millions acres of federal parks, including battlefields, cemeteries, historical sites, lakeshores, memorials, monuments, parkways, preserves, recreation areas, rivers, seashores, and trails. The NPS is supervised by both a director and the assistant secretary for fish and wildlife and parks, and serves as a Department of the Interior bureau funded by Congress. As its primary mission, the NPS is charged with the preservation of park lands for the enjoyment and education of current and future generations, incorporating measures such as pollution control to foster this preservation. The NPS advances its mission by serving as an environmental advocate of park lands, funding state and local governmental bodies in their efforts to develop park areas, and sponsoring educational activities to increase public awareness about parks. In addition, the NPS works in conjunction with the Environmental Protection Agency to enforce laws (e.g., Clean Air Act, Clean Water Act, Endangered Species Act, National Environmental Policy Act, Wild and Scenic Rivers Act, and Wilderness Act) intended to protect and preserve park lands. Comparable agencies in Argentina, Australia, and Germany have adopted some of the same strategies as the NPS. SEE ALSO ENVIRONMENTAL PROTECTION AGENCY.

Bibliography

Freemuth, John C. (1991). *Islands under Siege: National Parks and the Politics of External Threats.* Lawrence: University of Kansas Press.

Internet Resource

National Park Service Web site. Available from http://www.nps.gov.

Robert F. Gruenig

National Pollutant Discharge Elimination System (NPDES)

Under the Clean Water Act, the National Pollutant Discharge Elimination System (NPDES) helps control the discharge of pollutants into water bodies by regulating point sources. By definition, point sources are discrete conveyances such as man-made ditches, tunnels, channels, or pipes that directly discharge into surface waters. By regulating these forms of discharge, the NPDES hopes to protect the public health and assure the treatment of wastewater.

The main pollutants regulated by the NPDES include conventional pollutants (sanitary wastewater, which consists of domestic wastewater—what people flush down their kitchen sink, for example) and wastewater from commercial and industrial facilities, fecal coliform, oil and grease, toxic pollutants (organic and metals), and nonconventional pollutants (such as nitrogen and phosphorous). Industrial, municipal, or agricultural facilities discharging directly into surface water require NPDES permits. A household connected to a municipal or septic sewer system does not.

NPDES permits can be obtained at a state environmental protection office, or at an Environmental Protection Agency (EPA) regional office (in states without EPA approval to issue permits). The permits limit what can be discharged into the environment and provide established monitoring and reporting requirements. The EPA monitors NPDES compliance with on-site inspections and data review. Failure to comply with a permit's provisions can result in civil and criminal action against the violator.

By maintaining vigilant control of pollutants discharged into surface water, the NPDES helps to prevent harmful contamination of the public's water supply. SEE ALSO CLEAN WATER ACT; POINT SOURCE; WASTEWATER TREATMENT.

Internet Resources

Environmental Health & Safety Online. "NPDE—National Pollutant Discharge Elimination System." Available from http://www.ehso.com/npdes.htm.

U.S. Environmental Protection Agency. "National Pollutant Discharge Elimination System (NPDES)." Available from http://www.cfpub.gov/npdes.

Lee Ann Paradise

National Toxics Campaign

The National Toxics Campaign (NTC) was once a leading environmental organization, dedicated to helping local communities seek environmental justice. From its inception in the 1980s until it ended in 1993, this grassroots organization helped many citizen groups develop strategies to hold industry and government accountable for damages to human health and the environment.

The NTC's basic philosophy was that people have the right to a clean and healthy environment regardless of their race or economic standing. Unlike many of the larger environmental organizations that worked on national legislation and international issues, the NTC focused its efforts on empowering local groups and organizations to work together to solve local problems. The NTC succeeded in encouraging leaders of different ethnic groups to organize their own campaigns against polluters that affected residential areas. The NTC's leaders worked with many not-in-my-backyard (NIMBY) groups—groups of citizens trying to keep toxic-waste dumps out of residential areas.

In the beginning, the NTC's founder, John O'Connor, concentrated on local battles against chemical dumps and incinerators. Soon the organization started the only toxics analysis lab in the country that was run by the grassroots movement. As the organization grew, members were able to address more and more toxic-waste problems.

During its lifetime, the NTC was responsible for helping citizen groups bring many polluters to court and for strengthening environmental protection legislation. The NTC was instrumental in the expansion and reauthorization of the Superfund and in the passage of right-to-know legislation the Toxics Release Inventory, which required a limited set of industries to report a release of a limited set of chemicals. The organization played a central role in bringing environmental violations by U.S. military facilities to the attention of the public. Equally important, the NTC developed a network of leaders (including a significant number of organizations of people of color) to develop strategies for environmental justice.

Many people were surprised when the National Toxics Campaign ended in 1993. However, there are several other national organization that have been able to carry on similar grassroots campaigns. Groups such as the Center for Health, Environment and Justice (formerly called the Citizens Clearinghouse for Hazardous Waste), Highlander Center's STP Schools, Greenpeace, and People Against a Chemically Contaminated Environment (PACCE) support grassroots campaigns against toxic-chemical dumping. SEE ALSO ACTIVISM; CITIZEN INVOLVEMENT; CITIZEN SCIENCE; ETHICS; GIBBS, LOIS; NONGOVERNMENTAL ORGANIZATIONS (NGOS); PUBLIC PARTICIPATION.

Bibliography

Cohen, Gary, and O'Connor, John, eds. (1990). *Fighting Toxins: A Manual for Protecting Your Family, Community and Workplace.* Washington, D.C.: Island Press.

Corliss Karasov

Natural Resource Damage Assessment (NRDA)

Natural Resource Damage Assessment is the legal and technical process to pursue restoration for damages to natural resources caused by discharges of oil and releases of hazardous materials into the environment. Federal and state agencies, and Native American tribal governments are designated as NRDA trustees. They act on behalf of the public to restore injured natural resources under a number of laws such as the Comprehensive Environmental Response, Compensation and Liability Act (CERCLA), Oil Pollution Act of 1990 (OPA), and Federal Water Pollution Control Act (FWPCA). Typically, monetary damages are assessed against the polluter. Damages are compensatory, not punitive, and must be used for ecological restoration. The NRDA process is overseen by the Department of the Interior. SEE ALSO ARBITRATION; COMPREHENSIVE ENVIRONMENTAL RESPONSE, COMPENSATION AND LIABILITY ACT (CERCLA); CONSENSUS PROCESS; ENFORCEMENT; LITIGATION; MEDIATION.

Internet Resource

U.S. Department of Interior. "Training Module for NRDA." Available from http://www.doi.gov/oepc.

Susan L. Senecah

Nelson, Gaylord

U.S. SENATOR (D-WISCONSIN) AND FOUNDER OF EARTH DAY (1916–)

One of the first and most effective environmentalists elected to the U.S. Senate, Gaylord Nelson is considered the father of Earth Day and sponsored many of the important environmental laws passed by Congress in the 1960s and 1970s.

As governor of Wisconsin (1958 to 1962), he convinced the legislature to purchase **conservation easements** on private property of high natural and scenic value.

Nelson brought his environmental concerns to Washington when he was elected Wisconsin's Democratic U.S. Senator in 1962. He organized a nationwide conservation tour for President Kennedy in 1963 and, in 1965, introduced the first legislation to ban DDT, a chemical used to kill insects that proved harmful to many other species. In 1969, inspired by the effective student anti-Vietnam War **teach-ins**, Nelson hired Harvard law student Denis Hayes to organize a series of environmental teach-ins on college campuses nationwide. These teach-ins helped inspire a growing awareness of pollution and environmental degradation. This awareness eventually led an estimated twenty million Americans to participate in thousands of events organized across the United States to mark the first Earth Day on April 22, 1970. The mobilized public awareness of environmental problems resulting from Earth Day gave Nelson and other environmentalist members of Congress the support they needed to pass the many environmental acts of the 1970s. Nelson is best known for his work on the Environmental Protection Act (1969), the Clean Air Act (1970), the Safe Drinking Water Act (1974), and the Clean Water Act (1977). Nelson received two awards from the United Nations: the Environmental Leadership Award in 1982 and the Only One Earth Award in 1992. In 1995, he was awarded the Medal of Freedom, the nation's highest civilian honor. SEE ALSO ACTIVISM; EARTH DAY; HAYES, DENIS; LAWS AND REGULATIONS; UNITED STATES; POLITICS.

Bibliography

Mowrey, Mark, and Redmond, Tim. (1993). Not in Our Backyard: The People and Events that Shaped America's Modern Environmental Movement. New York: Morrow.

Internet Resource

Earth Day Network. Available from http://www.earthday.net.

Anne Becher and Joseph Richey

conservation easement legal agreement restricting a landowner's development rights to preserve long-term conservation and environmental values

teach-in educational forum springing from a protest movement (derived from sit-in protests)

NEPA   *See National Environmental Policy Act*

New Left

In the 1930s and through the 1950s, a political movement known later as the "Old Left" emerged in American politics. A liberal group of predominantly northern intellectuals, the Old Left shared a fascination with labor problems and frequently maintained an interest in communism as a solution to

America's economic troubles. The New Left, the successor to the Old Left, emerged in the 1960s and was heavily influenced by the early accomplishments of the civil rights movement. The New Left included many different groups, and was often dominated by middle-class college students disillusioned with life in America. Students for a Democratic Society (SDS) emerged as the best known of these groups, and pressed for a more democratic government, nuclear arms reduction, an end to the war in Vietnam, and better living conditions for the urban poor.

The New Left, in its widespread critique of American society, also included environmental and pollution reform in its agenda. Many New Left activists focused on the dangers of increased industrial production and increased consumption, leading to waste and pollution. One influence of the New Left was the development of the first Earth Day on April 22, 1970. Earth Day was originally planned by New Left activists as a teach-in and sit-in at university campuses, similar to earlier civil rights and antiwar activities to protest environmental degradation. Wisconsin Senator Gaylord Nelson developed and changed the idea for the event, hoping to organize a peaceful mass demonstration without the negative lawless image that public protest had acquired over the course of the turbulent 1960s. Approximately ten million people across the country participated in the original Earth Day, with even local and national polluters professing their support. Overall, though, the concept of Earth Day initiated by the New Left as a protest to industrial production bore little resemblance to the actual event, which was supported by the very polluters the New Left stood against.

New Left protest influenced the overall awareness of environmental issues, and helped lead to legislation, including the Clean Air Act in 1970. By the early 1970s, however, the New Left **counterculture** had become increasingly interested in the use of violence and associated with drug use and "free sex." This use of violence appeared in a small group of New Lefters called the Weathermen, or the Weather Underground, who advocated armed revolution against "American Imperialism," usually in the form of random bomb explosions. Other acts of New Left violence included the "liberation" of areas for public park space.

counterculture a culture with social ideas that stand in opposition to the mainstream culture

By the late 1970s, New Right conservatism had catapulted Ronald Reagan to the presidency, and before long a powerful backlash against many of the accomplishments of the New Left, the civil rights movement, and the 1960s in general took hold throughout the United States. SEE ALSO ACTIVISM; EARTH DAY; ENVIRONMENTAL MOVEMENT; POLITICS; PUBLIC PARTICIPATION; PUBLIC POLICY DECISION MAKING.

Bibliography

Gottlieb, Robert. (1993). *Forcing the Spring: The Transformation of the American Environmental Movement.* Washington, D.C.: Island Press.

O'Neill, William L. (2001). *The New Left: A History.* The American History Series. Wheeling, IL: Harlan Davidson.

Elizabeth D. Blum

NGOs *See Nongovernmental Organizations*

Nitrogen Oxides *See NO$_x$*

NO$_x$ (Nitrogen Oxides)

NO$_x$ is a common term for the more reactive nitrogen oxides and includes nitric oxide (NO) and nitrogen dioxide (NO$_2$), but excludes, for example, nitrous oxide (N$_2$O). NO$_2$ is a reddish brown, highly reactive gas that is formed in the air by the oxidation of NO. Anthropogenic emissions from the high-temperature combustion of coal, oil, gas, and gasoline can oxidize atmospheric nitrogen (N$_2$) to yield the majority of NO found in the environment. Natural sources of NO$_2$ are soil microbial processes. In the soil the **nitrification** and **denitrification** processes pass through compounds that can break down and release NO and N$_2$O into the atmosphere. This is a natural process that is enhanced when nitrogen fertilizers are used to improve crop yields.

Short-term exposure to NO$_2$ at concentrations found in the United States can increase respiratory illness in children. There is evidence that long-term exposure to NO$_2$ may lead to increased susceptibility to respiratory infection. The least reactive nitrogen oxide is N$_2$O, but it can affect both the ozone layer and global warming. Once in the atmosphere, it slowly diffuses into the **stratosphere** where it is destroyed by the shorter-wavelength UV radiation. The NO produced by this photodissociation is critical in establishing the amount of ozone in the stratosphere, so any increase in N$_2$O would decrease the ozone layer. The lifetime of N$_2$O is more than sixty years. Because it can absorb infrared radiation, the excess production of N$_2$O can contribute to global warming.

NO and NO$_2$ react with sunlight and unburned gasoline in a matter of hours to days to produce ozone that is critical in the development of photochemical smog. Atmospheric NO$_x$ also reacts to produce nitric acid. While it is stable in dry air, nitric acid is very soluble and, along with sulphuric acid, significantly contributes to acid rain. Because acid rain and smog involve the reactions of NO$_x$, restrictions on their emissions are a common approach to air quality management even though only NO$_2$ is classed as a criteria pollutant.

In most countries, smog control focuses on reducing ozone concentrations to the air-quality standard by controlling emissions of the precursors, including NO$_x$. In the United States the national ambient air quality standard (NAAQS) for NO$_2$ is 0.053 parts per million (ppm), and from 1988 to 1997, the average NO$_2$ concentration dropped 14 percent to 0.018 ppm. Each state prepares a state implementation plan (SIP) that describes how it will reduce pollutant levels, and presents that plan to the EPA for approval. The EPA, in turn, then supports state plans. The NO$_x$ SIP rule of 1998 is aimed at reducing summertime NO$_x$ emissions in order to cut down on the transport of ozone from power plants in the Midwest to eastern states. Other countries use similar approaches but rely on government and public pressure rather than statutory requirements to meet standards. SEE ALSO ACID RAIN; COAL; ELECTRIC POWER; GLOBAL WARMING; OZONE; PETROLEUM; SMOG; VEHICULAR POLLUTION.

Bibliography

Brimblecombe, Peter. (1996). *Air Composition and Chemistry*, 2nd edition. New York: Cambridge University Press.

Finlayson-Pitts, Barbara J., and Pitts, James N. (2000). *Chemistry of the Upper and Lower Atmosphere*. San Diego, CA: Academic Press.

nitrification the process whereby ammonia, typically in wastewater, is oxidized to nitrite and then to nitrate by bacterial or chemical reactions

denitrification the biological reduction of nitrate or nitrite to nitrogen gas, typically by bacteria in soil

stratosphere the portion of the atmosphere ten to twenty-five miles above the earth's surface

Graedel, Thomas E., and Crutzen, Paul J. (1995). *Atmosphere Climate and Change.* New York: Scientific American Library (distributed by W.H. Freeman).

Turco, Richard. (1997). *Earth under Siege.* Oxford: Oxford University Press.

Donald R. Hastie

NOAA *See National Oceanic and Atmospheric Administration*

Noise Control Act of 1972

In passing the Noise Control Act (NCA) of 1972, Congress hoped to "promote an environment for all Americans free from noise that jeopardizes health or welfare." The Office of Noise Abatement and Control (ONAC) of the Environmental Protection Agency (EPA) was charged with overseeing noise-abatement activities and coordinating its programs with those of other federal agencies that play an important role in noise control. The Noise Control Act was amended by the Quiet Communities Act of 1978 to promote the development of effective state and local noise control programs, to provide funds for noise research, and to produce and disseminate educational materials to the public on the harmful effects of noise and ways to effectively control it.

Throughout the 1970s ONAC issued reports identifying the products that are major sources of noise pollution and providing information on ways to control the noise they generate, for example, the regulation of noise emissions from aircraft. EPA publications included a public education and information manual for noise for schools and pamphlets on sound, sound measurement, and noise as a health problem. The EPA assisted communities in noise surveying, in designing local noise ordinances, and in the training of noise enforcement officers.

Faced with strong industry opposition, ONAC lost its funding in 1981 and the EPA's programs to control noise were halted. The Noise Control Act has never been rescinded, but it has also yet to be refunded. As of 2002, agencies such as the Department of Transportation, Department of Labor, and Federal Railroad Administration have developed their own noise control programs, with each agency setting its own criteria. In addition, states and cities have enacted noise ordinances, with some localities limiting noise more effectively than others.

Across the United States, antinoise groups are pressing local authorities to curb noise intrusions that have grown considerably over the past twenty years and are urging legislators to refund ONAC. Comprehensive federal oversight is needed to address transportation and product noises. With Europe and Japan working toward implementing modern noise-control policies (such as noise labeling of products), American manufacturers may find it difficult to meet foreign noise-emission standards. The European Noise Directive requires member nations to assess environmental noise exposure levels for their populations and to develop action plans to limit noise. SEE ALSO LAWS AND REGULATIONS, UNITED STATES; NOISE POLLUTION.

Bibliography

Bronzaft, Arline L. (1998). "A Voice to End the Government's Silence on Noise." *Hearing Rehabilitation Quarterly* 23, no. 1:6–12, 29.

Dallas, J.E. (1998). "The Quiet Communities Act of 1997: More than Meets the Ear." *Hearing Rehabilitation Quarterly*:16–22.

Shapiro, Sidney A. (1991). The Dormant Noise Control Act and Options to Abate Noise Pollution. Washington, D.C.: Report for the Administrative Conference of the United States.

Internet Resource

Noise Pollution Clearinghouse. Available from http://www.nonoise.org.

Arline L. Bronzaft

Noise Pollution

Noise pollution is the intrusion of unwanted, uncontrollable, and unpredictable sounds, not necessarily loud, into the lives of individuals of reasonable sensitivities. Using the "reasonable person" standard removes the notion that the judgment of sounds as unwanted is subjective. Unwanted sounds or noises can be traced back to Old Testament stories of very loud music and barking dogs as well as to ancient Rome where city residents complained about noisy delivery wagons on their cobblestone streets. The Industrial Revolution, the growth of cities, and the demand for transportation made the world even noisier. With the modern world so dependent on and enchanted with noise-producing and noise-related technology—automobiles, aircraft, helicopters, motorcycles, snowmobiles, jet skis, leaf blowers, amplified music, bass-driven car stereo systems—the **ambient** noise level is rapidly accelerating. This growth in noise has led to research examining the impact of noise on the lives and activities of reasonable people. The result has been a body of evidence that strongly suggests noise is hazardous to good mental and physical health.

To understand noise, one must know something about sound and how loudness is measured. Sound that travels through the air in waves has two major properties: the frequency or speed at which the waves vibrate and the intensity of each vibration. It is the intensity, or how many molecules are packed together with each vibration, that for the most part produces the sense of loudness, although frequency also contributes to the determination of loudness, with higher-pitched sounds sounding louder. Loudness is measured by a decibel scale (expressed as dB), but to reflect human hearing more accurately a modified version of this scale, known as the A scale, has been developed. On the A scale, loudness is measured in dBAs. The scale increases logarithmically so that an increase of 10 dB indicates a doubling of loudness, and an increase of 20 dB represents a sound that is four times louder. Whispers measure 20 dBA, normal conversation 50 to 60 dBA, shouting 85 dBA, and loud music over 120 dBA. Continuous exposure to sounds over 85 dBA may cause permanent hearing loss.

Exposure to very loud sounds that are enjoyable, and not technically noise to the listener, can lead to hearing impairment. Because many people, especially young children and teenagers, are not aware of the dangers of very loud sounds to their hearing, they should be warned that playing computer games with loud audio attachments, setting headsets at consistently high volume, or regularly playing ball in a loud gymnasium may affect their hearing over time. A survey of hearing threshold shifts among youngsters between the ages of six and nineteen found that one out of eight of them suffered a noise-related hearing problem. Children attending loud movies and sporting events, or visiting video arcades may be unwittingly exposing themselves to

ambient surrounding or unconfined; air: usually but not always referring to outdoor air

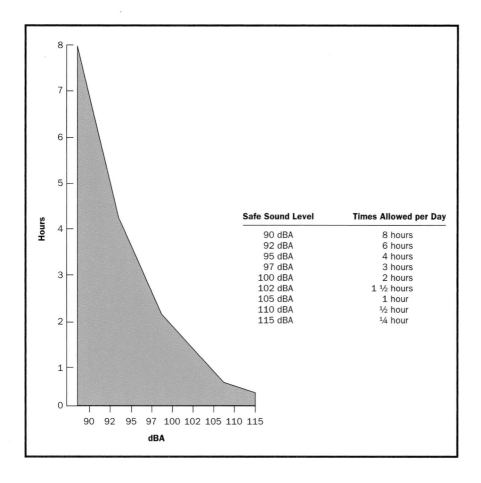

Safe Sound Level	Times Allowed per Day
90 dBA	8 hours
92 dBA	6 hours
95 dBA	4 hours
97 dBA	3 hours
100 dBA	2 hours
102 dBA	1 ½ hours
105 dBA	1 hour
110 dBA	½ hour
115 dBA	¼ hour

dangerously loud sounds. Teenagers are especially vulnerable as they are more likely to equip their cars with high-powered "boom boxes," attend loud dance clubs, and work in noisy fast-food restaurants.

Sounds need not be very loud to be deemed intrusive—for example, the drip of a faucet, an overhead jet, or a neighbor's stereo late at night. Noises are especially bothersome at night when one is trying to sleep, and a good night's sleep is vital to good health. Exposure to bothersome noises over time can be stressful, resulting in adverse health effects, such as hypertension. Although more research is needed to solidify a noise and health link, there is agreement that noise lessens the quality of life. Noises can be especially harmful to children. Scientific research indicates that noisy homes slow down cognitive and language development in young children. In addition, children living and attending schools near noisy highways, railroads, and airports have lower reading scores, and some children living or attending a school near a major airport have experienced elevated blood pressure.

In 1972 the U.S. government passed legislation recognizing the growing danger of noise pollution. It empowered the Office of Noise Abatement and Control (ONAC) within the Environmental Protection Agency (EPA) to curtail noise levels, but by 1982, during the Reagan administration, the office lost most of its funding. States and cities were no longer supported in their efforts to abate noise, and ONAC no longer published materials educating people on the dangers of noise. Recently, the federal government has passed legislation to lessen noise in national parks, for example, banning snowmobiles, but states and cities are on their own in controlling noise, with some cities more

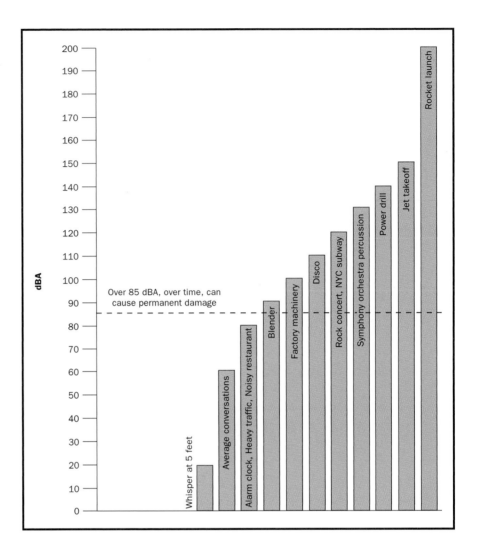

successful than others. Traffic noise, especially aircraft noise, is the major source of annoyance calling for better federal regulation within the United States. In contrast, the European Union is finalizing a noise directive that will require member states to produce noise maps and develop action plans to reduce noise levels.

Noise from snowmobiles, jet skis, and supersonic jets has also intruded on the environment, affecting animals' abilities to communicate, protect their young, and mate. Worldwide, antinoise groups believe their governments are doing too little to lessen the surrounding din, and groups from the United States, Europe, Canada, Australia, Africa, and Asia have joined together to educate both the public and governments about the long-term dangers of noise pollution, urging them to lower the decibel level. A quieter, healthier environment is within our grasp.

Bibliography

Bronzaft, A.L. (1998). "A Voice to End the Government's Silence on Noise." *Hearing Rehabilitation Quarterly* 23:6–12, 29.

Bronzaft, A.L., and Dobrow, S.B. (1988). "Noise and Health: A Warning to Adolescents." *Children's Environments Quarterly* 5:40–45.

Chen, A.C., and Charuk, K. (2001). "Speech Interference Levels at Airport Noise Impacted Schools." *Sound & Vibration* 35(7):26–31.

Evans, G.W., and Lapore, S.J. (1993). "Nonauditory Effects of Noise on Children. A Critical Review." *Children's Environments* 10:31–51.

Federal Interagency Committee on Aviation Noise (FICAN). (2000). *FICAN Position on Research into Effects of Aircraft Noise on Classroom Learning.* Washington, D.C.

Niskar, A.S.; Kiezak, S.M.; Holmes, A.; Esteban, E.; Rubin, C.; and Brody, D.J. (2001). "Estimated Prevalence of Noise Induced Hearing Threshold Shifts among Children 6 to 19 Years of Age. The Third Health and Nutrition Examination Survey. 1988–1994." *Pediatrics* 108:40–43.

Stansfeld, S.; Haines, M.; and Brown, B. (2000). "Noise and Health in the Urban Environment." *Reviews of Environmental Health* 15:43–82.

Internet Resources

League for the Hard of Hearing Web site. Available from http://www.lhh.org/noise.

Noise Pollution Clearinghouse Web site. Available from http://www.nonoise.org.

Arline L. Bronzaft

Nonaqueous Phase Liquids (NAPLs)

Nonaqueous Phase Liquids (NAPLs) are hazardous organic liquids such as dry cleaning fluids, fuel oil, and gasoline that do not dissolve in water. A significant portion of contaminated soil and groundwater sites contain NAPLs, and they are particularly hard to remove from the water supply. NAPLs are always associated with human activity, and cause severe environmental and health hazards.

Dense NAPLs (DNAPLs) such as the chlorinated hydrocarbons used in dry cleaning and industrial degreasing are heavier than water and sink through the water column. They can penetrate deep below the water table and are difficult to find when investigating sites for contamination.

Hydrocarbon fuels and aromatic solvents are described as light NAPLs (LNAPLs), which are less dense than water and float. These include lubricants and gasoline, pollutants often associated with leaking gasoline or oil storage tanks.

It is difficult or impossible to remove all of the NAPLs once they are released into the ground. Although many NAPL removal technologies are currently being tested, there have been few field demonstrations capable of restoring an NAPL-contaminated aquifer to drinking-water quality. NAPL contamination can affect aquifers for tens or hundreds of years.

Internet Resource

Newell, Charles J.; Bowers, Richard L.; and Rifai, Hanadi S. "Impact of Non-Aqueous Phase Liquids on Groundwater Remediation." In Environmental Expert.com Web site. Available from http://www.environmental-center.com/articles/article1079/article1079.htm.

Richard M. Stapleton

Nongovernmental Organizations (NGOs)

Collaborative efforts among the public have played an important role in shaping the political and social values and hence public policy of the United States. Organizing with others who share a similar vision enhances the potential for change. Nongovernmental organizations (NGOs) accomplish just that. Established outside of political parties, NGOs are aimed at advocating the public's

ORGANIZATION	FOUNDING	BUDGET	FOCUS	ACCOMPLISHMENTS
Conservation International	1987	$50,000,000	To preserve and promote awareness about the world's endangered biodiversity.	Working with the Cambodian government to create a one-million-acre protected area. Sponsored scientific research of coral reefs off Indonesia. Helped create the world's largest national rain forest.
Izzak Walton League of America	1922	$3,000,000	To protect and promote sustainable resource use.	Helped create the Land and Water Fund. Were instrumental in the protection of the Boundary Waters Canoe Area Wilderness, Everglades National Park, and Isle Royale National Park.
National Audubon Society	1905	$44,000,000	To restore and protect the natural habitat of birds and other wildlife for the benefit of human interest and biodiversity.	Involved the public in bird counts across the United States to track populations. Has opened nature centers to promote understanding of birds.
National Wildlife Federation	1936	$96,000,000	The largest member-supported conservation group working to protect wildlife and ecosystems.	Function in forty-six states to promote the protection of species and their environments. Worked in the western United States to prevent urban sprawl and sustainable forestry.
Natural Resources Defense Council	1970	$30,632,992	Using science and law to protect the planet's wildlife and wild places.	Worked with the EPA to restrict pesticide use, prevented the development of a large airport near the Florida Everglades, and have helped design a plan to restore Yosemite.
Nature Conservancy	1951	$245,000,000	To protect aquatic and terrestrial habitats for the survival of biodiversity.	Own over a thousand preserves and have protected more than fourteen million acres of land in the United States.
Wilderness Society	1935	$14,700,000	Protect the remaining wilderness in the United States by keeping roads, loggers, and oil drilling efforts out of wilderness areas.	Helped block oil exploration near Arches National Park, created the Wilderness Act, which was passed in 1964, and the Conservation Act which was passed in 1980.
Wildlife Conservation Society	1895	$95,000,000	Support international survival strategies as well as habitat conservation projects.	Formed Jackson Hole Wildlife Park in 1956, led the national campaign to reintroduce bison to the Kansas grasslands, and created the Bronx Zoo.
World Wildlife Fund	1961	$60,000,000	Protect and preserve endangered species.	Launched Wildlands and Human Needs projects to address the needs of people living in fragile ecosystems.
Sierra Club	1892	$43,000,000	To educate and enlist people to protect the environment through lawful means, and address key issues including commercial logging, urban sprawl, and water quality.	Assisted in preserving the North Grove Calaveras Big Trees, fought to return Yosemite to federal management, and worked to create the National Park Service.
Environmental Defense Fund	1967	$39,000,000	Create solutions to environmental problems including policies to reduce fossil fuels.	Won a ban on DDT use, prevented the development of a resort on former state park land that would endanger native species.
Greenpeace USA	1971	$19,266,530	Nonviolent direct action to expose environmental threats.	Drew attention to ocean incineration of toxic waste, resulting in a ban of the practice; also, won an end to sperm whale hunting, halted the testing of nuclear arms off Florida.
Friends of the Earth	1969	$3,000,000	To protect Earth from environmental disaster through toxic waste cleanup and groundwater protection.	Conducted lab tests proving that genetically altered food not approved for human consumption was being sold, won a federal court case that prevented Army Corps of Engineers from illegally issuing permits for developers to fill in wetlands.

concerns and pressuring governments to do a better job. These organizations may range from a handful of local citizens enacting recycling in their community to a million-member-strong organization with a budget of $20 million.

Agents of Information and Action

NGOs are often nonprofit groups that employ a variety of tactics for achieving awareness among the public and the government. The very nature and structure of NGOs has been advantageous in dealing with pollution issues for several reasons. First, membership within NGOs consists of people with a strong personal commitment to their cause. Second, the focused efforts of NGOs allow their leaders to become specialized. Third, the loose structure of NGOs enables them to respond with greater speed and flexibility than the government.

Throughout the forty years of the modern environmental movement, NGOs have been crucial in bringing visibility to pollution problems affecting both the local and international communities. According to Peter Willets, "Information is the currency of politics, and the ability to move accurate up-to-date information around the globe has been a key factor in the growing strength of environmental groups" (Willets, p. 114). The communication of information has been accelerated through the use of the Internet. In addition, NGOs also rely heavily on publications, media coverage, and conferences to collaborate with one another and to educate the public.

Although reformers of the Settlement House era of the late 1800s and early 1900s organized efforts for change within city neighborhoods, the formation of prominent mainstream organizations such as the Wilderness Society and Sierra Club are widely considered to be the first major environmental NGOs. Rooted in early-twentieth-century debates over the exploitation of land, these early NGOs lobbied the government by talking with local officials and publishing works on the importance of wilderness. One of the most notable efforts to drum up public support was a series of full-page advertisements taken out by the Sierra Club from 1965 to 1968 in the *New York Times* vilifying the prospects of building hydroelectric dams in and flooding the Grand Canyon.

Friends of the Earth and Greenpeace are two NGOs with international status that have fought to keep the public informed about pesticides and toxics pollution through direct action techniques. Their practices of physically obstructing or protesting industry has made them popular in the media since the groups' inception in the 1970s. In one particular instance, Friends of the Earth amassed a collection of Schweppes bottles and subsequently dumped them on the company's front steps. Their efforts to send a clear message to the beverage company about waste pollution attracted media coverage and brought about a rise in membership. Similarly, Greenpeace employed confrontational tactics by sailing the vessel *Phyllis McCormack* towards a French nuclear testing site to halt testing. In another campaign, Greenpeace members put themselves in small boats between whalers and whales.

The Rise of International Networks

By the mid-1980s there were thousands of NGOs. Their success across the globe was encouraging to environmentalists and it was encouraging to a public—both national and international—that had begun to see the importance of NGOs in environmental issues. Danish NGOs won a complete ban on throwaway beverage packaging while Australian NGOs won concessions on mining in their national parks. The use of phosphates in detergents was banned in Switzerland with the help of NGOs. But as pollution became a major factor in the global debate of acid rain, global warming, and ozone depletion, NGOs saw a great need to collaborate internationally.

The discovery of a hole in the ozone layer above Antarctica provoked furious action among American NGOs. Apparent disinterest shown toward the issue by European NGOs prompted several U.S. NGOs to send representatives to Europe to discuss the consequences of chlorofluorocarbons (CFCs) on the atmosphere. As a result of their meeting, the U.K. branch of Friends of the Earth drew up a campaign to publish its own guide to pollutants. In 1986 *Aerosol Connection* was a resounding success in communicating to the public how to support non-CFC products. Thousands of people were eager to get their hands on a copy. Raising public awareness weakened the position of the chemical companies in the United Kingdom, because they had controlled most public information about CFCs. The scientific information that NGOs supplied for the debate over CFCs helped speed negotiations on the Montréal Protocol, which called for a ban on CFC use. The experience clearly illustrated the power of NGOs to successfully lobby internationally.

NGOs experienced greater inclusion in the political arena throughout the 1990s. NGO pressure on World Bank policy set a precedent for collaboration by the World Bank with NGOs in the international realm. By challenging the World Bank to support environmentally viable water projects, NGOs exposed an array of existing problems to the media, to the U.S. government, and to congressional staff. Just a week after collaboration with the World Bank, NGOs from across the world gathered in Rio de Janeiro for the 1992 Earth Summit. Twenty years earlier, the 1972 UN Conference on the Human Environment in Stockholm was a major turning point for NGOs. Because only government officials were invited to the conference, NGOs gathered around the conference site to debate their own positions. To help clarify confusion surrounding conference issues, NGOs published a newspaper which they delivered to the media, embassy, and hotels where attendees were staying.

The 1992 Earth Summit

Having learned from the 1972 UN Conference, the planners of the 1992 Earth Summit coordinated a parallel conference for NGOs. Known as the Global Forum, this satellite conference enabled NGOs across the world to network, share research, and evaluate their collective role in protecting the environment. Together, NGOs drafted an extensive collection of treaties including the Earth Charter, a document meant to parallel the Summit's Rio Declaration, an agreement defining the rights and responsibilities of countries. Five years after the 1992 Earth Summit, five hundred NGOs met in New York to judge their progress and push for a redrafting of the Earth Charter. By 2000 a new draft was finalized to express the renewed vision NGOs hoped to fulfill.

By the mid-1990s NGOs had secured an important position in the environmental movement's crusade against pollution. Organizations large and small, striving to eradicate pollution, raised the public's level of awareness. Because pollution is at the same time a local and international problem, NGOs have been essential on all levels. Their dedication to issues and their multifaceted approaches to disseminating information makes them an important asset to the cause they represent and to the legislation they are hoping to influence. International NGO networks only serve to improve the environmental movement as receptivity to NGO work continues to expand worldwide.

Bibliography

Fox, Jonathan, and Brown, L. David. (1998). *The Struggle for Accountability: The World Bank, NGOs, and Grassroots Movements.* Cambridge, MA: MIT Press.

Gottlieb, Robert. (1993). *Forcing the Spring.* Washington, D.C.: Island Press.

Hays, Samuel P. (2000). *A History of Environmental Politics Since 1945.* Pittsburgh: University of Pittsburgh Press.

Hedblad, Alan, ed. (2003). *Encyclopedia of Associations*, 39th edition. Detroit: Gale Group.

Markham, Adam. (1994). *A Brief History of Pollution*. New York: St. Martin's Press.

Willets, Peter. (1982). *Pressure Groups in the Global System*. London: St. Martin's Press.

Internet Resources

CIESIN. "A Summary of the Major Documents Signed at the Earth Summit and Global Forum." Available from http://www.ciesin.org.

Citizens Campaign for the Environment. "Coalitions and Affiliations." Available from http://www.citizenscampaign.org.

Global Policy. "NGOs." Available from http://www.globalpolicy.org/ngos.

Environmental Defense Fund. "Notable Victories." Available from http://www.environmentaldefense.org.

Natural Resources Defense Council. "Environmental Legislation." Available from http://www.nrdc.org.

Transformational Movement. "Earth Charter." Available from http://www.transformworld.org.

United Nations. "UN Conference on Environment and Development (1992)." Available from http://www.un.org/geninfo/bp/enviro.html.

Worldwatch Institute. "WTO Confrontation Shows Growing Power of Activist Groups." Available from http://www.worldwatch.org.

Christine M. Whitney

Nonpoint Source Pollution

Nonpoint source pollution occurs when rainfall or snowmelt runs over land or through the ground, picks up pollutants, and deposits them into rivers, lakes, wetlands, and coastal waters or introduces them into groundwater. Some of the primary activities that generate nonpoint source pollution include farming and grazing activities, timber harvesting, new development, construction, and recreational boating. Manure, pesticides, fertilizers, dirt, oil, and gas produced by these activities are examples of nonpoint source pollutants. Even individual households contribute to nonpoint source pollution through improper chemical and pesticide use, landscaping, and other household practices.

After Congress passed the Clean Water Act in 1972, the water-quality community within the United States placed a primary emphasis on addressing and controlling point source pollution (pollution coming from discrete conveyances or locations, such as industrial and municipal waste discharge pipes). Not only were these sources the primary contributors to the degradation of U.S. waters at the time, but the extent and significance of nonpoint source pollution were also poorly understood and overshadowed by efforts to control pollution from point sources.

At the beginning of the twenty-first century, nonpoint source pollution stands as the primary cause of water-quality problems within the United States. According to the *National Water Quality Inventory* (published by the U.S. Environmental Protection Agency), it is the main reason that approximately 40 percent of surveyed rivers, lakes, and estuaries are not clean enough to meet basic uses such as fishing or swimming.

Leading Contributors to Nonpoint Source Pollution

States and other jurisdictions reported in the *National Water Quality Inventory* that agriculture and urban runoff are among the leading contributors to

A cow drinking in a dried-up riverbed. (U.S. EPA)

hydromodification any process that alters the hydrologic characteristics of a body of water

deteriorating water quality nationwide. The most common nonpoint source pollutants causing water-quality problems include nutrients (nitrogen and phosphorus), siltation (soil particles), metals, and pathogens (bacteria and viruses).

Agriculture is identified as the leading source of degradation of polluted rivers, streams, and lakes surveyed by states, territories, and tribes in the *National Water Quality Inventory*. Agricultural activities that result in non-point source pollution include concentrated animal feeding operations (CAFOs), grazing, plowing, pesticide spraying, irrigation, fertilizing, planting, and harvesting. A major nonpoint source pollutant from these activities is an excess of nutrients, which can occur through applications of crop fertilizers and manure from animal production facilities. Excessive nutrients may overstimulate the growth of aquatic weeds and algae, depleting the oxygen available for a healthy aquatic community.

Hydromodification that alters the flow of water is the second leading source of damage to U.S. rivers, streams, and lakes, according to the same *National Water Quality Inventory* report. Examples of hydromodification projects include channelization, dredging, and construction of dams. Excess sediment due to erosion caused by projects such as building dams can severely alter aquatic communities by clogging fish gills or suffocating eggs. Sediment may also carry other pollutants into water bodies (e.g., PCBs or mercury) which can accumulate in aquatic species, leading to fish consumption advisories.

Habitat modification is identified as the third-largest source of water pollution in surveyed rivers and streams in the *National Water Quality Inventory*.

Habitat modification occurs when the vegetation along stream banks is removed, diminishing buffers that help filter runoff and provide shade for the adjacent water body. These modifications can result in an increase in the water temperature (because of less shade) and an increase in quantity and velocity of runoff, making the river or stream less suitable for the organisms inhabiting it.

Runoff from urban areas is the fourth-largest source of water pollution in rivers and streams and the third-largest source of water pollution in lakes, according to the *National Water Quality Inventory*. Increased urban development brings additional roads, bridges, buildings, and parking lots, which can result in large amounts of runoff that quickly and easily drain into rivers and lakes. In contrast, the porous and varied terrain of natural landscapes like forests, wetlands, and grasslands traps rainwater and snowmelt and allows it to filter slowly into the ground. Urban runoff transports a variety of pollutants, including sediment from new development; oil, grease, and toxic chemicals from vehicles; and nutrients and pesticides from turf management and gardening. It can also carry pathogenic bacteria and viruses released from failing septic systems and inadequately treated sewage, which can result in closed beaches and shellfish beds, contaminated drinking water sources, and even severe human illness.

Programs for Nonpoint Source Control

The United States has made significant progress in addressing nonpoint source pollution since Congress amended the Clean Water Act in 1987 to establish a national program for controlling nonpoint source pollution. Under section 319 of the Clean Water Act, states adopted management programs to control nonpoint source pollution, and since 1990 the EPA has awarded grants to states to assist them in implementing those management programs. Other federal agencies also provide technical and financial support through grants and loans to states, local communities, and farmers and other landowners, to implement nonpoint source pollution controls. In addition, many state and local entities are dedicating increasing amounts of funding to control nonpoint source pollution.

GOLF AND THE ENVIRONMENT

The well-manicured deep green turfs of America's golf courses are often situated in pristine, water-rich environments. However, often the process of maintaining these golf courses involves heavy fertilization, pesticide treatments, and perpetual mowing and watering, which can lead to polluted groundwater and drinking water and damage to aquatic habitat and wildlife. Proper management of golf courses can reduce or prevent many of these problems, and a large coalition of public and private partners (including the EPA and a conglomeration of state and national golf associations) has developed and adopted voluntary guidelines that apply to the siting, development, and operation of golf courses. Through better site analysis and selection, better management and timing of pesticides applications, the use of slow-release fertilizers, employment of buffers to filter turf runoff, and other such practices, the golf industry is making considerable headway in managing the effects of nonpoint source pollution.

—Source: "Environmental Principles for Golf Courses in the United States," Second Conference on Golf and the Environment, Pinehurst, NC, Center for Resource Management. 1996.

Seventeen municipalities make up the Buzzards Bay watershed in the southeastern region of Massachusetts. Nonpoint source pollution from failing septic systems, farm animal wastes, and stormwater runoff were contributing to a decline in water quality in the bay, forcing the closing of many shellfish beds. Watershed partners, including various federal (e.g., U.S. Department of Agriculture), state (e.g., Massachusetts Department of Environmental Protection), local partners (e.g., Town of Marion), and area residents cooperated to support the construction of a wetland system to help filter the stormwater discharge into the bay. The success of this effort depended on a coordinated approach including all partners on a watershed basis.

—Source: *Watershed Success Stories* (2000). Interagency Watershed Coordinating Committee. Washington, D.C.

watershed the land area that drains into a stream; the watershed for a major river may encompass a number of smaller watersheds

State nonpoint source programs provide for the control of nonpoint source pollution primarily through best management practices (BMPs), which are on-the-ground technical controls used to prevent or reduce nonpoint source pollution. Common practices used to control nutrients from agriculture include altering fertilizer and pesticide application methods and storing and properly managing manure from confined animal facilities. Developing a buffer of vegetation between the land and the stream bank can help filter all types of nonpoint source pollutants from entering a receiving water body, including sediment transported by overland flow. Stream-bank protection and channel stabilization practices are also very effective in preventing sediment deposition in the water by limiting the bank erosion processes and streambed degradation. Urban runoff can be controlled by establishing trenches, basins, and detention ponds at construction sites to hold, settle, and retain suspended solids and associated pollutants. Basic pollution-prevention measures introduced around the home can also prevent nonpoint source pollutants from entering storm water. Practices include the proper storage, use, and disposal of household hazardous chemicals; proper operation and maintenance of onsite disposal systems; and even proper disposal of pet waste so that it does not wash into storm drains.

Watershed Approach to Managing Nonpoint Source Pollution

Nonpoint source pollution derives from many different sources over large geographic areas so regulating and controlling it are challenging. The **watershed** approach to managing nonpoint source pollution, however, is proving to be an effective technique. Everyone lives in a watershed, or an area of land in which all water drains. According to the U.S. Geological Survey, the nation can be divided into approximately 2,149 medium-sized watersheds, averaging about 1,700 square miles in each area. The watershed approach relies on coordinating all relevant federal, state, and local government agencies, and the stakeholders who live in a particular watershed, to help solve priority problems in that watershed. Historically, many water-quality problems were addressed piecemeal in individual water bodies by individual entities, usually limited by political, social, and economic boundaries. The watershed approach, however, relies on the coordination of all entities and stakeholders to help solve the watershed's most serious environmental problems, which in many instances are caused by nonpoint source pollution.

International Implications

Managing nonpoint source pollution is an international challenge. Like the United States, many developed countries initially directed resources toward controlling point source pollution. However, significant nonpoint source problems remain, especially resulting from an excess of nutrients and sediment in water bodies. The United Nations Environment Programme has identified increased nitrogen loadings, resulting mainly from agricultural runoff and wastewater, as one of the most serious water-quality issues affecting all countries. Sedimentation is a significant concern for other countries, frequently resulting from deforestation or clear cutting for fuelwood, or agricultural practices. One of the largest threats in developing countries relates to problems with sewage control, either through poor maintenance of sewage collection systems or a lack of it, leading to severe waterborne diseases.

The increasing world population promises even more challenges for managing nonpoint source pollution. Some international communities are embracing integrated solutions (like the watershed approach) to reduce it. Agenda 21 adopted at the United Nations Conference on Environment and Development in 1992 is but one example. SEE ALSO AGRICULTURE; CRYPTOSPORIDIOSIS; EDUCATION; HYPOXIA; PHOSPHATES; SEDIMENTATION; WATER POLLUTION; WATER POLLUTION: FRESHWATER; WATER TREATMENT.

Bibliography

United Nations Environment Programme. (1999). *Global Environment Outlook.* London: United Nations Environmental Program.

U.S. Environmental Protection Agency. (2000). *National Water Quality Inventory: 2000 Report.* Washington, D.C.: U.S. Environmental Protection Agency.

Internet Resources

Center for Watershed Protection Web site. Available from http://www.cwp.org.

Nonpoint Education for Municipal Officials (NEMO) Web site. Available from http://nemo.uconn.edu.

U.S. Department of Agriculture, Natural Resources Conservation Service Web site. Available from http://www.nrcs.usda.gov.

U.S. Environmental Protection Agency, Office of Wetlands, Oceans, and Watersheds Web site. Available from http://www.epa.gov/owow/nps.

Stacie Craddock

North American Free Trade Agreement   *See NAFTA*

NPDES   *See National Pollutant Discharge Elimination System*

NRC   *See Nuclear Regulatory Commission*

Nuclear Regulatory Commission (NRC)

The Nuclear Regulatory Commission's (NRC) primary mission is to protect public health and safety and to protect the environment from the effects of radiation from nuclear reactors, materials, and waste facilities. The NRC is empowered by the Atomic Energy Act of 1954 and its amendments to regulate source material (primarily uranium ore and processed uranium), special nuclear material, including material enriched in plutonium or the isotope uranium-235 above certain levels, and by-product material, and to regulate the uses of these materials. Primarily, this means it regulates nuclear power plants and civilian research reactors, the materials used to make fuel for these plants, the wastes produced, and other materials and uses of radioactive material that are derived from these sources.

The NRC is headed by a five-member commission that is appointed by the president (subject to Senate confirmation). The NRC does not regulate naturally occurring radioactive materials (NORM) that do not fall into one of these categories. Such naturally occurring materials include radioactive waste from oil and gas production. The NRC also does not regulate radiation-producing machines, such as x-ray machines, or radioactive materials produced in accelerators. SEE ALSO ANTINUCLEAR MOVEMENT; LAWS AND REGULATIONS; UNITED STATES.

Matthew Arno

PERVIOUS (PERMEABLE) CONCRETE

An increasing number of parking lots in California are being paved with pervious concrete to reduce runoff and allow water to drain through to underlying soil or groundwater. The concrete is made from Portland cement, gravel, and water and consists of up to one-quarter empty spaces that allow rainfall to penetrate at a rate of about three to five gallons per square foot, per minute. Beneficial soil microorganisms break down pollutants, such as oil and gasoline, trapped in the voids. In 2002 a Santa Barbara couple made the news as possibly the first homeowners in California to pave their driveway with pervious concrete.

Occupational Safety and Health Administration (OSHA)

The Occupational Safety and Health Administration (OSHA), the federal agency charged with protecting workers' health and safety, was created by Congress in 1971 to administer the Occupational Health and Safety Act of 1970. With few exceptions, including some state plans and specific industries, OSHA oversees all U.S. workers and their employers. OSHA's duty is to ensure that workplaces are free from hazards that are likely to cause serious harm or death to workers.

As part of that duty, OSHA establishes standards for workplace activities and exposures to hazardous materials. Working in conjunction with the National Institute for Occupational Safety and Health (NIOSH), OSHA uses scientific data to determine acceptable levels of risk for regulated materials and creates corresponding material safety data sheets (MSDS) for each. Levels are set forth in the Federal Code of Regulations (CFR), and employers must prevent workers from being exposed above the CFR's permissible exposure limits. Although employers must oversee their own programs, OSHA requires that records be kept for all workplace exposures, illnesses, injuries, and fatalities. The agency may only regulate the employer-employee relationship, but when individuals bring lawsuits against their employers, courts will generally find the employer negligent if there has been a failure to comply with OSHA standards. SEE ALSO HAZARDOUS WASTE.

Bibliography

Michaud, Patrick A. (1995). *Accident Prevention and OSHA Compliance*. Lewis Publishers.

2000 OSHA Handbook. (1999). PA: Chamber Educational Foundation.

Internet Resources

National Institute for Occupational Safety and Health Web site. Available from http://www.cdc.gov/niosh.

Occupational Health and Safety Administration Web site. Available from http://www.osha.gov.

Mary Elliott Rollé

Ocean Dumping

Ocean disposal of society's waste got its start indirectly long before the Agricultural Age when nearby streams, lakes, and estuaries were useful as waste repositories. As civilization moved to the coastal zone and navigation began in earnest, the oceans were viewed as even a larger waste repository. Early civilizations were located adjacent to bodies of water for sources of food, irrigation, drinking water, transportation, and a place to dispose of unnecessary items. Historically, the disposal of wastes into water by humans was universally practiced. It was a cheap and convenient way to rid society of food wastes (e.g., cleaned carcasses, shells, etc.), trash, mining wastes, and human wastes (or sewage). The advent of the Industrial Age brought with it the new problem of chemical wastes and by-products: These were also commonly disposed of in the water.

A trash-strewn beach.
(©Claude Charlier/Corbis.
Reproduced by permission.)

Early dumping started in rivers, lakes, and estuaries, whereas ocean dumping was simply not used because of the distance and difficulty in transporting waste materials. The wastes from ships, however, were simply dumped directly into the ocean. As civilization developed at river deltas and in **estuaries** adjacent to the ocean, and these areas soon began to display the effects of dumping, disposal in the ocean became a popular alternative. Over the past 150 years, all types of wastes have been ocean dumped. These include sewage (treated and untreated), industrial waste, military wastes (munitions and chemicals), entire ships, trash, garbage, dredged material, construction debris, and radioactive wastes (both high- and low-level). It is important to note that significant amount of wastes enter the ocean through river, atmospheric, and pipeline discharge; construction; offshore mining; oil and gas exploration; and shipboard waste disposal. Unfortunately, the ocean has become the ultimate dumping ground for civilization.

It has been recognized over the past fifty years that the earth's oceans are under serious threat from these wastes and their "witches' brew" of chemicals and nonbiodegradable components. Society has also come to understand that its oceans are under serious threat from overfishing, mineral exploration, and coastal construction activities. The detrimental effects of ocean dumping are physically visible at trashed beaches, where dead fish and mammals entangled in plastic products may sometimes be observed. They are additionally reflected in the significant toxic chemical concentrations in fish and other sea life. The accumulations of some toxins, especially mercury, in the bodies of sea life have resulted in some harvestable seafood unfit for human consumption.

estuary region of interaction between rivers and near-shore ocean waters, where tidal action and river flow mix fresh and salt water (i.e., bays, mouths of rivers, salt marshes, and lagoons). These ecosystems shelter and feed marine life, birds, and wildlife

Seriously affected areas include commercial and recreational fishing, beaches, resorts, human health, and other pleasurable uses of the sea. During the 1960s numerous groups (global, regional, governmental, and environmental) began to report on the detrimental impact of waste disposal on the ocean. Prior to this time, few regulatory (or legal) actions occurred to control or prevent these dumping activities.

Early U.S. Legislation

Late in the nineteenth century, the U.S. Congress enacted Section 10 of the River and Harbor Act of 1890, prohibiting any obstruction to the navigation of U.S. waters. The authority to implement the act through a regulatory permit program was given to the secretary of the army acting through the chief of the U.S. Army Corps of Engineers. In the late 1960s the corps enlarged the scope of its review of permit applications to include fish and wildlife, conservation, pollution, esthetics, ecology, and matters of general public interest. In addition, the National Environmental Policy Act of 1969 (NEPA) required the review of policy issues pertinent to the public interest and an environmental impact statement on activities that might significantly affect the quality of the environment.

In 1972 the U.S. Congress passed the Marine Protection, Research and Sanctuaries Act (Ocean Dumping Act or ODA) and the Federal Water Pollution Control Act amendments (Clean Water Act or CWA) that set a global standard for managing environmental restoration and protection, for maintaining the environment within acceptable standards, for prohibiting the disposal of waste materials into the ocean, and for regulating the discharge of wastes through pipelines into the ocean.

With the enactment of these laws, the corps's regulatory program became quite complex. The goal of the CWA is to restore and maintain the chemical, physical, and biological integrity of the nation's waters, with the corps responsible for regulating the discharge of dredged material into inland and coastal waters. The ODA regards oceans in a somewhat similar manner, requiring the review of all proposed operations involving the transportation or disposal of waste materials and their potential environmental impact. The corps also manages the ocean dumping permit program. Like the CWA, the ODA is concerned with the unregulated dumping of materials into ocean waters that endanger human health and welfare, the marine environment, and the earth's ecological systems, and that may have dire economic consequences. The corps implements these programs in full partnership with the U.S. Environmental Protection Agency and is subject to their oversight.

International recognition of the need to regulate ocean disposal from land-based sources on a global basis was the result of the UN Conference on the Human Environment in June 1972 and the Inter-Governmental Conference on the Convention of the Dumping of Wastes at Sea in November 1972. These conferences resulted in a treaty entitled Convention on the Prevention of Marine Pollution by Dumping of Wastes and Other Matter—London Convention 1972 (LC-72). The LC-72 came into effect in 1975 and currently has approximately eighty member nations. Another treaty addressing the issue of wastes disposed of from vessels, the International Convention for the Prevention of Pollution from Ships, 1973 (MARPOL), was adopted

MAJOR GLOBAL AGREEMENTS AND DOMESTIC LEGISLATION GOVERNING PROTECTION OF THE MARINE ENVIRONMENT

Key Global Agreements to Protect the Marine Environment from Dumping

Title	Description
1982 UN Convention on The Law of the Sea (UNCLOS 1982) (entry into force: November 1994);	Provides a framework for the determination of the rights and obligations of states relating to the oceans. Part XII contains provisions with regard to protection and preservation of the marine environment.
International Convention for the Prevention of Pollution from Ships, 1973, as modified by the protocol of 1978 relating thereto (MARPOL 73/78)	Provides measures for ships and national administrations to prevent pollution by oil (Annex I), noxious liquid substances in bulk (Annex II), harmful substances in packaged form (Annex III), sewage (Annex IV), garbage (Annex V), and air pollution from ships (Annex VI).
Convention on the Prevention of Marine Pollution by Dumping of Wastes and other Matter (London Convention 1972) (entry into force: August 1975)	Provides measures to limit the use of the oceans as disposal area for wastes generated on land.

Key Domestic Legislation to Protect the Marine and Coastal Environment

Title	Description
Federal Water Pollution Control Act Amendments of 1972 (CWA)	To restore and maintain the chemical, physical, and biological integrity of the nation's waters.
Marine, Protection Research, and Santuaries Act of 1972 (ODA)	To regulate the dumping of all types of materials into ocean waters and to prevent or strictly limit the dumping into ocean waters of any material which would adversely affect human health, welfare or amenities, or the marine environment, ecological systems, or economic potentialities.
National Environmental Policy Act of 1969 (NEPA)	To declare a national policy that will encourage productive and enjoyable harmony between people and the environment; to promote efforts that will prevent or eliminate damage to the environment and biosphere and stimulate human health and welfare.

in 1973. Countries signing MARPOL agree to enforce bans on dumping oil and noxious liquids into the ocean from ships, but the disposal of hazardous substances, sewage, and plastics remains optional. There are dozens of other international agreements dealing with ocean pollution, but the LC-72 and MARPOL are the most significant as far as dumping is concerned. The United States is an active member of both of these treaties.

The LC-72 and domestic ODA are similar in structure and requirements, with the U.S. regulation being more stringent. The dumping of industrial wastes, radioactive wastes, munitions (chemical or biological), sewage, and incineration at sea are directly prohibited. Moreover, the ocean disposal of other waste materials containing greater than trace amounts of certain chemicals (i.e., mercury, cadmium, petroleum hydrocarbons, chlorinated chemicals, and nondegradable plastics) is strictly prohibited. Allowed under strictly regulated conditions are the ocean disposal of dredged material (harbor sediments), geologic material, and some fish waste; burial at sea; and ship disposal.

The corps and the U.S. Environmental Protection Agency (EPA) implement the LC-72 and ODA in the United States. The Corps issues its permits after careful assessment using environmental criteria developed by the EPA. About 350 million tons of sediments are dredged annually in U.S. waters for the purpose of navigation for trade and national defense; approximately 20 percent of this total is disposed of in formally designated sites in ocean

bioassay a test to determine the relative strength of a substance by comparing its effect on a test organism with that of a standard preparation

waters. A small portion of sediments from major harbor areas (about seven to ten percent of the national total) is sufficiently contaminated that ocean placement is not allowed, and the sediment must be contained at regulated land sites. Proposed ocean disposal is assessed through the use of an effects-based approach, which evaluates the dredged material as a complex substance that may contain a wide variety of contaminants. The assessment will identify those sediments that may be detrimental to ocean biota and human health. The effects-based approach uses **bioassay** test organisms to integrate the potential effects of all the contaminants present in a combined impact assessment. This is done through the use of bioassays for acute toxicity and an estimate of contaminants' bioaccumulation potential. An assessment is also made on the potential of sediment contamination to impact water quality. A decision is then based on the suitability of a material for unrestricted or restricted ocean disposal, or not. For example, a dredged sediment from a contaminated portion of a harbor can be prohibited from ocean disposal and must be placed in a land containment facility.

In highly industrialized harbors such as those in New York or New Jersey, dredging and the disposal of dredged material are often controversial. Ocean placement is not allowed except in the case of the cleanest sediments and adding to the controversy, land disposal locations are very limited and very expensive. In contrast, world trade and shipping, which depend on navigation dredging for deep channels, are a vital component of regional and national economies. The long-term solution to contaminated sediments will depend on waste control from land sources and the cleanup of highly contaminated sediments that continue to impact the navigation channel.

The ocean placement of suitable dredged material or sediments at carefully selected ocean sites may be environmentally safe in relation to other alternatives. It might even be beneficial to the ocean through proper management. Eroding beaches, for instance, often receive clean dredged sand as a routine part of environmental improvement programs. Dredged material comprises 95 percent or more of all ocean disposal on a global basis. As navigable waterways and their role in world trade and defense continue to be important components of the economic growth and stability of coastal nations, the environmentally sound disposal of suitable dredged materials into the ocean will remain a necessary alternative. Moreover, the beneficial uses of these dredged sediments (when they are not contaminated with pollutants) for beach replenishment, wetlands, construction, aquatic and upland habitat improvement, and as construction materials will remain the highest priority in sediment and ocean disposal management. SEE ALSO BIOACCUMULATION; CLEAN WATER ACT; DREDGING; OCEAN DUMPING BAN ACT; RIVERS AND HARBORS APPROPRIATIONS ACT; WATER POLLUTION; WATER POLLUTION: MARINE.

Bibliography

Committee on Public Works, U.S. House of Representatives. (1973). *Laws of the United States Relating to Water Pollution Control and Environmental Quality, 93-1.* Washington, D.C.: U.S. Government Printing Office.

Engler, R.M. (1980). "Prediction of Pollution Potential through Geochemical and Biological Procedures: Development of Guidelines and Criteria for the Discharge of Dredged and Fill Material." In *Contaminants and Sediments*, edited by R.A. Baker. Ann Arbor, MI: Ann Arbor Science Publications.

Engler, R.M. (1990). "Managing Dredged Materials." *Oceanus* 33(2):63–69.

Engler, R.M.; Saunders, L.; and Wright, Thomas. (1991). "Environmental Effects of Aquatic Disposal of Dredged Material." *Environmental Professional* 13:317–325.

Engler, R.M.; Saunders, L.; and Wright, T. (1991). "The Nature of Dredged Material." *Environmental Professional* 13:313–316.

Huber, M.E., et al. (1999). "Oceans at Risk." *Marine Pollution Bulletin* 38(6):435–438.

International Maritime Organization. (1991). *The London Dumping Convention: The First Decade and Beyond.* London.

Nauke, M. (1985). "Disposal at Sea of Dredged Material under the London Convention." *Dredging and Port Construction* May:9–16.

Internet Resources

Greenpeace. Available from http://www.greenpeace.org/~odumping.

London Convention of 1972. Available from http://www.londonconvention.org/London_Convention.htm.

Robert M. Engler

Ocean Dumping Ban Act

The Ocean Dumping Ban Act, enacted in 1988, significantly amended portions of the Marine Protection, Research and Sanctuaries Act of 1972 and banned ocean dumping of municipal sewage sludge and industrial waste (with limited exceptions) by phased target dates. The disposal of sewage sludge in waters off New York City was a major motivation for its enactment. Eligible municipalities previously had been allowed to dispose of sewage sludge beyond the so-called 106-mile ocean waste dumpsite, but are now precluded from doing so. Ocean disposal of sewage sludge and industrial waste was totally banned after 1991. Narrow exceptions were created for certain Army Corps of Engineers dredge materials that are occasionally deposited offshore.

During the interim period from 1989, after the amendments were enacted, to 1991, when the total ban took effect, limited sewage and industrial waste dumping was allowed for businesses dumping under already existing permits. The U.S. Environmental Protection Agency (EPA) was directed to report to Congress on an annual basis regarding the effectiveness of compliance agreements, and the progress made by permitted parties toward developing alternative systems for managing sewage sludge and industrial waste. EPA also had to report on its own efforts in identifying and implementing alternative disposal systems and general progress toward the congressional goal of terminating the ocean dumping of sewage sludge and industrial waste. EPA has interpreted these 1988 amendments to include the ocean incineration of wastes so that they must be regulated in the same manner as ocean disposal. SEE ALSO BIOSOLIDS; LAWS AND REGULATIONS, INTERNATIONAL; MEDICAL WASTE; OCEAN DUMPING.

Internet Resource

U.S. Environmental Protection Agency. "Ocean Dumping Ban Act of 1988." Available from http://www.epa.gov/history.

Kevin Anthony Reilly

Ocean Pollution *See Water Pollution: Marine*

Oil Spills *See Disasters: Oil Spills; Petroleum*

breakdown degradation into
component parts

Organic Farming   *See Agriculture*

OSHA   *See Occupational Safety and Health Administration*

Oxygen Demand, Biochemical

Biochemical oxygen demand (BOD) is a measure of how much organic pollution is in water. The BOD test measures the amount of dissolved oxygen in water that is used up due to the **breakdown** of organic pollutants, such as sewage, in a certain number of days. Raw sewage has a BOD of forty to 150 milligrams per liter, whereas drinking water has a BOD of less than 0.5 milligrams per liter.

Engineers and scientists measure the BOD of a lake or river to see how healthy the water is. The lower the BOD, the healthier the water. Water needs to have oxygen in it to support aquatic life such as fish and plants. Oxygen in the water is replenished from the atmosphere through aeration, but if it is used up faster than it is replenished, the water becomes anaerobic (or hypoxic)—existing in the deficiency or absence of free oxygen. Anaerobic water cannot support life. SEE ALSO FRESH KILLS; HYPOXIA; WATER TREATMENT.

Bibliography

Peavey, Howard S.; Rowe, Donald R.; and Tchobanoglous, George. (1985). *Environmental Engineering*. New York: McGraw-Hill.

Julie Hutchins Cairn

Ozone

Ozone is a gas found in the atmosphere in very trace amounts. Depending on where it is located, ozone can be beneficial ("good ozone") or detrimental ("bad ozone"). On average, every ten million air molecules contains only about three molecules of ozone. Indeed, if all the ozone in the atmosphere were collected in a layer at Earth's surface, that layer would only have the thickness of three dimes. But despite its scarcity, ozone plays very significant roles in the atmosphere. In fact, ozone frequently "makes headlines" in the newspapers because its roles are of importance to humans and other life on Earth.

What Is Ozone?

Chemically, the ozone molecule consists of three atoms of oxygen arranged in the shape of a wide V. Its formula is O_3 (the more familiar form of oxygen that one breathes has only two atoms of oxygen and a chemical formula of O_2). Gaseous ozone is bluish in color and has a pungent, distinctive smell. In fact, the name ozone is derived from the Greek word *ozein*, meaning "to smell or reek." The smell of ozone can often be noticed near electrical transformers or nearby lightning strikes. It is formed in these instances when an electrical discharge breaks an oxygen molecule (O_2) into free oxygen atoms (O), which then combine with O_2 in the air to make O_3. In addition to its roles in the atmosphere, ozone is a chemically reactive oxidizing agent that is used as an air purifier, a water sterilizer, and a bleaching agent.

Where Is Ozone Found in the Atmosphere?

Ozone is mainly found in the two regions of the atmosphere that are closest to the earth's surface. About 10 percent of the atmosphere's ozone is in the lowest-lying atmospheric region, the troposphere. This ozone is formed in a series of chemical reactions that involve the interaction of nitrogen oxides, volatile organic compounds, and sunlight. Most ozone (about 90%) resides in the next atmospheric layer, the stratosphere. The stratosphere begins between 8 and 18 kilometers (5 and 11 miles) above the earth's surface and extends up to about 50 kilometers (30 miles). The ozone in this region is commonly known as the *ozone layer*. Stratospheric ozone is formed when the sun's **ultraviolet (UV) radiation** breaks apart molecular oxygen (O_2) to form O atoms, which then combine with O_2 to make ozone. Note that this formation mechanism differs from the one mentioned above for ozone in the lower atmosphere.

ultraviolet radiation high-energy, short-wavelength light beyond human vision

What Roles Does Ozone Play in the Atmosphere and How Are Humans Affected?

The ozone molecules in the stratosphere and the troposphere are chemically identical. However, they have very different roles in the atmosphere and very different effects on humans and other living beings, depending on their location.

A useful statement summarizing ozone's different effects is that it is "good up high, bad nearby." In the upper atmosphere, stratospheric ozone plays a beneficial role by absorbing most of the sun's biologically damaging ultraviolet sunlight (called UV-B), allowing only a small amount to reach the earth's surface. The absorption of ultraviolet radiation by ozone creates a source of heat, which actually defines the stratosphere (a region in which the temperature rises as one goes to higher altitudes). Ozone thus plays a key role in the temperature structure of the earth's atmosphere. Without the filtering action of the ozone layer, more of the sun's UV-B radiation would penetrate the atmosphere and reach the earth's surface. Many experimental studies of plants and animals and clinical studies of humans have shown that excessive exposure to UV-B radiation has harmful effects. Serious long-term effects can include skin cancers and eye damage. The UV-absorbing role of stratospheric ozone is what lies behind the expression that ozone is "good up high."

In the troposphere, ozone comes into direct contact with life-forms. Although some amount of ozone is naturally present in the lower atmosphere, excessive amounts of this lower-atmospheric ozone are undesirable (or bad ozone). This is because ozone reacts strongly with other molecules, including molecules that make up the tissues of plants and animals. Several studies have documented the harmful effects of excessive ozone on crop production, forest growth, and human health. For example, people with asthma are particularly vulnerable to the adverse effects of ozone. Thus, ozone is "bad nearby."

What Are the Environmental Issues Associated with Ozone?

The dual role of ozone links it to two separate environmental issues often seen in the newspaper headlines. One issue relates to increases in ozone in

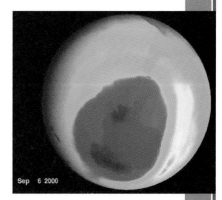

Earth, showing depletion of the ozone layer, over Antarctica. This graphic depicts the largest hole ever recorded, taken on September 6, 2000. (Goddard Space Flight Center, National Aeronautics and Space Administration.)

DISTRIBUTION OF OZONE IN THE ATMOSPHERE

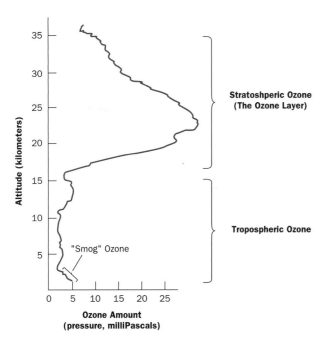

Stratoshperic Ozone (The Ozone Layer)

- Contains 90% of Atmospheric Ozone
- Beneficial Role: Acts as Primary UV Radiation Shield
- Current Issues:
  Long-Term Global Downward Trends
  Springtime Antarctic Ozone Hole Each Year
  Springtime Arctic Ozone Losses in Several Recent
  Years

Tropospheric Ozone

- Contains 10% of Atmospheric Ozone
- Harmful Impact: Toxic Effects on Humans and Vegetation
- Current Issues:
  Episodes of High Surface Ozone in Urban and Rural
  Areas

SOURCE: Adapted from the Introduction to World Meteorological Organization/United Nations Environment Programme report, *Scientific Assessment of Ozone Depletion: 1998* (WMO Global Ozone Research and Monitoring Project-Report No. 44, Geneva, 1999).

photochemical light-induced chemical effects

the troposphere (the bad ozone mentioned above). Human activities that add nitrogen oxides and volatile organic compounds to that atmosphere, such as the fossil fuel burning associated with power-generating plants and vehicular exhaust, are contributing to the formation of larger amounts of ozone near the earth's surface. This ozone is a key component of **photochemical** smog, a familiar problem in the atmosphere of many cities around the world. Higher amounts of surface-level ozone are increasingly being observed in rural areas as well. Thus, the environmental issue is that human activities can lead to more of the bad ozone.

The second environmental issue relates to the loss of ozone in the stratosphere. Ground-based and satellite instruments have measured decreases in the amount of stratospheric ozone in our atmosphere, which is called ozone-layer depletion. The most extreme case occurs over some parts of Antarctica, where up to 60 percent of the total overhead amount of ozone (known as the column ozone) disappears during some periods of the Antarctic spring (September through November). This phenomenon, which has been occurring only since the early 1980s, is known as the Antarctic ozone hole. In the arctic polar regions, similar processes occur that have also led to significant chemical depletion of the column ozone during late winter and spring in many recent years. Arctic ozone loss from January through late March has been typically 20 to 25 percent, and shorter-period losses have been higher, depending on the meteorological conditions encountered in the Arctic stratosphere. Smaller, but nevertheless significant, stratospheric ozone decreases have been seen at other, more populated latitudes of the earth, away from the polar regions. Instruments on satellites and on the ground have detected higher amounts of UV-B radiation at the earth's surface below areas of depleted ozone.

What Human Activities Affect the Stratospheric Ozone Layer?

Initially, theories about the cause of ozone-layer depletion abounded. Many factors were suggested, from the sun to air motions to human activity. In the 1970s and 1980s, the scientific evidence showed conclusively that human-produced chemicals are responsible for the observed depletions of the ozone layer. The ozone-depleting compounds contain various combinations of carbon with the chemical elements chlorine, fluorine, bromine, and hydrogen (the halogen family in the periodic table of the elements). These are often described by the general term *halocarbons*. The compounds include chlorofluorocarbons (CFCs which are used as refrigerants, foam-blowing agents, electronics cleaners, and industrial solvents) as well as halons (which are used in fire extinguishers). The compounds are useful and benign in the troposphere, but when they eventually reach the stratosphere, they are broken apart by the sun's ultraviolet radiation. The chlorine and bromine atoms released from these compounds are responsible for the breakdown of stratospheric ozone. The ozone destruction cycles are catalytic, meaning that the chlorine or bromine atom enters the cycle, destroys ozone, and exits the cycle unscathed and therefore able to destroy another ozone molecule. In fact, an individual chlorine atom can destroy as many as 10,000 different ozone molecules before the chlorine atom is removed from the stratosphere by other reactions.

What Actions Have Been Taken to Protect the Ozone Layer?

Research on ozone depletion advanced very rapidly in the 1970s and 1980s, leading to the identification of CFCs and other halocarbons as the cause. Governments and industry acted quickly on the scientific information. Through a 1987 international agreement known as the Montréal Protocol on Substances That Deplete the Ozone Layer, governments decided to eventually discontinue production of CFCs (known in the United States by the industry trade name "Freons"), halons, and other halocarbons (except for a few special uses). Concurrently, industry developed more ozone-friendly substitutes for the CFCs and other ozone-depleting halocarbons. If nations adhere to international agreements, the ozone layer is expected to recover by the year 2050. The interaction of science in identifying the problem, technology in developing alternatives, and governments in devising new policies is thus an environmental "success story in the making." Indeed, the Montréal Protocol serves as a model for other environmental issues now facing the global community.

What Actions Have Been Taken to Reduce the Amount of Ozone at Ground Level?

Ozone pollution at the earth's surface is formed within the atmosphere by the interaction of sunlight with chemical precursor compounds (or starting ingredients): the nitrogen oxides (NO_x) and volatile organic compounds (VOCs). In the United States, the efforts of the Environmental Protection

combustion burning, or rapid oxidation, accompanied by release of energy in the form of heat and light

Agency (EPA) to reduce ozone pollution are therefore focused on reducing the emissions of the precursor compounds. VOCs, a primary focus of many regulations, arise from the **combustion** of fossil fuel and from natural sources (emissions from forests). Increasingly, attention is turning to reducing the emissions of NO_x compounds, which also arise from the combustion of fossil fuels. The use of cleaner fuels and more efficient vehicles has caused a reduction in the emission of ozone precursors in urban areas. This has led to a steady decline in the number and severity of episodes and violations of the one-hour ozone standard established by the U.S. Environmental Protection Agency (EPA) (which is 120 parts per billion or ppb, meaning that out of a billion air molecules, 120 are ozone). In 1999 there were thirty-two areas of the country that were in violation of the ozone standard, down from 101 just nine years earlier. Despite these improvements, ground-level ozone continues to be one of the most difficult pollutants to manage. An additional, more stringent ozone standard proposed by the EPA to protect public health, eighty ppb averaged over eight hours, was cleared in early 2001 for implementation in the United States. For comparison, Canada's standard is sixty-five ppb averaged over eight hours. SEE ALSO AIR POLLUTION; ASTHMA; CFCs (CHLOROFLUOROCARBONS); ELECTRIC POWER; HALON; MONTRÉAL PROTOCOL; NO_x (NITROGEN OXIDES); SMOG; VEHICULAR POLLUTION; ULTRAVIOLET RADIATION; VOCs (VOLATILE ORGANIC COMPOUNDS).

Bibliography

World Meteorological Organization. (2003). *Scientific Assessment of Ozone Depletion: 2002.* Global Ozone Research and Monitoring Project, Report No. 47. Geneva: World Meteorological Organization.

Internet Resources

University Corporation for Atmospheric Research. "Cycles of the Earth and Atmosphere—Module Review." Available from http://www.ucar.edu/learn/1.htm.

U.S. Environmental Protection Agency. "Automobiles and Ozone." Available from http://www.epa.gov/otaq/04-ozone.htm.

U.S. Environmental Protection Agency. "Ozone Depletion." Available from http://www.epa.gov/docs/ozone.

Christine A. Ennis

Packaging   *See Waste Reduction*

Particulates

Particulates, or particulate matter (PM), refer to any mixture of solid particles or liquid droplets that remain suspended in the atmosphere for appreciable time periods. Examples of particulates are dust and salt particles, and water and sulphuric acid droplets. The length of time a particle survives in the atmosphere depends on the balance between two processes. Gravity forces the particles to settle to the earth's surface, but atmospheric turbulence can carry the particles in the opposite direction. Under normal conditions, only particles with diameters less than 10 micrometers (µm) remain in the atmosphere long enough to be considered atmospheric particulates. In quantifying particulate matter, it is typical to give the mass of particles less than a particular size per cubic meter of air. For example, 10 µg/m³ $PM_{2.5}$ means that in 1 cubic meter (m³) of air the mass of all particles with diameters less than 2.5 µm is 10 µg.

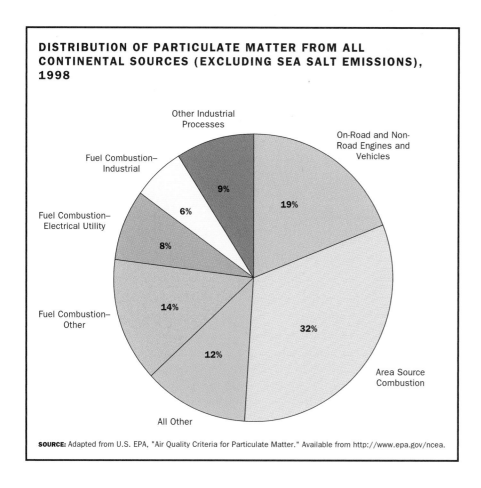

DISTRIBUTION OF PARTICULATE MATTER FROM ALL CONTINENTAL SOURCES (EXCLUDING SEA SALT EMISSIONS), 1998

Other Industrial Processes

Fuel Combustion– Industrial

Fuel Combustion– Electrical Utility

Fuel Combustion– Other

All Other

On-Road and Non-Road Engines and Vehicles

Area Source Combustion

9%
6%
19%
8%
14%
12%
32%

SOURCE: Adapted from U.S. EPA, "Air Quality Criteria for Particulate Matter." Available from http://www.epa.gov/ncea.

Most atmospheric particulate comes from natural sources and is mainly dust or sea salt from mechanical processes such as wind erosion or wave breaking. Although most of this material is of large size and so is lost from the atmosphere by gravitational settling, many of the smaller particles can travel very long distances. For example, dust from Saharan dust storms is carried across the Atlantic Ocean and can be detected in Florida. Similarly, dust from Asia is regularly detected in Hawaii and sometimes even continental North America. Adding to the naturally produced dust is a small but often locally important contribution from the **photochemical** oxidation of naturally occurring gas-phase hydrocarbons, such as alpha and beta pinene, emitted by trees. These particles frequently give forested areas a hazy atmosphere.

photochemical light-induced chemical effects

Although natural processes produce most of the atmospheric particulate on a global scale, **anthropogenic** processes are the source of most particulate in urban or industrial areas. The major anthropogenic sources are those that increase natural loading, such as extra dust due to agriculture or construction. However, a significant amount of particles are present in factory, power plant, and motor vehicle emissions, and produced from the reactions of anthropogenic gases present in those emissions.

anthropogenic human-made; related to or produced by the influence of humans on nature

Primary emissions are those that are produced before being released into the atmosphere or immediately afterward. They result from condensation that follows the rapid cooling of high-temperature gases. An example is the soot that comes from diesel engines. Secondary particles are produced over a longer time period and derive from gas-phase chemical reactions that produce low-vapor-pressure (condensable) products. This process is especially important, as it produces the ultrafine particles (0.01 µm) that have been shown to be closely related to human health effects. An example is the atmospheric oxidation of sulphur dioxide to sulphuric acid, in which sulphur dioxide is a gas but sulphuric acid exists in the form of droplets. Particles are an environmental concern because they lower visibility, contribute to acid rain, and adversely affect human health.

Particulate suspended in the atmosphere has diameters similar to the wavelengths of visible light, which makes it very good at scattering this light. In the presence of particulate, scattering reduces the light coming from distant objects, making it more difficult to see them. This loss of visibility is particularly important in areas that rely on clear vistas to attract tourists.

Sulphur dioxide emitted from fossil fuel combustion is oxidized to particulate sulphuric acid or sulphate, which is a major component of acid rain.

Particles can be a major irritant to the human bronchial and pulmonary systems. The body has natural mechanisms to limit the penetration of these particles into its sensitive areas. The nose is an effective filter for particles of greater than about three µm, and blowing the nose expels these. Smaller particles can penetrate deeper into the bronchial passages where mucous layers and small hairs called cilia catch the particles, which can then be expelled by coughing. The smallest particles, however, may penetrate all the way into the lungs. Irritation of the lung and bronchial tissue by particles prompts the body to produce mucous in self-defense, which can exacerbate existing respiratory problems such as bronchitis and asthma. There is also concern that harmful pollutants in, or attached to, the particles may be absorbed into the body. Heavy metals and carcinogenic polycyclic aromatic hydrocarbons (**PAHs**) from combustion can be introduced into the body in this way.

PAHs polyaromatic hydrocarbons; compounds of hydrogen and carbon containing multiple ring structures

24-hour standard in regulations: the allowable average concentration over twenty-four hours

Most jurisdictions have, and are continually updating, air-quality standards for particulate matter. In 1997 the U.S. Environmental Protection Agency (EPA) added a new annual $PM_{2.5}$ standard of 15 µg/m^3 and a new **24-hour standard** margin of 65 µg/m^3, while retaining the annual PM_{10} standard of 50 µg/m^3 and making minor technical changes to the 24-hour standard of 150 µg/m^3. Approximately 29 million U.S. citizens live in areas that do not meet the PM_{10} standards, but because of the need for three years of monitoring and the requirements of the clean air act, nonattainment areas for $PM_{2.5}$ have not yet been determined. The standards in most industrialized countries are similar to those in the United States.

Many countries have large areas that exceed the local air-quality standards, and thus they have instituted control programs to reduce particulate levels. Fortunately, many of the strategies in place to combat smog, acidic deposition, and smoke releases are also effective in reducing particle levels. Most countries now have integrated strategies to reduce common emissions (e.g., nitrogen oxides and hydrocarbons) that contribute to particulate matter, acid deposition, and smog. SEE ALSO AIR POLLUTION; ASTHMA; DIESEL; SCRUBBERS; SMOG; VEHICULAR POLLUTION.

Bibliography

Finlayson-Pitts, Barbara J., and Pitts, James N. (2000). *Chemistry of the Upper and Lower Atmosphere*. San Diego, CA: Academic Press.

Internet Resource

U.S. Environmental Protection Agency. "Air Quality Criteria for Particulate Matter." Available from http://www.epa.gov/ncea.

Donald R. Hastie

PBTs   *See Persistent Bioaccumulative and Toxic Chemicals*

PCBs (Polychlorinated Biphenyls)

PCBs, known to cause cancer in animals and believed to cause cancer in humans, are among the most widespread and hazardous synthetic pollutants. They comprise a group of 209 structurally similar compounds, so-called **congeners**. The individual congeners differ in the degree of chlorination and the positions of the chlorine atoms in the molecule. They are numbered from one to 209 according to a scheme proposed by Ballschmiter and Zell (hence, the term BZ numbers).

PCBs are obtained by the controlled reaction of biphenyl with elemental chlorine under the **catalytic** influence of iron or iron chloride. The commercial formulations were mixtures of various congeners. They were marketed under trade names such as Aroclor, Pyranol, and Clophen and were characterized by their average chlorine content, ranging from 21 to 68 percent. The pattern of congener composition is characteristic for each product and may serve to identify the source of a local contamination. PCBs are stable to heat, chemical, and biological decomposition. The mixtures are clear to yellow nonflammable thick liquids (i.e., of medium to high viscosity) or waxy solids. With an increasing degree of chlorination the low water **solubility** and **volatility** further decrease, while persistence, lipid solubility, and thus the capacity for **bioaccumulation** increase. Boiling limits are between 270 and 420°C (515 and 788°F) and water solubility ranges from 0.1 µg/L to 6 mg/L. The electrical conductivity is extremely low. Because of their excellent technical properties, PCBs were used extensively as insulating oils in electrical components such as transformers and capacitors. Additionally, they were used in hydraulic fluids, lubricating oils, plasticizers, paints, adhesive resins, inks, fire retardants, and various other products.

The industrial production of PCBs started in 1929. From that point, PCBs were produced in many countries, including the United States, China, France, Germany, Japan, the former Soviet Union, and the United Kingdom. When their bioaccumulation and adverse health effects became recognized, open uses were terminated in the early 1970s. During the late 1970s most countries completely stopped PCB production, for example, the United States outlawed its manufacture in 1977. Until production ceased, an estimated one million metric tons had been manufactured. A large proportion of this amount is still in use in closed applications, although some countries have set deadlines for the replacement of all PCB-containing transformers and capacitors. From production, open uses, leaks from closed systems, and improper disposal, large amounts of PCBs have entered the environment. As Dobson and van Esch note, a large part of these PCBs is believed to be located in aquatic sediments. Despite the termination of production, further emissions into the environment are expected from PCBs still in use, from dump sites, and by remobilization from contaminated soils and sediments.

congener a member of a class of chemicals having a of similar structure

catalytic of a substance that promotes reaction without being consumed

solubility the amount of mass of a compound that will dissolve in a unit volume of solution; aqueous solubility is the maximum concentration of a chemical that will dissolve in pure water at a reference temperature

volatility relating to any substance that evaporates readily

bioaccumulation build-up of a chemical within a food chain when a predator consumes prey containing that chemical

CHEMICAL STRUCTURE OF PCB 126

(3,3',4,4',5–pentachlorobiphenyl)

Chemical structure of PCB 126 (3,3',4,4',5-pentachlorobiphenyl).

deposit concentration of a substance, i.e., mineral ore

reevaporate return to the gaseous state

affinity physical attraction

neurology medical science relating to the nervous system

remediation cleanup or other methods used to remove or contain a toxic spill or hazardous materials from a Superfund site or for the Asbestos Hazard Emergency Response program

Sign warning people not to eat fish contaminated with PCB. (Courtesy of Richard Stapleton. Reproduced by permission.)

Once released into the environment, PCBs may continue to exist for years since they are very resistant to chemical and biological degradation. A constant process of **deposition** and **reevaporation** favors combined atmospheric and ocean-borne long-range transport to areas far from known PCB sources. In 2002 PCBs are found virtually everywhere on the planet: in deep-sea sediments as well as in arctic and antarctic environments. Their high **affinity** for organic material causes adsorption on particles and sediments, and accumulation in food webs. For the top members of the food chain, enrichment factors of twenty million and greater have been calculated. PCBs are stored in fatty tissue. They are transferred to offspring via eggs or mother's milk.

Although the acute toxicity of PCBs is rather low, their chronic effects are severe. They include chloracne, hepatocellular carcinoma, adenofibrosis, and damage to the nervous system. During the 1990s evidence was presented by Colborn, Dumanoski, and Myers that PCBs reduce reproductive capacity, and the numbers and viability of sperms and eggs. Maternal exposure to PCBs has been linked to **neurological** and cognitive problems in young children. Furthermore, there are indications that PCBs suppress normal immune responses. The toxicity of individual congeners varies significantly. The most toxic congeners are coplanar (i.e., flat) PCBs, which show a structural similarity to dioxin. This is taken into account by the system of toxicity equivalency factors (TEF), relating the toxicity of a compound to that of 2,3,7,8-tetrachlorodibenzo[1,4]dioxin. A certain percentage of the toxicity of PCB formulations can be attributed to a compound class closely related to dioxins known as furans, which are contained as impurities.

Although UV and biological processes are known to decompose PCBs to some extent, the most commonly chosen method is thermal deconstruction. Temperatures above 1,200°C (2,192°F) and an excess of oxygen are necessary to ultimately destroy PCBs. Temperatures of 600 to 900°C (1,112 to 1,652°F), as typically found in communal waste incinerators, favor the formation of the more toxic chlorinated furans. The **remediation** of PCB-contaminated sites usually consists of removing soil by excavation or sediments by dredging, and disposing of the contaminated material in hazardous waste landfills or incinerating them in approved facilities.

In the case of the upper Hudson River in New York State that is heavily contaminated with PCBs from former General Electric production sites, there has been much controversy over the benefits of dredging. Although opponents fear an elevated remobilization of PCBs from dredged sediments, supporters argue that modern dredging equipment minimizes resuspension and that PCBs would be slowly released from sediments over the course of the next several decades if not removed from the river system. Fishing has been banned in some parts of the Hudson River, and in 2002 the U.S. Environmental Protection Agency (EPA) drew up a $500 million plan to dredge its sediments.

Because of their high persistence, the large amounts deposited in the environment, and their bioaccumulative and toxic potential, PCBs will remain among the priority pollutants for decades. SEE ALSO BIOACCUMULATION; DREDGING; PERSISTENT ORGANIC POLLUTANTS (POPS); PESTICIDES; SUPERFUND; WATER POLLUTION.

Bibliography

Colborn, Theo; Dumanoski, Dianne; and Myers, John Peterson. (1996). *Our Stolen Future.* New York: Dutton.

Dobson, Stuart, and van Esch, G.J. (1993). *Polycholorinated Biphenyls and Terphenyls,* 2nd edition. Geneva: World Health Organization.

Erickson, Mitchel D. (1997). *Analytical Chemistry of PCBs,* 2nd edition. Boca Raton, FL: Lewis Publishers.

Hutzinger, Otto; Safe, Stephen; and Zitko, V. (1983). *The Chemistry of PCBs.* Melbourne, FL: Krieger.

Robertson, Larry W., and Hansen, Larry G., eds. (2001). *PCBs: Recent Advances in Environmental Toxicology and Health Effects.* Lexington: The University Press of Kentucky.

Safe, Stephen, and Hutzinger, Otto, eds. (1987). *Polychlorinated Biphenyls (PCBs): Mammalian and Environmental Toxicology.* New York: Springer-Verlag.

Waid, John. (1987). *PCBs and the Environment.* Boca Raton, FL: CRC Press.

Internet Resource

U.S. Environmental Protection Agency. "The PCB Home Page at EPA." Available from http://www.epa.gov/pcb.

Stefan Weigel

Persistent Bioaccumulative and Toxic (PBT) Chemicals

Persistent bioaccumulative and toxic (PBT) chemicals represent a group of substances that are not easily degraded, accumulate in organisms, and exhibit an acute or chronic toxicity. They may therefore pose serious concerns for human and environmental health. The effects of PBTs range from cancer, endocrine disruption, reproductive dysfunction, behavioral abnormalities, birth defects, disturbance of the immune system, damage to the liver and nervous system, to the extinction of whole populations.

The category PBT was defined by the United Nations Environmental Programme (UNEP). Persistent organic pollutants (POPs) are an integral part of the PBT group, which additionally includes trace metals and organo-metal compounds. A large proportion of PBTs are organohalogens—namely, organochlorine pesticides, polychlorinated biphenyls (PCBs), polychlorinated naphthalenes (PCN), chloroparaffins, and brominated flame retardants. Further PBTs are polycyclic aromatic hydrocarbons (PAH), metals and their compounds (e.g., the antifouling tributyltin TBT), and phthalates (plasticizers). In 1999 the Environmental Protection Agency (EPA) listed fourteen priority PBTs, most of which belong to the so-called dirty dozen identified by the UNEP: six pesticides, PCBs, hexachlorobenzene, octachlorostyrene, dioxins and furans, benzo(a)pyrene, alkyllead, and mercury and its compounds. The UNEP Stockholm Convention, signed in 2001, established control and phase-out measures for that initial set of twelve POPs. In response, some nations devised action plans to prevent the introduction of new PBTs into the marketplace, to identify further priority PBTs, and to phase out or reduce the emissions of priority PBTs. SEE ALSO BIOACCUMULATION; DIOXIN; MERCURY; PCBs (POLYCHLORINATED BIPHENYLS); PERSISTENT ORGANIC POLLUTANTS (POPs); PESTICIDES.

Bibliography

Bernes, Claes, Naylor, Martin. (1999). *Persistent Organic Pollutants: A Swedish View of an International Problem.* Stockholm: Almqvist and Wiksell Internation.

Lipnick, Robert L., ed. (2001). *Persistent, Bioaccumulative, and Toxic Chemicals.* Washington, D.C.: American Chemical Society.

Lipnick, Robert L., ed. (2001). *Persistent, Bioaccumulative, and Toxic Chemicals II: Assessment and New Chemicals.* Washington, D.C.: American Chemical Society.

Internet Resource

U.S. Environmental Protection Agency. "Persistent, Bioaccumulative and Toxic (PBT) Chemicals Program." Available from http://www.epa.gov/pbt.

Stefan Weigel

Persistent Organic Pollutants (POPs)

Persistent organic pollutants (POPs) are a subset of the more comprehensive term *persistent bioaccumulative and toxic chemicals* (PBTs). POPs commonly stands for organic (carbon-based) chemical compounds and mixtures that share four characteristics. They are semivolatile, stable under environmental conditions (**half-lives** of years to decades), fat-soluble, and possess the potential for adverse effects in organisms. Many POPs are organochlorine compounds. Among the twelve priority POPs defined by the United Nations Environmental Programme (and referred to as the "dirty dozen") are the pesticides aldrin, chlordane, DDT, dieldrin, endrin, heptachlor, mirex, and toxaphene (chlorobornanes); the industrial chemicals polychlorinated biphenyls (PCBs) and hexachlorobenzene; and the unintentional by-products dioxins and furans.

POPs' resistance to chemical and biological degradation and their propensity to evaporate led to their global distribution. By a constant process of deposition and reevaporation, POPs are transported by air and water currents to regions far from their sources until they ultimately gather in colder climates. Because of their **lipophilicity**, many POPs concentrate in organisms and accumulate to high levels in the top members of the food web such as predatory fish and birds, mammals and humans. Certain chemicals possess the ability to cross the placenta, while others are retained. Several contaminants present in the mother's body are thus handed down to the developing embryo in the womb—they are transferred to offspring across the placenta and through mother's milk. Adverse effects include cancer, endocrine disruption, reproductive dysfunction, behavioral abnormalities, birth defects, and interference with the immune and nervous systems. SEE ALSO BIOACCUMULATION; DIOXIN; PCBs (POLYCHLORINATED BIPHENYLS); PESTICIDES.

half-life the time required for a pollutant to lose one-half of its original concentration; for example, the biochemical half-life of DDT in the environment is 15 years

lipophilicity solubility in or attraction to waxy, fatty, or oily substances

Bibliography

Harrad, Stuart, ed. (2001). *Persistent Organic Pollutants: Environmental Behaviour and Pathways of Human Exposure.* Boston, MA: Kluwer.

Internet Resource

United Nations Environmental Programme. "Persistent Organic Pollutants." Available from http://irptc.unep.ch/pops.

Stefan Weigel

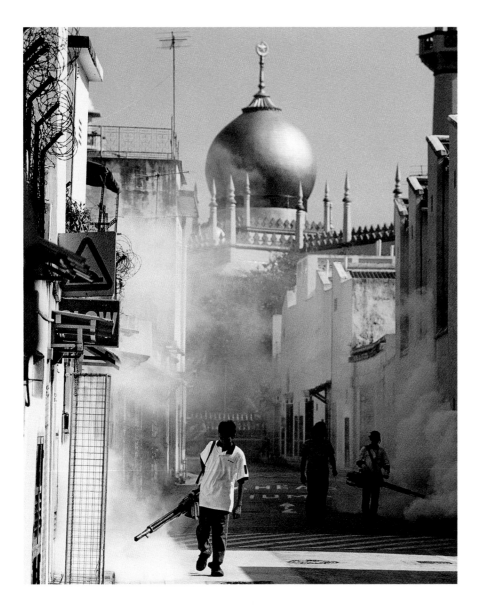

Fumigators walking down a street in the Sultan Mosque area of Singapore and spraying a pesticide to rid the area of mosquitoes. (©Steve Raymer/ Corbis. Reproduced by permission.)

Pesticides

Pesticides are substances or a mixture of substances, of chemical or biological origin, used by human society to mitigate or repel pests such as bacteria, nematodes, insects, mites, mollusks, birds, rodents, and other organisms that affect food production or human health. They usually act by disrupting some component of the pest's life processes to kill or inactivate it. In a legal context, pesticides also include substances such as insect attractants, herbicides, plant defoliants, desiccants, and plant growth regulators.

History of Pesticides

The concept of pesticides is not new. Around 1000 B.C.E. Homer referred to the use of sulfur to fumigate homes and by 900 C.E. the Chinese were using arsenic to control garden pests. Although major pest outbreaks have occurred, such as potato blight (*Phytopthora infestans*), which destroyed most potato crops in Ireland during the mid-nineteenth century, not until later that century were pesticides such as arsenic, pyrethrum, lime sulfur, and mercuric

A bird that died as a result of pesticide use. (U.S. EPA. Reproduced by permission.)

inorganic compounds not containing carbon

organochlorine chemical containing carbon and chlorine

vector an organism, often an insect or rodent, that carries disease; plasmids, viruses, or bacteria used to transport genes into a host cell: a gene is placed in the vector; the vector then "infects" the bacterium

bioconcentrate chemical buildup in an organism, i.e., fish tissue, to levels higher than in the surrounding environment

trophic related to feeding

organophosphate pesticide that contains phosphorus; short-lived, but some can be toxic when first applied

acetylcholine a chemical that transmits nerve signals to muscles and other nerves

systemic throughout the body

carbamate class of chemicals widely used as pesticides

chloride used. Between this period and World War II, **inorganic** and biological substances, such as Paris green, lead arsenate, calcium arsenate, selenium compounds, lime–sulfur, pyrethrum, thiram, mercury, copper sulfate, derris, and nicotine were used, but the amounts and frequency of use were limited, and most pest control employed cultural methods such as rotations, tillage, and manipulation of sowing dates. After World War II the use of pesticides mushroomed, and there are currently more than 1,600 pesticides available and about 4.4 million tons used annually, at a cost of more than $20 billion. The United States accounts for more than 25 percent of this market.

Older Insecticides

The first synthetic **organochlorine** insecticide, DDT (dichlorodiphenyltrichloroethane), discovered in Switzerland in 1939, was very effective and used extensively to control head and body lice, human disease **vectors** and agricultural pests, in the decades leading up to the 1970s. Benzene hexachloride (BHC) and chlordane were discovered during World War II and toxaphene (and heptachlor) slightly later. Shortly thereafter, two cyclodiene organochlorines, aldrin and dieldrin, were introduced, followed by endrin, endosulfan, and isobenzan. All these insecticides acted by blocking an insect's nervous system, causing malfunction, tremors, and death. All organochlorines are relatively insoluble, persist in soils and aquatic sediments, can **bioconcentrate** in the tissues of invertebrates and vertebrates from their food, move up **trophic** chains, and affect top predators. These properties of persistence and bioaccumulation led eventually to the withdrawal of registration and use of organochlorine insectices, from 1973 to the late 1990s, in industrialized nations, although they continued to be used in developing countries.

Organophosphate insecticides originated from compounds developed as nerve gases by Germany during World War II. Thus, those developed as insecticides, such as tetraethyl pyrophosphate (TEPP) and parathion, had high mammalian toxicities. Scores of other organophosphates including demeton, methyl schradan, phorate, diazinon, disulfoton, dimethoate, trichlorophon, and mevinphos have been registered. In insects, as in mammals, they act by inhibiting the enzyme cholinesterase (ChE) that breaks down the neurotransmitter **acetylcholine** (ACh) at the nerve synapse, blocking impulses and causing hyperactivity and tetanic paralysis of the insect, then death. Some are **systemic** in plants and animals, but most are not persistent and do not bioaccumulate in animals or have significant environmental impacts.

Carbaryl, the first **carbamate** insecticide, acts on nervous transmissions in insects also through effects on cholinesterase by blocking acetylcholine receptors. Other carbamate insecticides include aldicarb, methiocarb, methomyl, carbofuran, bendiocarb, and oxamyl. In general, although they are broad-spectrum insecticides, of moderate toxicity and persistence, they rarely bioaccumulate or cause major environmental impacts.

Botanical insecticides include nicotine from tobacco, pyrethrum from chrysanthemums, derris from cabbage, rotenone from beans, sabadilla from lilies, ryania from the ryania shrub, limonene from citrus peel, and neem from the tropical neem tree. Most, other than nicotine, have low levels of toxicity in mammals and birds and create few adverse environmental effects.

botanical derived from or relating to plants

Newer Insecticides

Synthetic **pyrethroid** insecticides, with structures based on the natural compound pyrethrum, were introduced in the 1960s and include tetramethrin, resmethrin, fenvalerate, permethrin, lambda-cyalothrin, and deltamethrin, all used extensively in agriculture. They have very low mammalian toxicities and potent insecticidal action, are photostable with low volatilities and persistence. They are broad-spectrum insecticides and may kill some natural enemies of pests. They do not bioaccumulate and have few effects on mammals, but are very toxic to aquatic invertebrates and fish.

pyrethroid chemicals derived from chrysanthemums and related plants

In recent years, new classes of insecticides have been marketed, none of which are persistent or bioaccumulate. They include juvenile hormone mimics, synthetic versions of insect juvenile hormones that act by preventing immature stages of the insects from molting into an adult, and avermectins, natural products produced by soil microorganisms, insecticidal at very low concentrations. *Bacillus thuringiensis* toxins are proteins produced by a bacterium that is **pathogenic** to insects. When activated in the insect gut, they destroy the selective permeability of the gut wall. The first strains were toxic only to *Lepidoptera*, but strains toxic to flies and beetles have since been developed. *B. thuringiensis* has been incorporated into plants genetically.

pathogenic causing illness

Nematicides

Soil **nematocides**, such as dichloropropene, methyl isocyanate, chloropicrin, and methyl bromide, are broad-spectrum soil fumigants. Others, aldicarb, dazomet, and metham sodium, act mainly through contact. All have very high mammalian toxicities and can kill a wide range of organisms from both the plant and animal kingdoms. Although **transient** in soil, they may have drastic ecological effects on soil systems.

nematocide a chemical agent which is destructive to nematodes

transient present for a short time

Molluscicides

Two **molluscicides**, metaldehyde and methiocarb, are used as baits against slugs and snails. Although of high mammalian toxicity, they cause few problems other than the occasional accidental death of wild mammals. Several molluscicides, used to control aquatic snails, *N*-trityl morpholine, copper sulfate, niclosamine, and sodium pentachlorophenate, are toxic to fish.

molluscicide chemical that kills mollusks

Herbicides

Hormone-type **herbicides** such as 2,4,5-T; 2,4-D; and MCPA; were discovered during the 1940s. They do not persist in soil, are selective in their toxicity to plants, are of low mammalian toxicity, cause few direct environmental problems, but are relatively soluble and reach waterways and groundwater.

herbicide a chemical pesticide designed to control or destroy plants, weeds, or grasses

Contact herbicides, which kill weeds through foliage applications, include dintrophenols, cyanophenols, pentachlorophenol, and paraquat. Most are nonpersistent, but triazines can persist in the soil for several years, are slightly toxic to soil organisms and moderately so to aquatic organisms. Herbicides cause few direct environmental problems other than their indirect effects, in leaving bare soil, which is free of plant cover and susceptible to erosion.

Fungicides

fungicide pesticide used to control, deter, or destroy fungi

Many different types of **fungicides** are used, of widely differing chemical structures. Most have relatively low mammalian toxicities, and except for carbamates such as benomyl, a relatively narrow spectrum of toxicity to soil-inhabiting and aquatic organisms. Their greatest environmental impact is toxicity to soil microorganisms, but these effects are short term.

Effects on the Terrestrial Environment

Pesticides are biocides designed to be toxic to particular groups of organisms. They can have considerable adverse environmental effects, which may be extremely diverse: sometimes relatively obvious but often extremely subtle and complex. Some pesticides are highly specific and others broad spectrum; both types can affect terrestrial wildlife, soil, water systems, and humans.

Pesticides have had some of their most striking effects on birds, particularly those in the higher trophic levels of food chains, such as bald eagles, hawks, and owls. These birds are often rare, endangered, and susceptible to pesticide residues such as those occurring from the bioconcentration of organochlorine insecticides through terrestrial food chains. Pesticides may kill grain- and plant-feeding birds, and the elimination of many rare species of ducks and geese has been reported. Populations of insect-eating birds such as partridges, grouse, and pheasants have decreased due to the loss of their insect food in agricultural fields through the use of insecticides.

Bees are extremely important in the pollination of crops and wild plants, and although pesticides are screened for toxicity to bees, and the use of pesticides toxic to bees is permitted only under stringent conditions, many bees are killed by pesticides, resulting in the considerably reduced yield of crops dependent on bee pollination.

The literature on pest control lists many examples of new pest species that have developed when their natural enemies are killed by pesticides. This has created a further dependence on pesticides not dissimilar to drug dependence. Finally, the effects of pesticides on the biodiversity of plants and animals in agricultural landscapes, whether caused directly or indirectly by pesticides, constitute a major adverse environmental impact of pesticides.

Effects on the Aquatic Environment

The movement of pesticides into surface and groundwater is well documented. Wildlife is affected, and human drinking water is sometimes contaminated beyond acceptable safety levels. Sediments dredged from U.S. waterways are often so heavily contaminated with persistent and other pesticide residues that it becomes problematic to safely dispose of them on land.

A major environmental impact has been the widespread mortality of fish and marine invertebrates due to the contamination of aquatic systems by pesticides. This has resulted from the agricultural contamination of waterways through fallout, drainage, or runoff erosion, and from the discharge of industrial effluents containing pesticides into waterways. Historically, most of the fish in Europe's Rhine River were killed by the discharge of pesticides, and at one time fish populations in the Great Lakes became very low due to pesticide contamination. Additionally, many of the organisms that provide food for fish are extremely susceptible to pesticides, so the indirect effects of pesticides on the fish food supply may have an even greater effect on fish populations. Some pesticides, such as pyrethroid insecticides, are extremely toxic to most aquatic organisms. It is evident that pesticides cause major losses in global fish production.

Effects on Humans

The most important aspect of pesticides is how they affect humans. There is increasing anxiety about the importance of small residues of pesticides, often suspected of being **carcinogens** or disrupting **endocrine** activities, in drinking water and food. In spite of stringent regulations by international and national regulatory agencies, reports of pesticide residues in human foods, both imported and home-produced, are numerous.

carcinogen any substances that can cause or aggravate cancer

endocrine the system of glands, hormones, and receptors that help control animal function

Over the last fifty years many human illnesses and deaths have occurred as a result of exposure to pesticides, with up to 20,000 deaths reported annually. Some of these are suicides, but most involve some form of accidental exposure to pesticides, particularly among farmers and spray operators in developing countries, who are careless in handling pesticides or wear insufficient protective clothing and equipment. Moreover, there have been major accidents involving pesticides that have led to the death or illness of many thousands. One instance occurred in Bhopal, India, where more than 5,000 deaths resulted from exposure to accidental emissions of methyl isocyanate from a pesticide factory.

Testing and Reclassification

New pesticides require extensive laboratory and field testing and may take about five years to reach market. A pesticide company has to identify uses, test effectiveness, and provide data on chemical structure, production, formulation, fate, persistence, and environmental impacts. The product is tested in the laboratory, greenhouse, and field under different environmental conditions. After several years of testing, the company submits a registration data package to the U.S. Environmental Protection Agency (EPA). Data include studies on acute, chronic, reproductive, and developmental toxicity to mammals, birds, and fish, the pesticide's environmental fate, rates of degradation, translocations to other sites, and ecological studies on its harmful effects to, and on, nontarget plants and animals.

After its review by government and other scientists, the EPA grants registration of the product for certain uses, with agreed label data and directions for use. About 1 in 35,000 chemicals survives from initial laboratory testing to the market, a process that generally takes several years, and involves more than 140 tests.

The continued use of a pesticide is supervised by the Federal Insecticide, Fungicide and Rodenticide Act (FIFRA), enacted in 1947 and modified many times since. A review may be called for when new evidence indicates possible unreasonable risks to human health or the environment, including toxicity or ill health to humans or animals, hazards to nontarget organisms, and risks to endangered species and suggests that the risks may outweigh the benefits of continued registration. After review, the EPA may take no action, alter the pesticide label to minimize risk, reclassify the approved uses or eliminate specific uses, or cancel or suspend the pesticide's registration entirely.

Pesticides and Food Safety

Pesticides are used on food crops and meat produced from domestic animals. The residues contained within domestically produced food are monitored closely by the EPA, whereas those for imported food are tracked by the Animal and Plant Health Inspection Service (APHIS) of the U.S. Department of Agriculture (USDA). Scientists determine the highest dose of a pesticide that might be ingested by animals (birds and mammals, including humans) to cause adverse health effects but not death; this is called the maximum tolerated dose (MTD). They also determine the no-observable-effect level (NOEL) and identify the amount of pesticides that may be safely consumed by humans, in terms of milligrams per kilogram of body weight, over a seventy-year lifetime. In calculating an acceptable exposure for a pesticide, scientists usually include a safety factor of one hundred below the NOEL, assuming a lifetime of exposure to the pesticide. Such calculations take for granted that a pesticide is applied to all labeled crops, at recommended rates, and that the treated food will be consumed daily for a lifetime. Pesticides that have been demonstrated to cause cancer in laboratory animals are not granted tolerance, or approved for application to food crops, based on legislation from Section 409, the so-called Delaney clause, of the federal Food, Drug and Cosmetic Act.

The Food and Drug Administration (FDA) and USDA, in addition to many states, have monitoring programs for pesticide residues in food. They sample approximately 1 percent of the national food supply. For every pesticide, the FDA conducts a total diet study (a market-based survey) to more accurately assess the exposure of the human population to pesticides. Similar calculations are made for exposure to pesticides that may reach drinking water through percolation into groundwater or runoff into waterways.

These adverse effects of pesticides on humans and wildlife have resulted in research into ways of reducing pesticide use. The most important of these is the concept of integrated pest management (IPM), first introduced in 1959. This combines minimal use of the least harmful pesticides, integrated with biological and cultural methods of minimizing pest losses. It is linked with using pesticides only when threshold levels of pest attacks have been identified. There is also a move toward sustainable agriculture which aims to minimize use of pesticides and fertilizers based on a systems approach. SEE ALSO AGRICULTURE; BIOACCUMULATION; CARSON, RACHEL; DDT (DICHLORODIPHENYL TRICHLOROETHANE); ENDOCRINE DISRUPTION; INTEGRATED PEST MANAGEMENT; PERSISTENT BIOACCUMULATIVE AND TOXIC CHEMICALS (PBTS); PERSISTENT ORGANIC POLLUTANTS (POPS); WATER POLLUTION.

Bibliography

Bohmart, B.L. (1997). *The Standard Pesticide Users Guide*, 4th edition. London: Prentice-Hall International.

Carson, Rachel. (1963). *Silent Spring*. London: Hamish Hamilton.

Ekstrom, C., ed. (1994). *World Directory of Pesticide Control Organizations*. Farnham, U.K.: British Crop Protection Council.

Leng, M.L.; Leovey, E.M.K.; and Zubkoff, P.L., eds. (1995). *Agrochemical Environmental Fate: State of the Art.* Boca Raton, FL: CRC Press.

Pimentel, D., Lehman, H., eds. (1993). *The Pesticide Question: Environment, Economics, and Ethics.* New York: Chapman and Hall.

Rand, G.M., ed. (1995). *Fundamentals of Aquatic Toxicology: Effects, Environmental Fate and Risk Assessment.* Washington, D.C.: Taylor and Francis.

Smith, R.P. (1992). *A Primer of Environmental Toxicology.* Philadelphia: Lea and Febiger.

Ware, G.W. (1994). *The Pesticide Book*, 4th edition. Fresno, CA: Thomson Publications.

Internet Resource

U.S. Environmental Protection Agency Web site. "Pesticides." Available from http://www.epa.gov/pesticides.

Clive A. Edwards

Petroleum

Petroleum is a naturally occurring liquid oil normally found in deposits beneath the surface of the earth. It is a type of oil composed of rock minerals, making it different from other kinds of oils that come from plants and animals (such as vegetable oil, animal fat, or essential oils). The word petroleum comes from the Latin words *petra* (rock) and *oleum* (oil), and so literally means *rock oil*. Despite this, petroleum is an organic compound, formed from the remains of microorganisms living millions of years ago. It is one of the three main fossil fuels, along with coal and natural gas.

Petroleum Economy

Petroleum, like all fossil fuels, primarily consists of a complex mixture of molecules called *hydrocarbons* (molecules containing both hydrogen and carbon). When it comes out of the ground, it is known as *crude oil*, and it may have various gases, solids, and trace minerals mixed in with it. Through refinement processes, a variety of consumer products can be made from petroleum. Most of these are fuels: gasoline, jet fuel, diesel fuel, kerosene, and propane are common examples. It is also used to make asphalt and lubricant grease, and it is a raw material for synthetic chemicals. Chemicals and materials derived from petroleum products include plastics, pesticides, fertilizers, paints, solvents, refrigerants, cleaning fluids, detergents, antifreeze, and synthetic fibers.

The modern petroleum industry began in 1859 in Pennsylvania, when a man named Edwin L. Drake constructed the first oil well, a facility for extracting petroleum from natural deposits. Since then, petroleum has become a valuable commodity in industrialized parts of the world, and oil companies actively search for petroleum deposits and build large oil-extraction facilities. Several deposits exist in the United States. However, around 1960 oil production in the country began to decline as oil in the deposits was being used up and fewer new deposits were being discovered. Demand for petroleum products continued to increase, and as a result the United States came to rely more and more on oil imported from other countries. In 2001 the amount of petroleum extracted from deposits in the United States was estimated to be only one-third of the amount demanded by U.S. consumers. A similar pattern exists in other industrialized countries, and some, like Japan and Germany, import almost all of the oil they use.

TEN LARGEST OIL SPILLS IN HISTORY (BY VOLUME)

Location	Date	Amount Spilled
1. Sea Island Installations, Persian Gulf, Kuwait	January 26, 1991	240,000,000 gallons (816,327 tons)
2. Ixtoc I exploratory well, Bahia del Campeche, Mexico	June 3, 1979	140,000,000 gallons (476,190 tons)
3. Production well, Fergana Valley, Uzbekistan	March 2, 1992	88,000,000 gallons (299,320 tons)
4. Nowruz No. 3 well, Persian Gulf, Nowruz Field, Iran	February 4, 1983	80,000,000 gallons (272,109 tons)
5. Tanker *Castillo de Bellver*, Table Bay, South Africa	August 6, 1983	78,500,000 gallons (267,007 tons)
6. Tanker *Amoco Cadiz*, off Portsall, Brittany, France	March 16, 1978	68,668,000 gallons (233,565 tons)
7. Tanker *Odyssey*, North Atlantic Ocean, off St. John's, Newfoundland, Canada	November 10, 1988	43,100,000 gallons (146,600 tons)
8. Tanker *Atlantic Empress*, Caribbean Sea, Trinidad and Tobago	July 19, 1979	42,704,000 gallons (145,252 tons)
9. Tanker *Haven*, Genoa, Italy	April 11, 1991	42,000,000 gallons (142,857 tons)
10. Production well D-103, 800 km southeast of Tripoli, Libya	August 1, 1980	42,000,000 gallons (142,857 tons)

SOURCE: Oil Spill Intelligence Report (1999). *International Oil Spill Statistics: 1998.* New York: Aspen Publishers. Available from *www.aspenpublishers.com/environment.asp*

However, on a per capita basis, the consumption in these countries is nowhere near the consumption in the United States.

The United States and Canada are unique in that, on average, an individual in these countries consumes about twice as much petroleum product as do individuals in most other industrialized nations. People in the United States and Canada rely more on personal vehicles for their transportation and tend to drive greater distances, making petroleum their major source of energy. In the United States, about two-thirds of the petroleum consumed is transportation fuel, and two-thirds of that (45% of the total) is gasoline for cars and trucks. About 40 percent of the energy used in the United States every year comes from petroleum.

Foreign Oil Dependence

Political leaders in the United States have long been gravely concerned about the country's growing dependence on foreign oil, which in many ways puts the country at the mercy of foreign governments, some of them hostile to the United States. The greatest production of crude oil in the world is in the Persian Gulf region of the Middle East, where about 65 percent of the world's known petroleum deposits are located. About half of U.S. imports come from members of the Organization of the Petroleum Exporting Countries (OPEC), a group of countries encompassing the Persian Gulf and certain parts of Africa and South America. Events in these often volatile regions can have a huge impact on oil prices in the United States and worldwide, and because of the crucial role oil plays in U.S. society any change in the price can precipitate uncontrollable shifts in the country's economy (see chart "World Oil Price 1970-2000"). The most famous example of this is the Arab Oil Embargo of 1973 to 1974, when U.S. support for Israel in a conflict in the Middle East led to a decision by OPEC to impose steep price increases on the sale of oil to the United States. One response by the U.S. government has been the establishment of the Strategic Petroleum Reserve, an emergency stockpile designed to sustain the country's oil needs for approximately three months in the event of a complete cutoff of imports. There is little doubt, however, that dependence on foreign oil is both a political liability for the United States as well as a risk to national security.

Workers using water hoses to clean oil from a beach following a spill. (United States Environmental Protection Agency. Reproduced by permission.)

Environmental Pollution

Petroleum-derived contaminants constitute one of the most prevalent sources of environmental degradation in the industrialized world. In large concentrations, the hydrocarbon molecules that make up crude oil and petroleum products are highly toxic to many organisms, including humans. Petroleum also contains trace amounts of sulfur and nitrogen compounds, which are dangerous by themselves and can react with the environment to produce secondary poisonous chemicals. The dominance of petroleum products in the United States and the world economy creates the conditions for distributing large amounts of these toxins into populated areas and ecosystems around the globe.

Smoke is pouring from a refinery burnoff vent. (© Royalty-Free/Corbis. Reproduced by permission.)

Oil Spills

Perhaps the most visible source of petroleum pollution are the catastrophic oil-tanker spills—like the 1989 *Exxon Valdez* spill in Prince William Sound, Alaska—that make news headlines and provide disheartening pictures of oil-coated shorelines and dead or oiled birds and sea animals. These spills occur during the transportation of crude oil from exporting to importing nations. Crude oil travels for long distances by either ocean tanker or land pipeline, and both methods are prone to accidents. Oil may also spill at the site where it is extracted, as in the case of a blowout like the Ixtoc I exploratory well in 1979 (see table "Ten Largest Oil Spills in History"). A blowout is one of the major risks of drilling for oil. It occurs when gas trapped inside the deposit is at such a high pressure that oil suddenly erupts out of the drill shaft in a geyser.

Accidents with tankers, pipelines, and oil wells release massive quantities of petroleum into land and marine ecosystems in a concentrated form. The ecological impacts of large spills like these have only been studied for a very

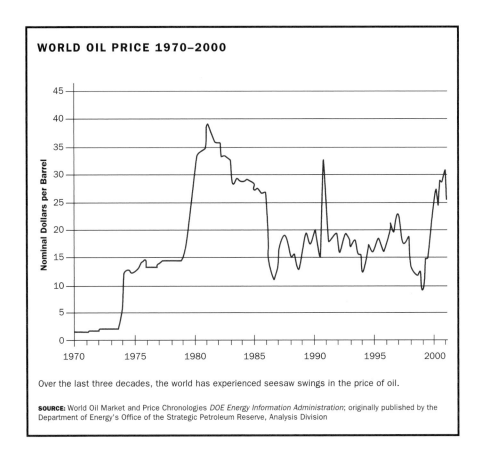

WORLD OIL PRICE 1970–2000

Over the last three decades, the world has experienced seesaw swings in the price of oil.

SOURCE: World Oil Market and Price Chronologies *DOE Energy Information Administration*; originally published by the Department of Energy's Office of the Strategic Petroleum Reserve, Analysis Division

few cases, and it is not possible to say which have been the most environmentally damaging accidents in history. A large oil spill in the open ocean may do less harm to marine organisms than a small spill near the shore. The *Exxon Valdez* disaster created a huge ecological disaster not because of the volume of oil spilled (eleven million gallons) but because of the amount of shoreline affected, the sensitivity and abundance of organisms in the area, and the physical characteristics of the Prince William Sound, which helped to amplify the damage. The *Exxon Valdez* spill sparked the most comprehensive and costly cleanup effort ever attempted, and called more public attention to oil accidents than ever before. Scientific studies of the effects of oil in Prince William Sound are ongoing, and the number of tanker accidents worldwide has decreased significantly since the time of the Valdez spill, due to stricter regulations and such required improvements in vessel design as double-hull construction.

Nonpoint Sources

Spills from tankers, pipelines, and oil wells are examples of *point sources* of pollution, where the origin of the contaminants is a single identifiable point. They also represent catastrophic releases of a large volume of pollutants in a short period of time. But the majority of pollution from oil is from nonpoint sources, where small amounts coming from many different places over a long period of time add up to large-scale effects. Seventy percent of the oil released by human activity into oceans worldwide is a result of small spills during petroleum consumption. These minor unreported spills can include routine discharges of fuel from commercial vessels or leakage from recreational boats.

OIL SEEPS

Almost half (45%) of the petroleum entering the marine environment is from natural seeps rather than anthropogenic sources. At seeps, oil and gas bubble out of cracks in the seabed creating special environments in which new organisms grow. These organisms survive through chemosynthesis rather than photosynthesis. They live in total darkness, more than four hundred meters below sea level, but survive by feeding directly off the hydrocarbons present in seeps or by eating carbon compounds resulting from chemosynthetic bacterial degradation of seep oil. Since 1984 oceanographers have discovered chemosynthetic communities of clams, mussels, tubeworms, bacterial mats, and other organisms on the seafloor of the Gulf of Mexico. United States Department of the Interior regulations protect these chemosynthetic communities from damage due to oil and gas drilling activities.

However, in North America, the majority of the release originates on land. Oil tends to collect in hazardous concentrations in the stream of wastewater coming out of cities and other populated areas. Runoff from asphalt-covered roads and parking lots enters storm drains, streams, and lakes and eventually travels to the ocean, affecting all of the ecosystems through which it passes. As cities grow, more and more people use petroleum products—lubricants, solvents, oil-based paint, and, above all, gasoline—and these are often improperly disposed of down drains and sewage pipes. Industrial plants also produce small, chronic spills that aren't noticed individually, but add up over time and enter waterways.

Taken together, land-based river and urban runoff sources constitute over half of the petroleum pollution introduced to North American coastal waters due to human activity, and 20 percent of the petroleum pollution introduced to ocean waters worldwide. When wastewater from these sources enters the marine environment it is usually by means of an estuary, an area where freshwater from land mixes with seawater. Estuaries are especially critical habitats for a variety of plants and animals, and are among the ecosystems most sensitive to pollutants.

Petroleum-Contaminated Soil

Not all oil released from land sources is quickly washed away to sea, however. Pipeline and oil-well accidents, unregulated industrial waste, and leaking underground storage tanks can all permanently contaminate large areas of soil, making them economically useless as well as dangerous to the health of organisms living in and around them. Removing or treating soil contaminated by petroleum is especially urgent because the hydrocarbons can leach into the underlying groundwater and move into human residential areas. The engineering field of *bioremediation* has emerged in recent decades as a response to this threat. In bioremediation, bacteria that feed on hydrocarbons and transform them into carbon dioxide can be applied to an affected area. Bioremediation has in many cases made cleaning up petroleum-contaminated sites a profitable real-estate investment for land developers.

Air Pollution

The U.S. Environmental Protection Agency (EPA) designates six criteria pollutants for determining air quality. These are: carbon monoxide (CO), nitrogen oxides (NO and/or NO_2, usually referred to as NO_x), sulfur dioxide (SO_2), ground-level ozone (O_3), particulate matter (including things like soot, dust, asbestos fibers, pesticides, and metals), and lead (Pb). Petroleum-fueled vehicles, engines, and industrial processes directly produce the vast majority of CO and NO_x in the atmosphere. They are also the principal source of gaseous hydrocarbons (also called volatile organic compounds, or VOCs), which combine with NO_x in sunlight to create O_3. Ozone, while important for blocking ultraviolet rays in the upper atmosphere, is also a key component of urban smog and creates human health problems when present in the lower atmosphere. Sulfur dioxide is a trace component of crude oil, and can cause acid rain when released into the air at oil refineries or petroleum power plants. Particulate matter is directly emitted in vehicle exhaust and can also form from the reaction of exhaust gases with water vapor and sunlight. Finally, leaded gasoline is a huge contributor of lead to the atmosphere, and the use of unleaded gasoline has decreased lead concentrations dramatically. The EPA and the World Bank are working to encourage the phaseout of leaded gasoline worldwide.

Petroleum-fueled transportation and coal-burning power plants are considered the chief causes of global warming. Excess amounts of carbon dioxide, methane, and NO_x, among other gases, trap heat in the atmosphere and create the greenhouse effect. Carbon dioxide (CO_2) is a main constituent of petroleum fuel exhaust, even though it is not toxic and therefore not classified as a pollutant. About one-third of the CO_2 emitted into the atmosphere every year comes from vehicle exhaust. Methane (NH_3), although usually associated with natural gas, is also emitted whenever crude oil is extracted, transported, refined, or stored.

The Future of Petroleum

The world's reliance on petroleum is expected to grow, despite widespread environmental, economic, and political consequences. The U.S. oil extraction industry continues to aggressively search for new oil deposits and lobby the federal government to open up restricted areas to drilling. The Arctic National Wildlife Refuge in Alaska has been on the oil industry agenda for several decades, creating a long-standing environmental controversy. Advances in oil well technology have allowed extraction in the deep ocean beyond the continental shelf, but these have not been enough to reverse the trend of declining production in the United States.

There are many compelling reasons to decrease society's dependence on petroleum for energy, and the most obvious place to begin is in the transportation sector. Energy-efficient engines and hybrid gas/electric cars can help to reduce some of the need for oil, providing higher gas mileage and less demand. A variety of alternative fuels have also been developed, such as ethanol, biodiesel (made from vegetable oil), and hydrogen. Each of these would produce little or no exhaust pollutants or greenhouse gases, and each derives from plentiful renewable resources. The United States is now in fact actively researching hydrogen as a viable alternative to gasoline, and the hydrogen fuel cell as a substitute for the internal combustion engine.

Petroleum is a useful chemical substance for many important purposes. But it is also a nonrenewable resource with a highly toxic composition, and it poses significant problems when used in huge volumes throughout the industrialized world. SEE ALSO AIR POLLUTION; ARCTIC NATIONAL WILDLIFE REFUGE; COAL; DISASTERS: OIL SPILLS; ECONOMICS; ELECTRIC POWER; ENERGY; FOSSIL FUELS; GLOBAL WARMING; OZONE; NO_x; RENEWABLE ENERGY; SULFUR DIOXIDE; UNDERGROUND STORAGE TANKS; VEHICULAR POLLUTION.

Bibliography

Oil Spill Intelligence Report. (1997). *Oil Spills from Vessels (1960–1995): An International Historical Perspective.* New York: Aspen Publishers.

Internet Resources

Committee on Oil in the Sea, National Research Council. (2003). *Oil in the Sea III: Inputs, Fates, and Effects.* Washington, D.C.: The National Academies Press. Available from http://www.nap.edu/catalog/10388.html.

Energy Information Administration. "Official Energy Statistics from the U.S. Government." Available from http://www.eia.doe.gov.

Exxon Valdez Oil Spill Trustee Council. "Restoring the Resources Injured by the Exxon Valdez Oil Spill and Understanding Environmental Change in the Northern Gulf of Alaska." Available from http://www.oilspill.state.ak.us.

National Biodiesel Board. "Need a Fill Up?" Available from http://www.biodiesel.org.

National Ethanol Vehicle Coalition. "National Ethanol Vehicle Coalition and E85." Available from http://www.e85fuel.com.

National Oceanic and Atmospheric Administration. "Office of Response and Restoration, National Ocean Service." Available from http://response.restoration.noaa.gov.

Schlumberger Excellence in Educational Development (SEED) Science Center. "Science Lab: Oil Well Blowout Simulator." Available from http://www.slb.com/seed/en/lab/blowout.

Trench, Cheryl J. (2001). "Oil Market Basics." Washington, D.C.: Energy Information Administration. Available from http://www.eia.doe.gov.

U.S. Department of Energy. "Energy Efficiency and Renewable Energy." Available from http://www.eere.energy.gov.

U.S. Department of Energy. "Fossil.energy.gov: A U.S. Department of Energy Web Site." Available from http://www.fossil.energy.gov.

U.S. Department of Energy. "Fossil Fuels: An Energy Education Website." Available from http://www.fossil.energy.gov/education.

U.S. Environmental Protection Agency. (1995). *Profile of the Petroleum Refining Industry*. Washington, D.C.: U.S. Government Printing Office. Available from http://www.epa.gov.

U.S. Environmental Protection Agency. (1999). *Profile of the Oil and Gas Extraction Industry*. Washington, D.C.: U.S. Government Printing Office. Available from http://www.epa.gov.

U.S. Environmental Protection Agency. "Air Quality Where You Live." Available from http://www.epa.gov/air/urbanair/index.html.

U.S. Geological Survey. Available from http://www.usgs.gov.

U.S. Geological Survey. (1997). "Bioremediation: Nature's Way to a Cleaner Environment." Available from http://water.usgs.gov/wid/html/bioremed.html.

Adrian MacDonald

Pharmaceutical Waste   *See Medical Waste; Resource Conservation and Recovery Act*

Phosphates

Pure phosphorus is rare in nature. It usually combines with oxygen to form phosphate ions or groups (PO_4^{3-}). Phosphates are considered organic when phosphate groups attach to carbon atoms or inorganic when phosphate ions associate with minerals such as calcium. Organic phosphates provide the energy for most chemical reactions in living cells.

The weathering of rocks releases inorganic phosphorus into the soil, and plants take this up and convert it to organic phosphate in their tissue. Humans and animals eat the plants, and when they die, phosphorus is returned to the soil by the action of bacteria and then again taken up by plants. This is the so-called phosphorus cycle.

Phosphates are normally a limiting factor for aquatic plant growth. When large amounts of phosphorus enter water, for instance, from farm runoff containing fertilizer, plants can grow out of control. Concentrations as low as 0.01 milligrams per liter (mg/L) can greatly impact a stream. This overfeeding is called eutrophication and may cause an algae bloom. The algae eventually die and sink to the bottom. Bacteria feeding on the algae remove oxygen from the water for respiration. As oxygen levels become lower, animals that need high oxygen levels such as fish will die. This is especially a problem at night when no photosynthesis occurs to replenish the oxygen.

If organic oxygen levels drop sufficiently, aerobic organisms can no longer survive and anaerobic bacteria take over. The end products of anaerobic respiration may smell like rotten eggs, fishy, or wormy. SEE ALSO AGRICULTURE; FISH KILLS; HEALTH, ENVIRONMENTAL; WASTEWATER TREATMENT; WATER POLLUTION.

Internet Resource

University of Maryland. "Impact of Phosphorus on Aquatic Life." Available from http://www.agnr.umd.edu/users.

Diana Strnisa

Photochemical Smog   *See Smog*

Phytoremediation   *See Bioremediation*

PIRGs   *See Public Interest Research Groups*

Plastic

Plastics are a subspecies of a class of materials known as polymers. These are composed of large **molecules**, formed by joining many, often thousands, of smaller molecules (monomers) together. Other kinds of polymers are fibers, films, elastomers (rubbers), and biopolymers (i.e., cellulose, proteins, and nucleic acids). Plastics are made from low-molecular-weight monomer precursors, organic materials, which are mostly derived from petroleum, that are joined together by a process called "polymerization." Plastics owe their name to their most important property, the ability to be shaped to almost any form to produce articles of practical value. Plastics can be stiff and hard or flexible and soft. Because of their light weight, low cost, and desirable properties, their use has rapidly increased and they have replaced other materials such as metals and glass. They are used in millions of items, including cars, bulletproof vests, toys, hospital equipment, and food containers. More than a hundred billion pounds of plastic were produced in 2000. Their increased use has resulted in concern with (1) the consumption of natural resources such as oil, (2) the toxicity associated with their manufacture and use, and (3) the environmental impact arising from discarded plastics.

molecule the smallest division of a compound that still retains or exhibits all the properties of the substance

Pollution Problems

Industrial practices in plastic manufacture can lead to polluting effluents and the use of toxic intermediates, the exposure to which can be hazardous. Better industrial practices have led to minimizing exposure of plant workers to harmful fumes; for example, there have been problems in the past resulting from workers being exposed to toxic vinyl chloride vapor during the production of polyvinyl chloride. Much progress has been made in developing "green processes" that avoid the use of detrimental substances. For example, phosgene, a toxic "war gas," was formerly used in the manufacture of polycarbonates. New processes, now almost universally employed, eliminate its

Cape fur seal lying on rock, dead of suffocation from a plastic wire wound around its neck, South Africa. (©Martin Harvey; Gallo Images/Corbis. Reproduced by permission.)

use. Also, the "just in time" approach to manufacture has been made possible by computer-controlled processes, whereby no significant amounts of intermediates are stored, but just generated as needed. In addition, efforts are ongoing to employ "friendly" processes involving enzyme-catalyzed low-temperature methods akin to biological reactions to replace more polluting high-temperature processes involving operations like distillation.

Spillage of plastic pellets that find their way into sewage systems, and eventually to the sea, has hurt wildlife that may mistake the pellets for food. Better "housekeeping" of plastic molding facilities is being enforced in an attempt to address this problem. Most plastics are relatively inert biologically, and they have been employed in medical devices such as prosthetics, artery replacements, and "soft" and interocular lenses. Problems with their use largely result from the presence of trace amounts of nonplastic components such as monomers and plasticizers. This has led to restrictions on the use of some plastics for food applications, but improved technology has led to a reduction in the content of such undesirable components. For example, the use of polyacrylonitrile for beverage bottles was banned at one time because the traces of its monomer, acrylonitrile, were a possible carcinogen. However, current practices render it acceptable today. There has been concern about **endocrine disruption** from phthalate-containing plasticizers used for plastics such as polyvinyl chloride (PVC). The subject of this possible side effect is controver-

endocrine disruption disruption of hormone control systems in the body

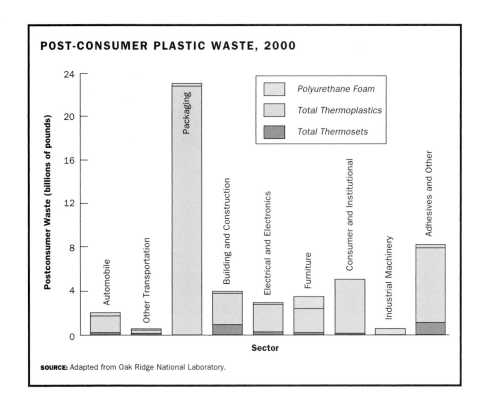

POST-CONSUMER PLASTIC WASTE, 2000

Legend:
- Polyurethane Foam
- Total Thermoplastics
- Total Thermosets

Y-axis: Postconsumer Waste (billions of pounds)

X-axis: Sector

Sectors: Automobile, Other Transportation, Packaging, Building and Construction, Electrical and Electronics, Furniture, Consumer and Institutional, Industrial Machinery, Adhesives and Other

SOURCE: Adapted from Oak Ridge National Laboratory.

sial, but caution in use is warranted pending further study. Plastics may also result in problems resulting from their improper use, and there is need of better education concerning limitations of use, for example, precautions that should be taken with items such as frying pan coatings and microwavable containers. When exposed to high temperatures, some plastics decompose or oxidize and produce low molecular weight products that may be toxic.

Reduced Use and Recycling

There is growing concern about the excess use of plastics, particularly in packaging. This has been done, in part, to avoid the theft of small objects. The use of plastics can be reduced through a better choice of container sizes and through the distribution of liquid products in more concentrated form. A concern is the proper disposal of waste plastics. Litter results from careless disposal, and decomposition rates in landfills can be extremely long. Consumers should be persuaded or required to divert these for recycling or other environmentally acceptable procedures. Marine pollution arising from disposal of plastics from ships or flow from storm sewers must be avoided. Disposal at sea is prohibited by federal regulation.

Recycling of plastics is desirable because it avoids their accumulation in landfills. While plastics constitute only about 8 percent by weight or 20 percent by volume of municipal solid waste, their low density and slowness to decompose makes them a visible pollutant of public concern. It is evident that the success of recycling is limited by the development of successful strategies for collection and separation. Recycling of scrap plastics by manufacturers has been highly successful and has proven economical, but recovering discarded plastics from consumers is more difficult. It is well recognized that separated plastics can be recycled to yield more superior products than possible for mixed ones.

Labeling plastic items with symbols has been employed, which enables consumers to identify them easily for placement in separate containers for curbside pickup. However, success depends on how conscientious consumers are in employing such standards and the ability of collectors to keep various types of plastic separate. Even a small amount of a foreign plastic in recycling feedstock can lead to the appreciable deterioration of properties, and it is difficult to achieve a high degree of purity. Manual sorting at recycling centers helps, but even trained sorters have difficulty identifying recyclables. Furthermore, manual sorting is an unattractive task and retaining labor willing to be trained for this is problematic. Automatic sorting techniques have been developed that depend on various physical, optical, or electronic properties of plastics for identification. Such methods prove difficult because of the variety of sizes, shapes, and colors of plastic objects that are encountered. Although in principle it is possible to create devices that can separate plastics with varying degrees of success, the equipment generally becomes more expensive with increasing efficiency. Technology for this continues to improve, and it is becoming possible to successfully separate mixed plastics derived from curbside pickup using such equipment.

To separate plastics, it is first necessary to identify the different types as indicated in the table. One must also distinguish between thermoplastics and thermosets. The latter, as found in tires and melamine dishes, has molecules that are interconnected by "crosslinks" and cannot be readily melted for recycling unless they are chemically reduced to low-molecular-weight species. For tires, recycling has not proved economical so disposal has involved grinding them up as asphalt additives for roads or burning in cement kilns.

Over 1.5 million pounds of plastic bottles were recycled in 2000, representing a four-fold increase in the amount of plastic recycled the previous decade. Nonetheless, the capacity to recycle bottles appreciably exceeds their supply by about 40 percent, so local governments and environmental groups need to encourage greater participation in this practice among consumers.

Profitable operations are currently in place for recycling polyethylene terephthalate (PET) from bottle sources and converting it into products such as fibers. One persistent problem, though, is obtaining clean enough feedstock to avoid the clogging of orifices in spinnerets by foreign particles. This has limited the ability to produce fine denier fibers from such sources. PET recycling is also constrained by regulations limiting its use to produce items in contact with food because there had been concern about contamination in consideration of improved recycling techniques.

A leading candidate for recycle feedstock is carpets because replacement carpets are usually installed by professionals able to identify recyclables and who serve as a ready source for recycling operations. They face the problem, however, of separating the recyclable carpet components from other parts such as jute backing and dirt. Such recycling operations have been only marginally profitable.

Polystyrene (PS) is another potentially recyclable polymer, but identifying a readily collectable source is problematic. One had been the Styrofoam "clamshells" fast-food chains use to package hamburgers. Recyclers were able to profitably collect polystyrene from such sources and produce salable products. However, largely because of public pressure, this use of polystyrene has

MAJOR TYPES OF PLASTICS BY S.P.I. CODES

SPI Code	Type of Resin	Example Products	% of Plastic
1	PET - Polyethylene terephthalate	Soft drink bottles, medicine containers	0.5%
2	HDPE - High-density polyethylene	Milk and water bottles, detergent bottles, toys	21%
3	PVC - Polyvinyl chloride	Pipe, meat wrap, cooking oil bottles	6.5%
4	LDPE - Low-density polyethylene	Wrapping films, grocery bags	27%
5	PP - Polypropylene	Syrup bottles, yogurt tubs, diapers	16%
6	PS - Polystyrene	Coffee cups, "clamshells"	16%
7	Other		8.5%

TYPES OF PLASTIC PACKAGING

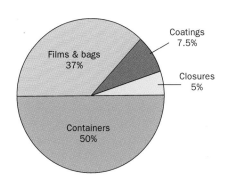

By Type of Use

By Type of Resin

- A one gallon plastic milk container that weighed 120 grams in 1960 now weighs just 65 grams.
- The average 1992 American car contains 300 pounds of plastic made from about 60 different resins.
- Every year, we make enough plastic film to shrink-wrap the state of Texas.
- 10% of the average grocery bill pays for packaging (mostly paper and plastics)—that's more than goes to the farmers.
- In 1993, plastics accounted for 11.5% of the U.S. municipal waste stream by weight (23.9% by volume). In 1994, plastics comprised 9.5% (by weight) of the waste stream.
- The rate of plastic soda bottle recycling rose from 33% in 1990 to 50% in 1994.
- 0.9 million tons of plastics (4.7%) were recycled in the U.S. in 1994.
- Products made from recovered plastic bottles include drainage pipes, toys, carpet, filler for pillows and sleeping bags, and cassette casings.

SOURCE: Modern Plastics, January, 1992

declined, so related recycling practices have largely disappeared too. Cafeteria items from school lunchrooms are another potential, but the collection of such objects involves the development of an infrastructure, often not in place. In these cases, it is necessary to separate the polystyrene from paper and food waste, but washing and flotation techniques have been developed for this purpose.

Increasing amounts of plastic components appear in automobiles, and their recovery from junked cars is a possibility. Its success depends on the ability of a prospective "junker" to identify and separate the plastic items. Three efforts may aid in this accomplishment:

1. The establishment of databases to enable junkers to learn what kinds of plastic are used in what parts of what model cars.

2. A reduction in the number of different plastics used for car construction.

3. The design of cars such that plastic parts may be removed easily (this would require special types of fasteners).

This illustrates a general need—the design of plastic-containing products with the ability to recycle in mind. As a consequence of public concern about the environmental problems arising from plastic use, industry is responding to these needs. The effort continues to use fewer different kinds of plastics and to adopt designs that allow for easier recycling but still retain desirable properties.

There are, however, some worthwhile products that can be produced from mixed plastic, such as "plastic lumber" used for picnic benches and marine applications such as docks and bulkheads that successfully replace wooden lumber which often contains toxic preservatives and arsenic. But, the market for such a product is limited, so efforts to obtain separated plastics are preferred.

Degradable Plastics

Discarded plastics are hard to eliminate from the environment because they do not degrade and have been designed to last a long time. It is possible to design polymers containing monomer species that may be attacked by chemical, biological, or photochemical action so that degradation by such means will occur over a predetermined period of time. Such polymers can be made by chemical synthesis (as with polylactic acid) or through bacterial or agricultural processes (as with the polyalkonates). Although such processes are often more expensive than conventional ones, cost would undoubtedly drop with increased production volume. One success story was the introduction of carbonyl groups into polyethylene by mixing carbon monoxide with ethylene during synthesis. These carbonyl groups are chomophores that lead to chain breaking upon the absorption of ultraviolet light. The polymer is then broken down into small enough units that are subject to bacterial attack. This approach has been successful, for example, in promoting the disappearance of rings from beverage cans, which are potentially harmful to wildlife.

A problem with the degradation of plastics is that it is probably undesirable in landfills because of the leachants produced that may contaminate water supplies. It is better in these instances to ship the plastics to composting facilities. This requires the separation of degradable plastics from other materials and the availability of such facilities. In most cases, the infrastructure needed for such an approach is not in place. This has discouraged its use for disposable diapers that are said to constitute 1 to 2 percent of landfill volume.

Degradable polymers may have limited use in the reduction of litter and production of flushable plastics, for example, feminine hygiene products, but it seems unlikely that the use of such materials will be a viable means of disposal for large amounts of plastic products. Degradation leads to the loss of most of the potential energy content of plastics that might be recovered by trash-to-energy procedures.

Trash to Energy

A method of plastic disposal with more positive environmental implications is burning and recovering the energy for power generation or heating. Plastics contain much of the energy potential of the petroleum from which they are made, and they, in a sense, are just borrowing this energy that may be recovered when the plastic is burned. Environmentalists and the public have objected to this procedure, leading to legislative restrictions. This has arisen, in part, because of the image of "old-fashioned" incinerators polluting the air with toxic fumes and ash. However, it is possible to construct a "high-tech" incinerator designed to operate at appropriate temperatures and with sufficient air supply that these problems are minimized. Remaining toxic substances in fumes may be removed by scrubbing, and studies have shown that no significant air pollution results. Toxic ash, for the most part, does not arise from the polymer components of the feedstock, but rather from other materials mixed with the polymers as well as from fillers, catalyst content, and pigments associated with the polymers. Proper design of the polymers and crude separation of the incinerator feedstock can reduce this problem. Furthermore, if the feedstock was not incinerated but placed in landfills, contaminants would ultimately enter the environment in an uncontrolled way. Incineration reduces the volume, so that the ash, which may contain them, can be disposed of under more controlled conditions. Also, it is possible to insolublize the ash by converting it into a cementlike material that will not readily dissolve.

Facilities for converting trash to energy in an environmentally acceptable way are expensive and at present not cost-effective when considering short-range funding. However, in the long run, they are environmentally desirable and reduce the need for alternative means for plastic waste disposal. It is imperative that legislators and taxpayers soon adopt this long-range perspective. SEE ALSO ENDOCRINE DISRUPTION; RECYCLING; SOLID WASTE; WASTE.

Bibliography

American Plastics Council. (2001). "2000 National Post Consumer Plastics Recycling Report." Arlington, VA: Author.

Gerngross, T.U., and Slater, S.C. (2000). "How Green Are Green Plastics." *Scientific American* August.

Hocking, M.B. (1991). "Paper vs. Polystyrene, a Complex Choice." *Science* 251.

Limbach, B.M. (1990). *Plastics and the Environment, Progress and Commitment.* Washington, D.C.: Society of the Plastics Industry.

Piaecki, B.; Rainry, D.; and Fletcher, K. (1998). "Is Combustion of Plastics Desirable?" *American Scientist* 86: 364.

Stein, R.S. (1992). "Polymer Recycling: Opportunities and Limitations." *Proceedings of the National Academy of Sciences of the United States of America* 89: 835.

Stein, R.S. (2002). "Plastics Can Be Good for the Environment." *NEACT Journal* 21: 10–12.

Vesilind, P.A. (1997). *Introduction to Environmental Engineering.* Boston, MA: PWS Publishing.

Richard S. Stein

Point Source

Point source pollution is contamination that enters the environment through any discernible, confined, and discrete conveyance, such as a smokestack, pipe, ditch, tunnel, or conduit. Point source pollution remains a major cause of pollution to both air and water. Point sources are differentiated from non-point sources, which are those that spread out over a large area and have no specific outlet or discharge point. Point source pollution in the United States is regulated by the Environmental Protection Agency (EPA).

Point Sources of Water Pollution

Point sources of water pollution include municipal sewage treatment plant discharges and industrial plant discharges. Municipal sewage treatment plant point sources can contribute pollution in the form of oxygen-depleting nutrients and in the form of pathogens that cause serious health hazards in drinking water and swimming areas. Industrial point sources can contribute pollution in the form of toxic chemicals and heavy metals. Examples of non-point source water pollution include agricultural and urban runoff, and runoff from mining, and construction sites.

The Clean Water Act (CWA), passed by Congress in 1972, provides the basic structure for regulating the discharge of pollutants from point sources to waters of the United States. The CWA gives the EPA the authority to establish effluent limits. Effluent is the outflow from a municipal or industrial treatment plant. The CWA also requires the acquisition of a National Pollution Discharge Elimination System (NPDES) permit prior to the discharge of pollutants. States may be authorized to implement CWA programs, but the EPA retains oversight responsibilities.

The EPA manages effluent limits for point sources in two ways: through technology-based controls and through water quality-based controls. Industry-wide effluent limits are established on a technology basis. These are minimum standards based on available treatment technology and pollution prevention measures. Effluent limits are also established on a water-quality basis. Water quality-based criteria are scientifically defensible standards that ensure protection of designated uses of a receiving water. Either standard may be superceded by the more stringent standard, as determined by the control authority.

Municipal point sources are the result of community sewage treatment systems. At the sewage treatment plant, wastewater is treated to remove solid and organic matter, disinfected to kill bacteria and viruses, and then often discharged to a surface water. Not all solids and organic matter are removed during treatment, resulting in degraded receiving water quality, due to a reduction in **dissolved oxygen**. Nutrients such as phosphorus that are not removed during treatment can cause overgrowth of algae and other organisms, also leading to lower dissolved oxygen. Many toxic substances can pass through conventional municipal treatment systems. Improperly treated sewage can be released as a result of upsets to the treatment process or as a result of operator error.

During heavy rain, discharges from sewage treatment systems can be a serious problem. In many municipalities, storm-water runoff is combined with municipal sewage in a common system. The increased water volume leads to reduced treatment. Combined sewer overflows occur when water flow exceeds treatment plant capacity, resulting in untreated sewage being discharged directly to rivers, lakes, or the ocean.

Industrial point sources are the result of industries using water in their production processes, and then treating the water prior to discharge. Some of the industries requiring process waters include pulp and paper mills, food processors, electronic equipment manufacturers, rare metal manufacturers, textile manufacturers, pharmaceutical manufacturers, forest product producers, leather tanners, and chemical manufacturers.

The National Pretreatment Program is charged with controlling the 126 **priority pollutants** from industries that discharge into sewer systems. These

dissolved oxygen (DO) the oxygen freely available in water, vital to fish and other aquatic life and for the prevention of odors; DO levels are considered a most important indicator of a water body's ability to support desirable aquatic life; secondary and advanced waste treatment are generally designed to ensure adequate DO in waste-receiving waters

priority pollutant a designated set of common water pollutants

pollutants fall into two categories: metals and toxic organics. The metals include lead, mercury, chromium, and cadmium. The toxic organics include solvents, pesticides, dioxins, and polychlorinated biphenyls (PCBs).

Unlike municipal treatment methods, which are similar across the country, industrial treatment methods are industry-specific. For example, electroplating wastewater may require cyanide removal through oxidization. In general, physical processes may be used to remove solids and biological processes to remove organics. Chemical treatment, such as precipitation and neutralization, is also widely used.

The *National Water Quality Inventory: 2000 Report* is compiled based on the water quality reports required to be submitted to the EPA by states every two years. The report identifies "impaired" waters: water that cannot support its designated use, such as fishing or swimming, due to contamination. According to the report, municipal point sources contributed to 37 percent and industrial discharges contributed to 26 percent of reported water-quality problems in the impaired portion of estuaries. Municipal point sources were the leading cause of contamination in 21 percent of the impaired ocean shorelines, and industrial discharges were the leading cause in 17 percent. Municipal point sources were a leading source of contamination in 10 percent of the impaired **river miles** and 12 percent of the impaired **lake acres**. These figures are improved over the percentages recorded in the *1992 Report* when municipal point sources were a leading contamination source in 15 percent of the impaired river miles and 21 percent of the impaired lake acres.

river mile one mile, as measured along a river's centerline

lake acre an acre of lake surface

The NPDES permit program can be credited with achieving significant improvements to the water quality of the United States. Immediately following passage of the CWA, efforts focused mainly on regulating traditional point sources, such as municipal sewage plants and industrial facilities. In the late 1980s, efforts to address "wet weather point sources," such as urban storm sewer systems, began. Currently, there is a greater focus on nonpoint source pollution. The EPA is moving away from a source-by-source and pollutant-by-pollutant approach to a watershed-based approach. A watershed, or "place-based," approach is a process that emphasizes addressing all stressors within a hydrologically defined boundary or drainage basin. Equal emphasis is placed on protecting healthy waters and restoring impaired waters.

Point Sources of Air Pollution

Point sources of air pollution include stationary sources such as power plants, smelters, industrial and commercial boilers, wood and pulp processors, paper mills, industrial surface coating facilities, refinery and chemical processing operations, and petroleum storage tanks. Examples of nonpoint sources of air pollution include: on-road mobile sources such as cars and trucks; nonroad mobile sources such as construction and recreation equipment engines; and natural sources such as windstorms and fires. Exposure to air pollution is associated with adverse effects on human health including respiratory problems and lung diseases. Air pollution can also significantly affect ecosystems.

The Clean Air Act (CAA) was passed by Congress in 1970 and amended in 1990. Under the CAA, EPA sets limits on how much of a pollutant is allowed in the air anywhere in the United States. Each state is required to develop a state implementation plan (SIP) to explain how it will do its job

Smoke is pouring from the smokestack of an incinerator. (U.S. EPA. Reproduced by permission.)

under the CAA. A permit must be obtained for large sources that release pollutants into the air. The permits require information on which pollutants are being released, how much pollutant is released, what steps are being taking to reduce pollution, and plans for monitoring.

The EPA has set national air quality standards for six principal air pollutants (also known as criteria pollutants): carbon monoxide (CO), lead (Pb), nitrogen dioxide (NO_2), ozone (O_3), particulate matter (PM), and sulfur dioxide (SO_2). CO, Pb, NO_2, and SO_2 result from direct emissions from a variety of sources, including point sources. PM can result from direct emissions or can form when emissions and other gases react in the atmosphere. Ozone is not emitted directly, but forms when nitrogen oxides (NO_x) and volatile organic compounds (VOCs) react in the presence of sunlight. The EPA refers to chemicals that cause serious health and environmental impacts as hazardous air pollutants (HAPs) or air toxics. Currently, 189 air toxics have been identified, including chemicals such as benzene, chloroform, and mercury.

The EPA tracks air pollution in two ways: (1) emissions form all sources going back thirty years and (2) air quality measured from monitoring stations around the country going back twenty years. The EPA summarizes its most recent evaluations in the report *Latest Findings on National Air Quality: 2000 Status and Trends*. Since 1970, the total emissions for the six criteria pollutants have been reduced 29 percent. National air quality levels measured at monitoring stations across the country have also shown improvements over the past twenty years for all six criteria pollutants. Over 160 million tons of pollution (from both point sources and non-point sources) are emitted into the air each year in the United States.

In *2000 Status and Trends*, the EPA reports an increasing focus on tracking and controlling ground-level ozone and fine particles, key components of smog and haze. Progress has been slowest for ground-level ozone. In some regions of the United States, ozone levels have actually increased in the past ten years. The ozone increase correlates to the increase in NO_x emissions from power plants and other sources. NO_x emissions also contribute to acid rain, haze and particulate matter. Sulfates, formed mainly from coal-fired power plant emissions, are the main source of particles in the eastern United States. The emissions also contribute to the formation of acid rain. The EPA's emissions trading program successfully reduced these air pollutants, resulting in improved visibility in the eastern United States.

While point source pollution is declining in the United States, it remains a global environmental concern. According to the UN report *Global Environment Outlook 2000*, rapid urbanization and industrialization in many developing countries is creating high levels of air and water pollution. SEE ALSO AIR POLLUTION; CWA; CAA; CUYAHOGA RIVER; DISASTERS; DONORA, PENNSYLVANIA; NATIONAL POLLUTANT DISCHARGE ELIMINATION SYSTEM (NPDES); NONPOINT SOURCE POLLUTION; THERMAL POLLUTION; TOXIC RELEASE INVENTORY; WASTEWATER TREATMENT; WATER POLLUTION.

Bibliography

U.S. Environmental Protection Agency. (2000). *National Water Quality Inventory: 2000 Report*. Washington, D.C.: U.S. Government Printing Office.

U.S. Environmental Protection Agency, Office of Wastewater Management. (1999). *Introduction to the National Pretreatment Program*. Washington, D.C.: U.S. Government Printing Office.

Vigil, Kenneth M. (1996). *Clean Water: The Citizen's Complete Guide to Water Quality and Water Pollution Control*. Portland, OR: Columbia Cascade Publishing Company.

Internet Resources

U.S. Environmental Protection Agency. National Pollutant Discharge Elimination System. Available from http://www.epa.gov/npdes.

U.S. Environmental Protection Agency, Office of Science and Technology. Available from http://www.epa.gov/OST.

United Nations Environment Programme. *Global Environment Outlook 2000*. Available from http://www.unep.org/geo2000.

Denise M. Leduc

Politics

Beginning in 1970, the "environmental decade," a swift and sweeping transformation in American law radically reshaped U.S. pollution control policies. This regulatory revolution was mounted on three political foundations: skillful pressure-group politics, effective legislative advocacy, and aroused public concern about environmental degradation. These traditional American political techniques promoted, and continue to shape, contemporary pollution control through U.S. political governmental institutions.

The Political Foundations: Pressure-Group Politics

Americans and their public officials paid scant attention to growing evidence of environmental degradation across the nation until the late 1960s. Air and water pollution control was considered the responsibility of state and local

Antoine Waechter (at center), French Green Party member, participating in a demonstration against the building of the Serre de la Fare dam, along the Loire headwaters. (© Bernard Bisson/Corbis. Reproduced by permission.)

governments. Most states did little more than set drinking water standards to protect public health from a few contaminants like bacterial diseases, fearing that more aggressive control of air and water pollutants would inhibit economic growth and drive resident business and industry to other states. Such mounting environmental degradation as the Cuyahoga River fire and Love Canal focused national attention on the need for environmental restoration. This was translated into bold new governmental policies largely by environmental pressure groups during the 1960s and 1970s.

The strength of the new environmental movement lay in organized political activism, coalition building, and legislative advocacy—the fundamentals of effective group politics. The focus of this political pressure was primarily the federal government with its vast authority and resources for creating nationwide pollution control. No single event dramatized the environmental movement's rise to national importance more than the first Earth Day in April 1970—a nationally televised Washington rally witnessed by 35 million Americans—that swiftly elevated public awareness of environmental degradation while advertising, especially for public officials, environmentalism's newly acquired political clout.

Pressure-Group Politics Old and New

Environmentalism's political strength depends on its leadership's skill in creating a broad and diverse alliance of interests to support environmental advocacy. The environmental movement embraces a great diversity of influential organizations, including traditional conservation groups like the Sierra Club and the National Wildlife Federation, established public health advocates like the American Cancer Society, newly formed environmental pressure groups like the Environmental Defense Fund and Friends of the Earth, major labor unions, public interest science organizations, and countless local organizations. Additionally, environmentalists are proficient recruiters. After the first Earth Day, environmentalist organizations multiplied and enriched their political resources, often creating innovative new organizational forms and strategies. Prior to 1970, fewer than twenty-five significant national environmental groups existed with a combined membership approaching 500,000—of these, perhaps a half-dozen organizations were important participants in national policymaking. Several hundred influential national environmentalist groups are politically active; five of the most important—the Audubon Society, Sierra Club, Environmental Defense Fund, National Wildlife Federation, and Wilderness Society—alone have a combined membership exceeding seven million. Although all the major organizations use the sophisticated resources of pressure-group politics—mass-mailing technology, skilled media specialists, and full-time legislative lobbyists—the environmental movement has also benefited by developing specialized legal advocacy groups, like the National Resources Defense Council, staffed primarily with lawyers and scientific experts committed exclusively to litigation that establishes important legal precedents and enforces pollution-control regulations for environmental protection.

Creating and Mobilizing Public Opinion

The radical transformation of U.S. pollution-control laws would have been impossible without strong, consistent public pressure on federal and state governments, especially on the Congress and state legislators. Current public opinion polls suggest that more than 80 percent of Americans agree with the goals of the environmental movement. The strength of this support is suggested by other polls consistently reporting since 1980 that more than two-thirds of the public believe environmental protection should be a major government priority, even at the risk of reducing economic growth. The breadth and depth of this ecological consciousness are remarkable, considering that few Americans understood the implications of ecology or the nature of domestic environmental pollution only a few decades ago. The most important political impact of this vigorous public environmentalism is on the electoral system: Candidates for major federal and state office are now customarily expected to support strong pollution controls and other ecologically protective policies, at least in principle. While Americans often disagree vigorously over pollution control methods, air and water pollution regulation itself is now an enduring component of the "American political consensus"—those policies Americans overwhelmingly view as the essential responsibility of their government.

A Regulatory Revolution: The Environmental Decade

The design of U.S. air and water pollution control was crafted in federal law during the "environmental decade" between 1970 and 1980. Responding to dramatic media revelations of ecological deterioration, growing environmental group pressure, and voter concerns, Congress laid the legislative foundation for all contemporary regulation through six statutes: The Clean Air Act Amendments (1970), the Federal Water Pollution Control Act Amendments

(1972), the Safe Drinking Water Act (1974), the Toxic Substances Control Act (1976), and the Comprehensive Environmental Response, Compensation and Liability Act (Superfund) in 1980. Altogether, the Congress wrote or amended nineteen major environmental laws in this remarkable decade. And by changing the law, Congress also reordered its political underpinning.

Federal Leadership

The laws listed above radically recast the U.S. approach to pollution management. Most important, the federal government assumed the primary responsibility for air and water pollution regulation; Washington set national pollution standards and supervised their implementation, thus defining pollution control priorities and prescribing acceptable control methods. The Clean Air Act, for example, now requires all states to control at least six dangerous pollutants (sulfur oxides, nitrogen oxides, carbon monoxide, lead, particulates, and volatile organic compounds) and a rapidly growing list of other substances currently believed to be "air toxins." The act additionally mandates that car manufacturers install pollution-control devices on all new automobiles. The new pollution laws also extended federal protection to the natural environment instead of exclusively to human health and safety. The Toxic Substances Control Act, for example, authorizes the federal government to regulate the manufacture, sale, or use of any chemical presenting "an unreasonable risk of injury to health or the environment."

Regulatory Federalism

Regulatory federalism has become a fundamental regulatory principle. This means that Washington prescribes national pollution standards and control procedures, but allocates the appropriate resources to states so they assume the primary responsibility for implementing and enforcing these requirements. States are then said to exercise "delegated authority." Using delegated authority, for instance, thirty-eight states as of 2002 issue permits for water pollution discharges required by the Federal Water Pollution Control Act Amendments and forty-nine states certify pesticides for local use as required by the Federal Environmental Pesticides Control Act (1972). Thus, the states assume an essential and highly influential role in national pollution regulation; pollution policymaking continually requires negotiation, conflict, and cooperation between the states and Washington.

New Regulatory Agencies

New federal agencies were created, and others reorganized, to implement these new control programs. The most important federal pollution control entity is now the U.S. Environmental Protection Agency (EPA), created in 1972. The EPA is the nation's largest regulatory agency with 18,000 employees, a 2002 budget exceeding $7.5 billion, and responsibility to fully or partially implement all the nation's important pollution control laws. In 1970 the President's Council on Environmental Quality (CEQ), a much smaller agency, was created within the White House to advise the President on environmental affairs. At the same time, the National Oceanic and Atmospheric Administration (NOAA) was created within the U.S. Department of Commerce to conduct research on and monitoring of ocean and atmospheric pollution. The authority and staff of many other federal agencies concerned with environmental quality, such as the Department of the Interior, were also vastly expanded to implement new pollution control programs. These agencies also provide research support and grants to the states to facilitate the enforcement of pollution control laws. The EPA, for instance, has distributed more than $150 billion in grants to state and local governments to upgrade their sewage treatment systems.

New Policymaking Procedures

Federal pollution laws created new, often controversial, regulatory procedures. The most contentious of these is *risk assessment*—the process used by regulatory agencies to determine if a substance constitutes a sufficient threat to human health and safety, or to the environment, to require control. Federal pollution laws, including the Toxic Substances Control Act, the Safe Drinking Water Act, and Superfund, require the EPA or other responsible agencies to conduct such risk assessments—usually focused on the risk of cancer—on thousands of chemicals never previously evaluated according to the rigorous new standards. Risk assessments proceed slowly due to the huge number of substances involved, a lack of basic information about their distribution and impact, and intense controversy about the appropriate procedures for the assessments.

Federal pollution legislation has also vastly increased opportunities for the public, and particularly environmental advocacy groups, to become informed and involved in federal environmental decision making. Major federal pollution laws such as the Clean Air and Clean Water Acts removed a major legal impediment to public involvement in pollution control by granting individuals and organizations **standing** to sue federal and state agencies for failure to enforce pollution control laws. Almost all federal environmental laws also require the responsible federal and state agencies to actively inform the public and to provide numerous opportunities for public comment and review of contemplated regulations.

standing the legal right to pursue a claim in court

At the beginning of the twenty-first century, it is apparent that the environmental movement permanently and comprehensively altered the law and politics of U.S. pollution regulation. Pressure-group politics, public opinion, and congressional legislation were the powerful driving forces in this change. The result was unprecedented, aggressive federal leadership in an active national program of pollution control based on federally mandated pollution standards and pollution controls. By promoting new national pollution control laws and agencies, expanded opportunities for public involvement in pollution regulation, and vigorous public concern for environmental degradation, the environmental movement has created a continuing "environmental era." SEE ALSO ACTIVISM; BROWER, DAVID; CARSON, RACHEL; CITIZEN SUITS; EARTH DAY; ENVIRONMENTAL IMPACT STATEMENT; GOVERNMENT; INDUSTRY; LAWS AND REGULATIONS, UNITED STATES; LEGISLATIVE PROCESS; NATIONAL ENVIRONMENTAL POLICY ACT (NEPA); NEW LEFT; PROGRESSIVE MOVEMENT; PUBLIC PARTICIPATION; PUBLIC POLICY DECISION MAKING; RISK.

Bibliography

Buck, Susan J. (1996). *Understanding Environmental Administration and Law*, 2nd edition. Washington, D.C.: Island Press.

Cohen, Richard E. (1995). *Washington at Work: Back Rooms and Clean Air*, 2nd edition. Boston: Allyn and Bacon.

Graham, Mary. (1999). *The Morning after Earth Day: Practical Environmental Politics.* Washington, D.C.: Brookings Institution Press.

Marzotto, Toni; Moshier Burnor, Vicky; and Bonham, Gorden Scott Bonham. (2000). *The Evolution of Public Policy: Cars and the Environment.* Boulder, CO: Lynne Rienner Publishers.

Rosenbaum, Walter A. (2002). *Environmental Politics and Policy*, 5th edition. Washington, D.C.: CQ Press.

Internet Resource

Project on Teaching Global Environmental Politics Web site. Available from http://webpub.alleg.edu/employee.

Walter A. Rosenbaum

Pollution Prevention

One key to achieving a sustainable society and tackling the complex environmental challenges of the twenty-first century is pollution prevention (P2), reducing or eliminating pollution before it is created. The idea has been discussed since 1976, but has only lately gained widespread support from both the private and public sectors. It is an environmentally sound and cost-effective practice.

In 1990 Congress passed a federal statute, the Pollution Prevention Act of 1990. The act defined pollution prevention (i.e., source reduction) as a practice that

1. Reduces the amount of any hazardous substance, pollutant, or contaminant entering any waste stream or otherwise released into the environment (including fugitive emissions) prior to recycling, treatment, or disposal; and

2. Reduces the hazards to public health and the environment associated with the release of such substances, pollutants or contaminants. The term includes equipment or technology modifications, process or procedure modifications, reformulation or redesign of products, substitution of raw materials, and improvements in housekeeping, maintenance, training or inventory control.

Since the Industrial Revolution, U.S. environmental policy has focused on end-of-pipe environmental remediation, control, and disposal. The end-of-pipe approach involves combatting pollution, regardless of what form (solid or hazardous waste, air emissions, or water discharge), only after it has been created.

To control end-of-pipe pollution, society issues permits. These permits set threshold limits for how much pollution a facility is allowed to create, taking into consideration the ecosystem in which the company operates. The more fragile the environment, the more consideration, presumably, is given to the allowable level of pollutant discharge. The result is that a company may obtain a permit to emit a certain amount of carcinogenic chemicals into the air or water as a by-product of its operations. The same system also holds true for communities. A community, for instance, acquires a permit to operate a landfill. The permit will stipulate certain types of waste for disposal, as well as place limits on the quantity that may be dumped on a daily basis.

DECADE OF P2 RESULTS, FROM 1990 TO 2000
(QUANTITATIVE PRELIMINARY DATA FROM SELECTED STATES)*

State	Pollution prevented (lb unless otherwise noted)	Total cost savings
Illinois	3.73×10^9 overall pollutant reduction.	$27 million
Maine	220 toxic-use companies reduced 105 million lb. toxic chemicals/5.6 million lb of waste reduced at 167 hazardous waste generators—63 toxic-release companies have reduced releases by 12.7 million lb.	Hazardous waste reduction overall has generated $50 million in cost savings to Maine businesses.
New Jersey	50% reduction in toxic/waste/nonproduct output for TRI chemicals.	$8 million
Massachusetts	Reduction in toxic chemical waste generation by 60 million lb or 57%, reduction in total chemical use by 317 million lb or 41%, reduction in toxic releases to the environment by 18 million lb or 87%.	
New Hampshire (7 years)	P2 assistance provided to 43 companies resulted in cost savings of $2.8 million per year for a total of $19,600,000.	
Vermont	Reduction in 38% in the total amount of hazardous waste generated through the end of 1998.	
Rhode Island	RI DEM performed more than 250 site assessments that resulted in elimination of more than $40 million tons of industrial waste.	
North Carolina	Since 1993 air pollutants reduced by 122,000 lb, water pollutants reduced by 11,836,500 lb, waste reduced by approx. 64 million lb. In 1999 energy conserved: 344,000 kW; water conserved: approx. 77,296,000 gallons.	$55,318,400 estimated total cost savings from P2 efforts.
Iowa	Iowa Waste Reduction Center at the Univ. of Northern Iowa has conducted more than 2,100 on-site reviews at Iowa small businesses since its inception in 1988. Approx. 87 million lb of hazardous and solid waste have been reduced as a result.	
Alaska	Since 1994 the Alaska Dept. of Environmental Conservation's Compliance Assistance Office has helped businesses reduce waste by 201,500 lb.	Cost savings are estimated at $1,752,000.
Virginia	From 1998–2001 Virginia Department of Environmental Quality reports that more than 1.5 million lb of air pollutants, 488 million lb of water pollutants, and 710 million lb of waste have been reduced as a result of P2 efforts in the state.	
Kentucky	Has achieved a 50% reduction in hazardous waste generation/releases based on TRI and hazardous waste data using the state's voluntary 1989 goals.	Estimated cost savings of $500,000 annually since 1991.
Maryland	Maryland Dept. of Environment reported a reduction of 17,780,109 million lb of waste for the period of 1997–2000.	Cost savings during same period estimated to be $125,863,000.

*As of 2002, the information listed above is being compiled into a comprehensive study and evaluation of P2 efforts over the past decade. The data listed here were reported by individual state agencies and are only intended to present baseline data, by which future P2 efforts can be measured. These numbers were the result of a survey conducted by NPPR from 1990 through 2000. Some programs submitted surveys, others reports. In no way should these numbers be used to compare programs. Unless otherwise noted, results date from the period when the state commenced its P2 program. Some states have had assistance programs much longer than others, and some may have smaller operating budgets for their programs.

Even recycling efforts, important as they are, focus attention on the back end of the pollution process, after waste has been produced. Recycling is an end-of-pipe solution.

Another outdated aspect of U.S. environmental policy is the single-medium approach to environmental problems. Single-medium approaches focus on one specific environmental medium (i.e., land, water, or air) at a time, generally to the exclusion of other media. Air pollution experts, for example, do not typically investigate other facets of a facility, such as its overall operation, waste generation, or water discharges. They view their one medium in isolation and may recommend new procedures or remedies that can adversely impact other media. It is not uncommon to see an inspector recommend measures to improve air quality that affect water quality or waste generation—thus simply transferring pollution from one medium to another.

The United States takes the single-medium approach because major environmental statutes are single-medium in scope. The Clean Air Act, Clean Water Act, and Resource Conservation and Recovery Act (RCRA) each focus on individual media. They contain strong requirements that focus on end-of-pipe approaches to meet them. These statutes are at the core of

persistent bioaccumulative toxics a group of substances that are not easily degraded, accumulate in organisms, and exhibit an acute or chronic toxicity

green choice a product that is not harmful for the environment

organic referring to or derived from living organisms; in chemistry, any compound containing carbon

U.S. environmental protection strategy. They have produced admirable results over the years, but are now facing the law of diminishing returns in the face of new complex environmental challenges such as global climate change, energy and water shortages, and **persistent bioaccumulative toxics** that pass easily from one medium to the next. Today's challenges demand the more innovative and vigorous approach of pollution prevention.

Low-Hanging Fruit

There are many ways pollution can be prevented. Some of the simplest, the "low-hanging fruit" involve basic housekeeping and maintenance modifications that do not include major capital investments, but may produce significant dividends in terms of cost savings for compliance and operations.

In an industrial setting, low-cost options can involve simply changing the filters on equipment more frequently, improving the maintenance of machinery, or replacing a solvent with a water-based alternative that performs just as well. In an office setting, it may involve requiring that all documents are printed on both sides of paper and that mugs are used instead of disposable cups. Less toxic alternatives, whether they be cleaners or office paper produced without chlorine, are **green choices**. A farming operation can reduce its use of toxic pesticides or explore the economic feasibility of becoming an **organic** operation.

Energy efficiency is a major component of pollution prevention and an increasingly important issue as we face shortages throughout the United States and global climate change. Again, low-hanging fruit opportunities abound. Options exist for more energy-efficient lighting and computer equipment. Simple business practices like turning equipment off at night can have a positive net environmental and cost outcome.

Even choosing an office building or a plant location can have dramatic environmental implications. Is the facility located near mass transit? If it is, it gives employees the option of using public transportation and reduces the emissions of greenhouse gases from automobiles.

Every state offers some type of pollution prevention assistance to aid companies and communities in identifying P2 opportunities. Because P2 is often not intuitive, government programs help provide a menu of available options to develop comprehensive programs. Many state agencies have engineers and planners on staff who have a wealth of expertise in working with a wide variety of industries. They provide training to company and community officials and disseminate technology, the sharing of information on technical issues and equipment. See the table for the results of P2 efforts in selected states over the past decade.

Identifying Systemic Pollution Prevention Opportunities

The next phase of pollution prevention is to focus on more systemic changes. These may involve more capital investment and a major cultural change on the part of an organization—none of which can happen without the support of senior management.

This is one reason why many companies are making sure that their innovative programs are integrated into their core business decisions. The lone environmental officer who focuses a company on complying with regulations still exists, but he or she is in many cases more actively involved in the daily business decisions being made by that company. This is crucial if serious process and operational changes are going to be adopted to help reduce pollution.

Companies are investigating the use of pollution prevention equipment and comprehensive process changes that are less toxic and generate less waste. Utilizing equipment that is more efficient in its use of materials is a common pollution prevention practice.

As stated, effective prevention will not occur without the backing of senior management, whether it be in the public or private sector. Many organizations create an official policy document, or expand their mission statement to incorporate innovative and cleaner production initiatives. Some organizations go as far as making a senior budget officer responsible for their company's P2 efforts. That way, there is a commitment from top management, particularly those who control the company's purse strings.

Regulatory and Public Information Right-to-Know Programs

TRI and other types of right-to-know programs publicly highlight chemical releases that industrial facilities release to the environment. These public disclosure programs force a company to evaluate its production process and the pollution it generates. The public component of the program helps put the spotlight on these firms, making it more likely that they will try to reduce future releases.

Some environmentalists have also advocated reforming environmental report and permit programs so that reporting facilities essentially perform a pollution prevention audit—identifying waste streams and exploring opportunities to reduce them—in the process of complying with regulatory requirements.

P2 Partners

The public and private sectors play different, but equally important, roles in the effort to promote P2. Government regulatory drivers (statutes and regulations) provide incentives for companies to minimize pollution and thus avoid requirements in the first place. An example of an excellent regulatory measure is the use of P2 and a Supplemental Environmental Project (SEP). A SEP essentially means that an agency can require a company to implement a P2 program as part of their settlement. A state agency can also stipulate that a P2 program be part of an operating permit. There are a number of states conducting this kind of green permit program.

State and local governments also offer critical technical assistance to companies and communities in identifying P2 options tailored to their needs. There are numerous tools available, including public information clearinghouses, on-site assessments, and a score of publications featuring case studies and guidebooks. Government can also offer market-based incentives, including low-interest loans for P2 equipment, reduction in reporting requirements, and public recognition programs that promote a company's environmental performance.

The private sector plays the unique role of being the laboratory. Companies are able to experiment with different P2 practices and techniques within their facility. Given the proper flexibility and support, they can provide some of the major technical and cost data necessary for P2 to expand.

Nongovernmental organizations such as community councils and environmental groups play an important advocacy role in the world of prevention. In the past, they have frequentlyprovided visionary leardership, helping, for instance, to shepherd the Pollution Prevention Act into reality in 1990.

Future Legislative Action

As stated earlier, the single-medium approach to environmental protection is an impediment to progress. Many attempts have been made to change laws or regulations on the federal, state, and local levels to leverage more opportunities for prevention and cleaner production without dismantling the current regulatory framework. The U.S. Environmental Protection Agency (EPA) has overseen several initiatives designed to allow more flexibility within the current system, in the hope of attaining more creativity and innovation. The Common Sense Initiative, 1994 to 1998, was an industry-based approach involving the automobile manufacturing, computers and electronics, iron and steel, metal finishing, petroleum refining, and printing industries. The program initiated more than forty-five projects, half of which, according to the EPA, are still ongoing. Similarly, the no-longer-funded Project XL allowed communities or businesses to test alternate ways of reducing environmental pollution. The National Environmental Performance Track is the current (2003) EPA program that encourages environmental solutions. This program recognizes and gives incentives to more than three hundred business members that go beyond regulatory environmental compliance and develop economically sound initiatives that further increase environmental protection. At the state level, environmental agencies can apply to the Performance Partnership Grants Program, authorized by Congress in 1996. It allows states to combine funds from up to sixteen environmental program grants into a single grant, for example, to address issues such as sprawl. In addition, state voluntary programs have proliferated and included recognition and environmental management system programs.

The Pollution Prevention Act of 1990 provided a good foundation for pollution prevention in the United States. It established much-needed definitions, contained provisions to set up an information clearinghouse and awards programs, and most important, provided start-up funds for states and the EPA to work on dedicated P2 programs. Unfortunately, many provisions of the act were never fully implemented and appropriations were insufficient to orchestrate a comprehensive program. For example, less than one percent of federal grant monies to states for other media programs such as air, waste, and water goes to P2.

Real change will come only by modifying key single-medium statutes. One idea, long proposed, is a unified organic statute. The existing statutes would be woven into a more holistic law, which is multimedia in scope, with prevention as the foundation. Others advocate the consolidation of only specific aspects of existing legislation.

As of 2002, the National Pollution Prevention Roundtable (NPPR) is undertaking a major study to help quantify the results of pollution prevention efforts over the past ten years. Although the study is not complete, its raw data indicating significantly reduced or eliminated pollution and cost savings are impressive, considering the minimal resources that are available nationwide for prevention efforts. The table highlights some of the data provided by state programs. SEE ALSO ABATEMENT; ENERGY, ALTERNATIVE; POLLUTION SHIFTING; RECYCLING; REUSE; TECHNOLOGY, POLLUTION PREVENTION; WASTE REDUCTION.

Bibliography

Hirschhhorn, Joel S.; and Oldenburg, Kirsten U. (1997). *Prosperity without Pollution: The Prevention Strategy for Industry and Consumers.* New York: John Wiley & Sons.

Theodore, Louis; Dupont, Ryan; and Ganesan, Kumar. (1999). *Pollution Prevention: The Waste Management Approach to the 21st Century.* Boca Raton, FL: CRC Press.

Marcus, Alfred A.; Sexton, Ken; and Geffen, Donald A. (2002). *Reinventing Environmental Regulation: Lessons from Project XL.* Washington, D.C.: Resources for the Future.

Internet Resources

Canadian Center for Pollution Prevention. Available from http://www.c2p2online.com.

National Pollution Prevention Roundtable. Available from http://www.p2.org.

U.S. Environmental Protection Agency. National Environmental Performance Track. Available from http://www.epa.gov/performancetrack.

U.S. Environmental Protection Agency. Pollution Prevention Home Page. Available from http://www.epa.gov/p2.

Natalie Roy

Pollution Shifting

Pollution shifting is defined as the transfer of pollution from one medium (air, water, or soil) to another. Early legal efforts to control pollution focused on single media. For example, in the United States, the Clean Air Act covers air and the Clean Water Act covers water. However, pollution is not constrained by statute; it can shift between media by both natural and human action. Pollution management is improved when all media are considered.

Intentional pollution shifting may occur to destroy a pollutant, convert it to a safer form, or reduce its quantity or concentration. Examples of intentional pollution shifting include combustion, **air stripping**, **air scrubbers**, and **adsorption**. Intentional pollution shifting is accomplished by chemical reaction and/or mass transfer. Chemical reactions can convert reactants in one media into products in a different media. In mass transfer shifting, differences in concentration are used to transfer pollutants from one media to another. For example, volatile compounds will transfer from relatively contaminated water to relatively clean air.

Combustion, Air Stripping, and Adsoprtion

Combustion is the process of burning, a chemical reaction. It involves combining combustible material with oxygen under conditions that produce light and heat in addition to by-products. The combustion of wastes, such as municipal solid waste, sludge, or hazardous waste, results in gaseous emissions and a solid ash residue. It significantly reduces the volume and mass of waste requiring disposal, by shifting some wastes to gaseous form. Although carbon dioxide has been implicated in global warming, many of the gaseous emissions have no negative health impact, such as nitrogen gas, carbon dioxide, and

air stripping a treatment system that removes volatile organic compounds (VOCs) from contaminated groundwater or surface water by forcing an airstream through the water and causing the compounds to evaporate

air scrubbers pollution-control devices that remove pollutants from waste gases before release to the atmosphere

adsorption removal of a pollutant from air or water by collecting the pollutant on the surface of a solid material; e.g., an advanced method of treating waste in which activated carbon removes organic matter from wastewater

water vapor. However, pollutants can also be present, including nitrogen oxides, sulfur dioxide, carbon monoxide, particulate matter, metals, acid gases, dioxins, and furans. Contaminants in exhaust gases are minimized by optimization of the combustion process, for example, maintaining proper temperature and oxygen levels. They can also be captured with pollution-control equipment, such as air scrubbers and filters. In addition, the ash may contain hazardous compounds, such as heavy metals.

In air stripping, contaminates dissolved in water are transferred to gaseous form by contact with relatively clean air, an example of mass transfer. Air stripping works best with volatile organic compounds (VOCs) and dissolved gases. VOCs are compounds with high vapor pressures, that is, compounds that tend to evaporate quickly. A common application of air stripping is the cleanup of groundwater contaminated by leaking fuel storage tanks. Air stripping is optimized by maximizing the surface area between the contaminated water and clean air, accomplished by creating fine water droplets in air or small air bubbles in water. Systems can be located away from the contamination (e.g., a system cleaning groundwater that is located on the earth's surface), or located within the contaminated zone (e.g., a system located in wells installed in contaminated groundwater). In some cases, the contaminated air from air stripping is released to the atmosphere, where the pollutants are destroyed by sunlight or reaction with other chemicals, adsorbed into soil or water, or diluted. Preferably, the organics in the exhaust from air stripping are destroyed by incineration or oxidation, or captured by adsorption.

The air stripping process may also be reversed. In air scrubbing, pollutants are transferred from contaminated air to clean water. However, a chemical reaction is often incorporated into air scrubbing, converting pollutants to a safer form. For example, sulfur dioxide produced during coal combustion can be removed from exhaust gas by mass transfer to water containing sodium hydroxide or carbonate, which converts the sulfur dioxide to calcium carbonate. Natural air stripping and air scrubbing also occur. Surface waters, such as lakes and oceans, serve as sinks for pollutants released to the atmosphere. Contaminated water left exposed to the atmosphere will release VOCs.

off-gas control control of gases released into the air

The final pollution shift considered here is adsorption, in which a contaminant in water or air is adsorbed onto a solid material. Adsorption is used for **off-gas control**, groundwater remediation, landfill leachate treatment, industrial wastewater treatment, and water treatment for drinking or industrial purposes. The most commonly used adsorbent is granular activated carbon (GAC). GAC has a tremendous amount of surface area per mass, on the order of one thousand square meters per gram. Its surface attracts many organic compounds; thus, a small amount of GAC can adsorb a significant amount of organic material. GAC may be regenerated, during which contaminants are destroyed.

Multimedia Approach

The multimedia approach to environmental management considers all media. It can be applied to single facilities, entire companies, and regions. According to the U.S. Environmental Protection Agency Multimedia Enforcement Division, it can result in:

- Improved detection and resolution of environmental compliance problems

- achievement of optimal enforcement results

- more effective enforcement

- more efficient use of resources

- fundamental changes in the regulated community's perceptions and behavior regarding environmental compliance

Such benefits are realized by considering an entire pollution system, that is, all media. SEE ALSO AIR POLLUTION; TECHNOLOGY, POLLUTION PREVENTION; WASTE; WATER POLLUTION.

Bibliography

LaGrega, M.; Buckingham, P.; and Evans J. (1994). *Hazardous Waste Management*, New York: McGraw-Hill.

Tchobanoglous, G.; Theisen, H.; and Vigil, S. (1993). *Integrated Solid Waste Management*. New York: McGraw-Hill.

Internet Resources

Canadian Centre for Pollution Prevention Web site. Available from http://www.C2P2online.com.

Reshkin, K. (2002). EPA Student Center Web site. Available from http://www.epa.gov/students.

Jess Everett

Polychlorinated Biphenyls   *See PCBs*

POPs   *See Persistent Organic Pollutants*

Popular Culture

Popular culture can be thought of as a composite of all the values, ideas, symbols, material goods, processes, and understandings that arise from mass media, such as the advertising and entertainment industries, as well as from other avenues, such as games, food, music, shopping, and other daily activities and processes.

Understanding Circumstances

For many people, popular culture may be the primary way of understanding, reinforcing, and modifying the circumstances of their lives. Most of the everyday knowledge and experiences that are shared by people (in the form of reading, watching, wearing, using, playing, working, talking, and so forth) make up the concept of popular culture. Popular culture, however, is distinguished from such traditional institutions as education, politics, and religion, although the distinction often becomes hazy. Over time, and with repeated exposure to societal norms (through, for instance, mass media), people form conscious and unconscious impressions of various aspects of life, including attitudes about pollution.

Chronicling the Good Life

Popular culture in the United States and much of the Western world has concentrated on the reoccurring major theme of the search for "the good life."

The recycle symbol.

Since the establishment of the United States, there have been two opposing themes of popular culture. The first theme, a materialistic one, emphasized a belief in happiness and success through technology, material wealth, and upward social mobility, while the second theme, a simpler one, sought happiness and success in a life of simplicity, one with few possessions, and a spiritual connection. Over the 230-plus years that these two themes permeated American society, they have alternated between being the majority and minority views. During years of prosperity, the materialistic theme dominated, whereas during more modest times the simpler theme was emphasized.

Social Values, Awareness, and Preferences

As the world's population continues to increase dramatically, and as issues such as global warming, ozone depletion, and the extinction of species garner worldwide attention, popular culture becomes more intertwined with people's environmental beliefs and values. The social values, awareness, and preferences of people are enmeshed in the fundamental moral and religious views between nature and humanity: Is it right to manipulate nature? What is the responsibility of society to future generations? Are the rights of other species more or less important than human rights, or are they equally important? These and many other questions are fundamental to the cultural beliefs and values that guide how people live.

Attitudes about Pollution

Popular culture helps to shape people's general understanding about pollution and the environment. Poll results released in the 1990s have consistently shown that from 50 to 75 percent of all Americans consider themselves to be "environmentalists." Moreover, from extensive survey results analyzed by Riley Dunlap and Rik Scarce, three major conclusions have been made about Americans: (1) they have become much more proenvironment since the 1960s; (2) since the 1980s, their environmentalism extends beyond opinions into their basic values and fundamental beliefs; and (3) their attitude about the environment affects the way they interact, consume, and vote.

Images of Pollution in Popular Culture

Images of the natural environment have been prominent in American popular culture since the ecology movement of the 1970s and 1980s. Music and art focusing on human interaction with the environment became popular beginning in the 1960s. Some popular early images of pollution that are now rooted into popular culture:

- A public service TV advertisement, which features a Native American with a tear running down his cheek (sometimes called "the crying Indian"). After paddling his canoe up a polluted river with dirty smokestacks crowding the shores, he comes ashore to a littered riverbank only to have more trash tossed carelessly out of a car and land at his feet. The narrator for the Keep America Beautiful television public service advertisement then declared, "People start pollution, people can stop it." (It premiered on the second Earth Day in 1971.)

- The song "Calypso," by John Denver, which was about French explorer and environmentalist Jacques Cousteau's ship, the *Calypso*. It included the lines "To light up the darkness and show us the way / For though we are strangers in your silent world / To live on the land we must learn from the sea."

- The song "Mercy Mercy Me," by Marvin Gaye, which laments: "Oh mercy mercy me / Oh, things ain't what they used to be no, no / Where did all the blue sky go? / Poison is the wind that blows from the north and south and east."

- The song "Big Yellow Taxi," by Joni Mitchell, released on her 1970 album *Ladies of the Canyon*. The song's lyrics include, "Don't it always seem to go / That you don't know what you've got 'til it's gone / They paved paradise and put up a parking lot."

- The Smokey the Bear advertisement campaign by the U.S. Forest Service. Over the years (starting in the 1940s), the campaign reminded people: "Remember—Only YOU can prevent forest fires."

- The recycling symbol, with the familiar three colored arrows that represent three recycling-related actions: (1) The red arrow stands for separating recyclables from garbage and recycle them, (2) the blue arrow stands for manufacturing new products from the recyclables, and (3) the green one represents purchasing products made from recycled materials ("green products").

Iron Eyes Cody, the teary-eyed Native American man that was for many years a part of the Keep America Beautiful campaign. (Keep America Beautiful, Inc.)

The relationship between popular culture and popular opinion is circular. Nowhere is this more apparent than in the movie business. Hollywood needs good stories and bad guys. Awareness of environmental issues provided it with a wealth of both.

In what was arguably Hollywood's first environmental thriller, life mimicked theater. In *The China Syndrome* (1979), a TV reporter (played by Jane Fonda) and her cameraman (Michael Douglas) collaborate with a whistle-blower (Jack Lemmon) to expose the risk of a meltdown at a California nuclear power plant. Within weeks of its release, reactor number two at Pennsylvania's Three Mile Island nuclear plant suffered a partial meltdown.

It did not take long for Hollywood to find drama involving real-life whistle-blowers. *Silkwood* (1983), starring Meryl Streep, Kurt Russell, and Cher, told the story of Karen Silkwood, a chemical technician at the Kerr-McGee plutonium fuels production plant in Crescent, Oklahoma, and a member of the Oil, Chemical and Atomic Workers Union. Silkwood was an activist critical of plant safety who was inexplicably exposed to plutonium. She was gathering evidence to support her claim that Kerr-McGee was negligent in maintaining plant safety when she was killed in a suspicious one-car crash. The movie was a box-office success; Kerr-McGee settled out of court with Silkwood's family for $1.3 million.

Two later blockbuster movies focused on legal fights against corporate bad guys:

- *A Civil Action* (1999) (based on the book of the same name), starring John Travolta and Robert Duvall, portrayed the true story of a dedicated—some would say obsessed—lawyer, Jan Schlichtmann, who took on a case involving drinking water contaminated by industrial pollution from two highly regarded corporations, which caused the deaths of innocent children in Woburn, Massachusetts.

- *Erin Brockovich* (2000), starring Julia Roberts and Albert Finney, tells the story of an unlikely real-life heroine, Erin Brockovich, who built a powerful case based on suspicious connections between a powerful electric utility, its abuse of toxic chromium, and the poisoned water supply of Hinkley, California, whose residents had suffered a legacy of death and disease.

Language

The increase in environmental awareness is reflected in the common vernacular: What were once called swamps are now called wetlands; what were once called jungles are now called rain forests; and what was once called a round globe is now called Mother Earth. A shift of perception from insignificant pieces of land to valuable components of an overall ecosystem has shown a fundamental change in cultural awareness. Language, though, is only one example of how a rising awareness of the effects of pollution and a greater understanding of ecosystems has been reflected in U.S. society. An average day contains many small examples of how the environment crisscrosses American lives.

A Typical Day of Enviro-Culture

A day in the life of an average American is filled with popular culture's representations of pollution and the environment. A person makes breakfast with cereal from a company that touts itself as environmentally conscious. Flipping channels while eating breakfast, an individual learns from CNN that an oil spill has occurred overnight near a sensitive coastline, while the Weather Channel reports that beach erosion caused by a hurricane off the coast of North Carolina is harming the natural resources of the sensitive Outer Banks. This average American drives to work in a sport utility vehicle (SUV), which was bought on its ability to drive up rugged mountain roads, but declined to buy a compact car that was advertised to help save the environment because of its fuel economy. This individual arrives in a crowded, concrete parking lot that surrounds a multiple-story office building, as do the other thousands of employees who also drive up singly and sometimes in pairs. The person stops by the grocery store on the way home from work in order to pick up prepared food that has been processed in a factory, but that is heralded as the right way to feed oneself in a wholesome and nutritious manner. And so it goes.

The American individual is exposed daily to images and ideas from popular culture (oftentimes unknowingly) in prepackaged advertisements on television, in newspapers and magazines, on the side of food products, on the Internet, and from hundreds of other sources. Certainly, most people's understanding of pollution issues and policies is formed from such brief tidbits—news reports, literature, and entertainment they encounter throughout their busy day.

American Lore: The Ecology of Images

The use of environmental images in popular culture has figured distinctly in American lore. Included in a paper titled "Ecology of Images," cultural theorist Andrew Ross calls the use of environmental images in popular culture the "ecology of images." The negative images of the natural environment included within the popular culture since the ecology movement emerged in the 1970s have included burning rivers, oil-slick waterfowl, and dirty smokestacks. The positive images include a green planet, rushing, clear waters, and white-peaked mountains. The negative images are often used by activists, who often direct blame onto the industrial sector of the community. The positive images are often shown by the business sector, in an effort to demonstrate how well they get along with nature and the environment. Nature and the environment are used as the means to produce the material goods that are needed and desired in society, but they are often abused as a result of in this materialistic way of living.

Commercialism

Popular culture is a world in which everything is for sale one way or another—a world of commercialism. The environment is often thought of as a product to be consumed, and, as a result, pollution becomes one facet of an ever-growing concern of the American popular culture. Companies involved in the capitalization and industrialization of the United States increasingly promote their products, and themselves, as being in tune with nature.

Greenwashing. D.C. Kinlaw states in *Competitive and Green: Sustainable Performance in the Environmental Age*, published in 1993, that businesses increasingly associate themselves with nature (sometimes called the "greenwashing" of the environment). Kinlaw continues by saying that only "by making the environment an explicit part of every aspect of the organization's total operation, can the leaders of an organization expect to maintain its competitive position and ensure its survival." By associating themselves with a good environmental policy (even though they may have a poor environmental record), companies can incorporate these advertised ideals into the popular culture for economic gain and for a supposed improvement in the quality of life. Major department stores and name brands promise the "good life" when they advertise a seemingly endless array of clothes, electronics, home furnishings, kitchen appliances, or whatever other material goods they offer. Similarly, Arkansas officials advertise that their state is "the Natural State," Texans can say "Don't Mess with Texas," and Midwesterners can say their states are "America's Breadbasket," but in reality these lands must be used (and often they are environmentally abused) to produce the lumber, oil, wheat, corn, cattle, and pigs necessary to support the economy and economic standards of the United States.

Two Sides of Nature. Nature must be used to fulfill the needs of people, as they endlessly demand new and better products with which to live the good life. Sometimes called "eco-pornography," the pollution that results from manufacturing is not always evident in everyday life, in the blue skies and clear waters of the images seen in popular culture in the form of television commercials, greeting cards, corporate promotions, and in books, magazines, calendars, travelogues, and videos.

The perspective of the environment as a commodity is found throughout the domain of popular culture. The cultural realm shapes and reflects the values, awareness, and preferences concerning pollution. Whether the vehicle is advertising, music, slogans, symbols, or mascots, the power of popular culture to shape society's behaviors and thoughts with respect to pollution is significant.

Bibliography

Anderson, Alison. (1997). *Media, Culture, and the Environment.* New Brunswick, NJ: Rutgers University Press.

Dunlap, Riley E.; and Scarce, Rik. (1991). "The Polls—Poll Trends: Environmental Problems and Protection." *Public Opinion Quarterly* 55:713–734.

Grossberg, Lawrence; Wartella, Ellen; and Whitney, D. Charles. (1998). *Media Making: Mass Media in a Popular Culture.* Thousand Oaks, CA: SAGE Publications.

Kempton, Willett; Boster, James S.; and Hartley, Jennifer A. (1995). *Environmental Values in American Culture.* Cambridge, MA: MIT Press.

Kinlaw, D.C. (1993). *Competitive and Green: Sustainable Performance in the Environmental Age.* San Diego, CA: Pfeiffer & Company.

Rushkoff, Douglas. (1994). *Media Virus! Hidden Agendas in Popular Culture.* New York: Ballantine Books.

Ross, Andrew. (1994). *The Chicago Gangster Theory of Life: Nature's Debt to Society.* New York: Verso.

Internet Resources

America Remembers. "Iron Eyes Cody: The 'Crying Indian.'" Available from http://www.americaremembers.com/FI09100-2.htm.

Dyer, Judith C. "The History of the Recycling Symbol: Gary Anderson, Recycling Dude Extraordinaire." Available from http://home.att.net/~DyerConsequences/recycling_symbol.html.

Earth Odyssey. "Recycling Symbols." Available from http://www.earthodyssey.com/symbols.html.

Federal Trade Commission. "Part 260: Guides for the Use of Environmental Marketing Claims." Available from http://www.ftc.gov/bcp/grnrule/guides980427.htm.

FOX.com "The Simpson's: Official Web Site." Available from http://www.thesimpsons.com.

Keep America Beautiful. "Public Service Announcements." Available from http://www.kab.org/psa1.cfm.

Snopes.com. "Urban Legends Reference Pages: Movies (Iron Eyes Cody)." Urban Legends Reference Pages. Available from http://www.snopes.com/movies/actors/ironeyes.htm.

STLyrics. "Friends—Soundtrack Lyrics (Mitchell, Joni—Big Yellow Taxi [Traffic Jam Mix])."Available from http://www.stlyrics.com/lyrics/friends/bigyellowtaxitrafficjammix.htm.

U.S. Department of Agriculture Forest Service, the National Association of State Foresters, and the Advertising Council. "Smokey's Vault: History of Campaign." Available from http://www.smokeybear.com/vault/history.asp.

Wood, Harold. "Earth Songs." Available from http://www.planetaryexploration.net/patriot/earth_songs.html.

Yamhill County Building and Planning Department, McMinnville, OR. "Yamhill County Solid Waste." Available from http://www.ycsw.org/index.asp.

William Arthur Atkins

Population

Throughout most of human history, the world's population has grown gradually. It took thousands of years for the global population to reach one billion people (around 1800). Then, in a little more than a century, the population jumped to two billion (by 1960), and to three billion by 1980. In just twenty years—between 1980 and 2000—the world's human population doubled from three billion to six billion people.

PLAN YOUR FAMILY FOR A BETTER FUTURE

Billboard promoting birth control, China, 1984. (UPI/Corbis-Bettmann. Reproduced by permission.)

The human population explosion during the past century was the result of several factors. Fertility rates remained high, while medical and agricultural advances such as antibiotics, immunizations, clean water, and improved food availability reduced mortality rates—especially among infants and children.

It is difficult to predict how rapidly the human population will continue to increase, due to the many factors that affect population growth. Another important question that scholars ask is "How many people can the earth support?" While the human population grows, the earth's size and resources remain the same. Technology can increase the amount of food that can be produced on a piece of land, but it cannot increase the amount of land and water on the planet. Many people regard population growth as the single most serious global issue, because population size is closely linked to environmental and human health conditions.

Environmental problems are aggravated by population explosions. More people means more resources and energy are consumed and more pollution is created and more waste is sent to landfills. More land is needed to grow crops and build houses. More trees are cut down for new homes. More cars are built, more fossil fuels are used, and more gases are released into the environment. More natural wilderness areas or beautiful landscapes are destroyed to provide resources and cropland. In short, population growth makes other environmental problems harder to solve.

Projecting Population Change

Scholars have spent centuries trying to find reliable ways to predict population change. One of the most famous population researchers was Thomas Malthus, a British clergyman who studied population growth in the 1770s. In his famous 1798 *Essay on the Principle of Population*, Malthus argued that human populations tend to grow exponentially, while food production is limited by land available for agriculture. In short, human populations tend to increase faster than food supply, leading to an imbalance.

Malthus projected that population increases in England would quickly outstrip the available food supplies, leading to famine and misery. Malthus's predictions for England never occurred in his lifetime. England's population did increase, but advances in science and technology enhanced food production. Malthus's theory also failed to take into account colonial growth as a result of other factors. Still, scholars use Malthus's concepts of geometric population growth today, though new models of population change are far more complex.

Researchers who study population change consider many factors for each country. Population change for any group of people is determined by fertility, mortality, and migration rates. What is the average number of children per family? What is the life expectancy? Are people migrating into or out of a country? Each of these is, in turn, affected by other factors.

It is important to remember that population projections are just estimates based on past information; they do not account for unknowns such as future wars, epidemics, or the effects of climate change. However, the scholars who make the projections attempt to improve their accuracy by revising projections as new information is collected. The United Nations Population Division is one of the organizations responsible for making population projections. After considering the potential impact of the current AIDS epidemic, the United Nations recently lowered its population projection for 2050 by more than one billion people.

United Nations Projections

At the beginning of the twenty-first century, the world population is still growing at a rate of 1.2 percent annually. This is the same as adding 77 million people (roughly the population of France) to the world each year. A world population projection published by the United Nations in 2002 estimates that the world's human population will reach 8.9 billion by 2050.

This population increase is not expected to occur evenly across the globe. The populations of some nations are shrinking while those of other nations are swelling. During the past few decades, reproduction rates have decreased in countries where the standard of living has improved; these improved living standards are generally associated with higher education levels across a population and access to birth control. Today, as many as thirty-three countries are witnessing population declines due to lower birthrates. Japan, Bulgaria, Italy, Bulgaria, Estonia, and the Russian Federation are among the countries that have achieved negative population growth.

Population explosions tend to occur in regions already struggling with hunger. Africa is expected to undergo the most rapid growth, increasing from 784 million people in 2000 to nearly 1.8 billion in 2050. Eight countries—India, Pakistan, Nigeria, the United States, China, Bangladesh, Ethiopia, and the Democratic Republic of Congo—are expected to account for half of the world's population increase during the next fifty years. India may overtake China as the most populous country, rising from just over one billion to more than 1.5 billion between 2000 and 2050. Birthrates are not the only reason

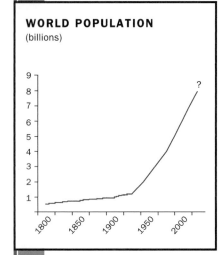

WORLD POPULATION
(billions)

for the anticipated rises. The United States has a low birthrate, but a high immigration rate.

How Many People Can the Earth Support?

Is there a limit to the number of people the world can support? Some people contend that new technologies will make it possible for the earth to support ever-larger human populations. They describe the earth's resources as virtually inexhaustible, due to the potential of technology. They point to the scientific advances that helped increase crop yields across India and China as an example of the human ability to adapt through technology.

Other scholars believe that there are limits to how much technology can accomplish. They argue that the earth's capacity to support human population growth is finite—because natural resources can be damaged or depleted. For example, India's increased crop production has not keep pace with its growing population. India's per-person food production is actually dropping as the food supply is shared among more and more people.

Water shortages may be the most insurmountable obstacles for human survival, as populations continue to grow. On every continent (including North America), rising demands for water are already causing water tables to drop to dangerously low levels, depleting future water supplies. Several of the world's major rivers are being drained dry before running their courses. Most of this water is used for irrigation (to grow food); less is used for industry and domestic use.

Water scarcity is already a serious survival problem for people living in the more populous and arid regions of the world. Scholars predict that most of the world will face water scarcity as human demands on the earth's resources continue to rise. Despite hope for technologies such as desalinization to solve the world's water shortages, the prospects to solve global problems are unlikely. So far, desalinization is too expensive for most nations.

A second challenge the world faces is food production. There is hope that breakthroughs in plant genetics and other sciences will continue to improve food production. Yet many scholars argue that even the most remarkable advances in agricultural technology, aquaculture, and ranching could not raise food production enough to meet the world's growing needs. Food production is also limited by the availability of fresh water and land that can be farmed—two finite resources.

Malnutrition is already a growing problem in many regions that depend on grains. Likewise, countries that depend on fish as a primary protein source are also faced with shrinking food supplies as the world's fish populations are further depleted.

Impact on Human Health and the Environment

Population growth affects almost every element of human health and the environment by exacerbating preexisting problems. For example, if a nation is already struggling to provide food, education, and healthcare to its people, the needs of an even larger population may exhaust the nation's ability to provide for anyone. As a result, the rate of poverty, homelessness, and disease are likely to rise. In most cases, rapid population growth results in a decline in human living standards.

"Population, when unchecked, increases in a geometrical ratio. Subsistence increases only in an arithmetical ratio. A slight acquaintance with numbers will show the immensity of the first power in comparison of the second."

—Thomas Robert Malthus, *An Essay on the Principle of Population,* (1798)

The impact of human population on the environment is complex. A popular theory is that the degree of human impact on the environment is determined by three factors: population size, how much each person consumes, and how much waste each person produces. India may have a much larger population than the United States, but people in United States tend to consume and waste far more goods than people in any other part of the world. According to this theory, a rise in the U.S. population would have a greater impact on the environment than would a similar increase in India's population.

What Is Being Done?

There are many views on what to do about global population growth. Several advocacy groups, such as Negative Population Growth, Zero Population Growth, Planned Parenthood, and the Carrying Capacity Network, focus on raising public awareness about birth control and the need to lower fertility rates. At least one group (Negative Population Growth) advocates that the U.S. government should provide incentives for smaller families and should limit immigration in the United States. The world's most populous country, China, has been exploring a variety of laws and incentives to limit urban families to one child per family, with the goal of reversing the country's unsustainable population growth. However, due to the government's inability to restrict family size in rural areas, where the overwhelming majority of China's population lives, and other factors, China's population growth is not expected to turn around until at least 2020.

Slowing population growth is also a priority for many environmental organizations, including the National Audubon Society, the Sierra Club, the Wilderness Society, the National Wildlife Federation, and the Environmental Defense Fund. Most of these groups have policy statements and/or education programs that deal with population issues. SEE ALSO EARTH SUMMIT; EHRLICH, PAUL; HISTORY; LIFESTYLE; MALTHUS, THOMAS; POPULAR CULTURE; POVERTY; ZERO POPULATION GROWTH.

Bibliography

Brown, Lester R.; Gardner, Gary; and Halweil, Brian. (1999). *Beyond Malthus: Nineteen Dimensions of the Population Challenge*. New York: Norton.

Cohen, Joel E. (1995). *How Many People Can the Earth Support?* New York: Norton.

Internet Resources

United Nations Population Division. *World Population Prospects: The 2002 Revision Population Database*. Available from http://esa.un.org/unpp/sources.html.

Corliss Karasov

Poverty

Continuing industrialization and technological advances benefit many (though not all) of the people in the developed countries, but the gap between the rich and poor countries is significant and increasing. In general, poverty deprives people of adequate education, health care, and of life's most basic necessities—safe living conditions (including clean air and clean drinking water) and an adequate food supply. The developed (industrialized) countries today account for roughly 20 percent of the world's population but control about 80 percent of the world's wealth. Poverty and pollution seem to operate in a vicious cycle that, so far, has been hard to break. Even in the developed nations, the gap between the rich and the poor is evident in their respective social and environmental conditions.

Poverty, the Environment, and Pollution

Regardless of the reason or the area of the world in which a poor population lives, certain reciprocal elements will act on the population and its environment. Lack of education, oppression, lack of appropriate infrastructure—from water-treatment facilities to better roads and communication—all exacerbate the twin problems of poverty and environmental degradation. One cannot ask people to heal the environment, or even just mind it, if they can barely sustain themselves. For example, tropical fish are considered to be either delicacies or exotic pets by people who can pay for them and people in tropical regions can earn good money for catching these fish. But to catch the fish more easily they use cyanide or dynamite to stun the fish. The former pollutes (and moves up the food chain) and the latter destroys the reef environment. Agricultural practices that tax the soil lead to soil erosion, which lowers crop yields and pollutes rivers and streams with silt. The accumulation of the silt—from the loose eroded soil—kills the fish in the river and streams. Another cause of soil erosion is the cutting down of trees, in massive numbers, either for use as firewood (because the winters are harsh and there is no other way to stay warm) or to sell for much needed cash. Eventually, not only will the soil erode to a point where it can no longer sustain agriculture, but the trees would be gone too. The above examples show that practices that fail to consider environmental health perpetuate the poverty cycle, thereby further destroying the environment.

The environment as a whole tends to be jeopardized more in the poorer areas. In the United States, Louisiana is a poor state in which there is an area known as "Cancer Alley." It is a stretch on the lower Mississippi River that is home to 125 companies, many of which manufacture products that result in highly hazardous waste. Cancer rates in the area are higher than the national average, and respiratory illnesses, as well as incidents of liver and kidney toxicity, are rampant. In one typical area, Ascension Parish, environmental justice activist Robert Bullard points out, "eighteen petrochemical plants are crammed into a nine and a half square mile area" (Bullard, p. 106).

Poor people tend to be less well educated (because they do not have the time and resources to obtain an education), and less politically powerful. Many people in Louisiana's Cancer Alley were never aware of the dangers of hazardous waste as industries started moving in. Many of them, after years of discrimination, are distrustful of politicians and public officials. Their land is cheap, and Louisiana provides the big industries with tax breaks, which appeal to companies looking at the bottom line.

Globally, the large industries find the same advantage in poor nations. Pollution controls and hazardous-waste-disposal regulations are stricter, and more expensive, in the developed nations. Many companies find it cheaper to export their waste to the developing countries, which are starving for cash. The hazardous waste disposal in those countries is unsafe and dangerously polluting. The people handling the waste are poorly educated, and therefore may suffer severe health consequences as a result of their work. However, if they are paid a salary they are better off than many others. In addition, the developing countries themselves, eager to grow economically, may develop heavy industry but not the controls or infrastructure necessary to contain the pollution. It is easy to see, therefore, that there is a huge divide, economically and ideologically, between the developed and developing countries.

The North–South Divide

Economists talk about the North–South divide when referring to the economic growth and development of nations. The developed, or industrialized, countries, most of which are in the northern hemisphere, are referred to as the North. The developing countries, which are economically underdeveloped to varying degrees, are referred to as the South. When it comes to pollution and environmental preservation, the North and South have different priorities that seem to put them at odds with each other.

The concept of *sustainable development* is crucial to understanding the conflict between the North and South. The United Nations, in a 1987 report of its World Commission on Environment and Development, defines sustainable development as the ability to grow economically and improve quality of life in such a way that "meets the needs of the present without compromising the ability of future generations to meet their own needs." (Nebel, p. 16)

As mentioned above, the most pressing priority for the southern hemisphere nations is economic growth: the poverty rate in the developing countries can reach 90 percent (by comparison, the North has a poverty rate, on average, of 15 percent). Environmental conservation and pollution control are far less a priority in the South. The priority in the North is sustainable development—the ability to continue on the course of consumption and energy use while ensuring a healthy environment. The developing countries feel this attitude is elitist, even racist (most poor nations or groups are not white). They contend that the developed countries' demands for environmental regulations place an undue burden on the developing nations. Worse yet, the largest polluters are the developed countries, which also consume the most global resources. Many of the problems of environmental destruction in the poor countries are a direct result of consumption levels in the developed countries (poaching for ivory in Africa is but one example, albeit extreme).

Historically, European colonization disrupted those societies that normally lived in balance with their environment. Mostly hunting, agricultural, or fishing in nature, the people grew or consumed enough to sustain themselves, never taking more than they needed. The European settlers diverted the native agriculture to grow certain target crops (sugarcane and tobacco, for example) that were valuable in Europe. Not rotating the crops depleted the soil and reduced crop yields. It also made the colonized countries' economies wholly dependent on the fluctuations in cash-crop prices. The settlers also mined and deforested the environment, causing heavy damage. To this day, developing nations are in the ironic position of exporting a big percentage of their agricultural yield, while having to import food. Even after gaining their independence, many of these countries were unable to build an economy independent of European and U.S. consumption patterns. The developing nations are heavily in debt to the developed countries, and their cash crops and other commodities (such as diamonds in Africa) are controlled by international corporations. The entire set of circumstances creates severe tension between the North and the South and is getting renewed attention with the emphasis now being given to environmental justice.

Environmental Equity

In 1997 a study by the Harvard Center for Population and Development Studies found that life expectancy for people living in poor communities in the United States was markedly lower than life expectancy for people living in wealthier communities, sometimes by as much as fifteen years. While many factors contribute to this alarming discrepancy, it has become clearer since the 1980s that poor communities, which are also predominantly non-white, bear the brunt of adverse pollution affects. In 1983, for example, a U.S. General Accounting Office report found that in eight southeastern states that were studied, "Blacks make up the majority of the population in three out of four communities where landfills are located." (U.S. GAO, p. 1) Worldwide, the trend is similar. Big corporations find it easier and cheaper to export trash and to build polluting factories in poor developing nations.

Environmental justice is, to use the U.S. Department of Energy's definition, "the fair treatment and meaningful involvement of all people regardless of race, color, national origin, or income with respect to the development, implementation, and enforcement of environmental laws, regulations, and policies" (http://www.epa.gov/compliance/environmentaljustice/index.html). In the United States, the 1980s saw the beginning of an environmental justice movement that started focusing attention on the undue burdens placed on poor communities when it comes to living in a polluted environment. Fighting what some refer to as environmental racism, the grassroots environmental justice movements at times clashed with older environmental groups, who formed around the idea of conservation, and whose concern for the natural environment seemed elitist. There was a perception that organizations such as the Sierra Club concerned themselves with the conservation of the natural environment but did not care about pollution in inner cities and poor rural communities. Much more research is being done on the connection between hazardous living conditions and poverty—not only on the effects, but also on the causes. Among the environmental justice group's many victories was Executive Order 12898, signed by President Bill Clinton on February 11, 1994, directing federal agencies to correct the "disproportionately high and adverse human health or environmental effects" that their operations have on the minorities and low-income populations.

Earth Summit and Agenda 21

Environmental justice and the connection between poverty and pollution have been gaining increased attention globally, both from governmental and non-governmental organizations (NGOs). In 1992 the United Nations Conference on Environment and Development (UNCED) met in Rio de Janeiro, Brazil, in what came to be known as the Earth Summit (June 3 to 14). Unprecedented in size, the meeting focused on sustainable development, and its main result was a document of goals and plan of action known as Agenda 21. The document was adopted by over 170 governments represented at the conference. One of the principles on which Agenda 21 is based is recognizing that "[a]ll States and all people shall cooperate in the essential task of eradicating poverty as an indispensable requirement for sustainable development" (Rio Declaration, Principle 5). Chapter 3 of Agenda 21 is dedicated to the issue of poverty. In it, the document acknowledges that sustainable development is not possible

A study prepared for the California state Waste Management Board, known as the Cerrell Report, concluded that trash incinerators should not be built within five miles of "middle and higher socioeconomic strata neighborhoods." The report, "Political Difficulties Facing Waste-to-Energy Conversion Plant Siting," says that plans to build such plants will face less opposition if placed in poor neighborhoods instead of wealthy ones. The report provides personality profiles of people most likely and least likely to fight an incineration plant.

without a sweeping, global effort to eradicate poverty, and certain recommendations are made as to how to achieve this goal.

Following the Earth Summit, the United Nations noted that poverty was in fact increasing. In a follow-up meeting to the Earth Summit, the UN General Assembly in its 1997 Programme for the Further Implementation of Agenda 21 called for refocusing sustainable development efforts on the eradication of poverty as an overriding priority. In 1995 the United Nations also declared 1997 to 2006 to be "the First United Nations Decade for the Eradication of Poverty" (http://www.un.org/esa/socdev/poverty/poverty.htm).

The disappointing decade that followed the Earth Summit, especially with the increase in poverty, led to the World Summit on Sustainable Development, a meeting in Johannesburg, South Africa, that took place between August 26 and September 4, 2002. While the document resulting from the meeting was only fifty pages long, it contained concrete goals, and as such has more practical value than Agenda 21, in the opinion of many participants. In addition to the goals, over three hundred international partnerships were formed to launch an initiative that would improve access to safe drinking water, improve sanitation, address toxic-waste management, and address many other sustainable development issues.

According to the World Bank, at the start of the twenty-first century 1.2 billion people lived in absolute poverty, a condition defined by the United Nations as "characterized by severe deprivation of basic human needs" (UN Report of the World Summit for Social Development, p. 44), including access to safe water, sanitation, food, and appropriate health care. Economically, the World Bank defines absolute poverty as living on less than one dollar per day. An additional 2.8 billion people lived on less than two dollars per day. Eight out of one hundred children didn't live to see their fifth birthday. While strides have been made in the fight against poverty, these advances were not uniformly distributed around the globe. It is well understood that the ecological crisis our planet is facing—one that includes pollution, scarcity of resources, environmental degradation, and loss of biodiversity—cannot be addressed without addressing, and alleviating, the problem of poverty. To do that, an integrated approach, one that addresses the entire poverty cycle, is needed. Such an approach would have to include the eradication of gender bias in community participation and access to education; equal representation to all citizens, regardless of economic status; access to safe drinking water, proper sanitation, and proper health care (including family-planning resources); universal access to education; and improved employment opportunities. It is obvious from this partial list that only committed international cooperation can bring about these changes. In maintaining the status quo, we pay a tremendous price in human suffering and in an environmental crisis that will affect generations to come. The good news is that more people are refusing to pay the price these days, and are taking steps to form partnerships that will bring about a positive change. SEE ALSO AGENDA 21; CANCER; CANCER ALLEY, LOUISIANA; CHÁVEZ, CÉSAR E.; DISASTERS: CHEMICAL ACCIDENTS AND SPILLS; DISASTERS: ENVIRONMENTAL MINING ACCIDENTS; DISASTERS: NATURAL; DISASTERS: NUCLEAR ACCIDENTS; DISASTERS: OIL SPILLS; EARTH SUMMIT; ENVIRONMENTAL RACISM; HEALTH, HUMAN.

Bibliography

Bullard, Robert D. (2000). *Dumping in Dixie: Race, Class, and Environmental Quality.* Boulder, CO: Westview.

Ehrlich, Paul R.; Ehrlich, Anne H.; and Daily, Gretchen C. (1995). *The Stork and the Plow*. New York: G.P. Putnam's Sons.

Nebel, Bernard J.; and Wright, Richard T. (2000). *Environmental Science: The Way the World Works*, 7th edition. Upper Saddle River, NJ: Prentice Hall.

Ponting, Clive. (1992). *A Green History of the World*. New York: St. Martin's Press.

Olden, Kenneth. (1998). "The Complex Interaction of Poverty, Pollution, Health Status." *The Scientist* 12(4):7.

United Nations Report of the World Summit for Social Development (Copenhagen, March 6-12, 1995).

U.S. General Accounting Office. (1983). "Citing of Hazardous Waste Landfills and Their Correlation with Racial and Economic Status of Surrounding Communities." RCED-83-168, June 1, 1983.

Internet Resources

Online Ethics Center for Engineering and Science at Case Western Reserve University. *Ethics and Values in Pre-College Science Instruction*. "Case #6: Love Canal." Available from http://onlineethics.org/edu/precol/classroom/cs6.html.

U.S. Environmental Protection Agency Office of Environmental Justice. Available from http://www.epa.gov/compliance/environmentaljustice/index.html

United Nations Department of Economic and Social Affairs. "Economic and Social Development." Available from http://www.un.org/esa.

United Nations Department of Economic and Social Affairs, Division for Social Policy and Development. "First United Nations Decade for the Eradication of Poverty 1997-2006." Available from: http://www.un.org/esa/socdev/poverty/poverty.htm.

United Nations Rio Declaration on Environment and Development. Available from http://www.unep.org/documents/default.asp?documentid=78&articleid=1163.

World Bank Group. "World Bank Poverty Net." Available from http://www.worldbank.org/poverty/index.htm.

Adi R. Ferrara

Precautionary Principle

The precautionary principle, also referred to as the precautionary approach, justifies the use of cost-effective measures to prevent environmental degradation even in the absence of full scientific certainty. This principle has obvious applications to various forms of environmental pollution. The principle can be traced to German national law in 1976, which states, "[e]nvironmental policy is not fully accomplished by warding off imminent hazards and the elimination of damage which has occurred. Precautionary environmental policy requires furthermore that natural resources are protected and demands on them are made with care."

The principle's first applications beyond national boundaries came in 1987. It was quickly adopted into numerous **multilateral treaties** and international declarations, including the 1987 Montréal Protocol on Substances that Deplete the Ozone Layer, the 1990 Bergen Declaration on Sustainable Development, the 1992 Convention on Biological Diversity, and the 1999 Treaty of Amsterdam, which has broadened and redefined the goals and institutions of the European Union.

multilateral treaty treaty between more than two governments

The principle's scope varies dramatically in these documents as well as in national legislation that contains it. In some, it is limited to toxic substances that are persistent and can **bioaccumulate**. In others, like the Bergen Declaration, it covers all government policies with the potential to degrade the environment, even when some causal relationships have not been fully

bioaccumulation buildup of a chemical within a food chain when a predator consumes prey containing that chemical

145

hormone a molecule released by one cell to regulate development of another

phthalate particular class of complex carbon compounds

polyvinyl chloride (PVC) class of complex carbon compounds containing chlorine

established scientifically. Some critics contend that the Principle restricts technology. It has been a focus of U.S.–European Union (EU) trade disputes, as Europeans have argued for its application to genetically modified foods, animal-growth-promoting **hormones**, and **phthalates** (softeners) in **polyvinyl chloride (PVC)** children's toys. The U.S. government also contends that the principle is a nontariff barrier, that is a policy that interferes with exports or imports other than a simple tariff such as quota. SEE ALSO LAWS AND REGULATIONS, INTERNATIONAL; LAWS AND REGULATIONS, UNITED STATES; TREATIES AND CONFERENCES.

Bibliography

Goklany, Indur M. (2001). *Precautionary Principle: A Critical Appraisal of Environmental Risk Analysis.* Washington, D.C.: Cato Institute.

Michael G. Schechter

President's Council on Environmental Quality

The Council on Environmental Quality (CEQ) was created by the National Environmental Policy Act (NEPA) in 1969 during the first term of President Richard Nixon. The primary role of the council is to advise the President on environmental policy. Because it is limited to an advisory role, CEQ does not have a highly visible public profile. It is composed of three members, including a chairperson, who are appointed by the president with the **advise and consent** of the Senate. CEQ's importance in environmental policy has fluctuated significantly over the years of its existence.

advise and consent the formal responsibility of a government body to provide counsel and approval for the actions of another body, especially the Senate to the President

The NEPA is the federal law that requires federal agencies to prepare environmental impact statements (EISs) prior to undertaking or approving any action that might have a significant effect on the quality of the environment. In adopting NEPA, Congress realized that a wide range of federal activity had an impact on environmental quality. In practice, one of the most important functions of CEQ is to oversee the implementation of the EIS process by other federal agencies. Initially, the oversight took the form of guidelines for implementing the EIS process; the guidelines were advisory and not mandatory. In 1979, at the request of President Jimmy Carter, the CEQ issued mandatory regulations that had to be followed by all agencies. Since there had been many court cases interpreting the language of NEPA, the CEQ regulations essentially **codified** the case law created by the courts. Generally, government regulations interpret and explain confusing statutory language, but unfortunately they themselves are often very confusing. CEQ's regulations under NEPA are an exception to this rule; they are written in clear and concise language. The extensive and clearly written regulations are most likely a factor in the reduced number of court cases filed under NEPA since 1979.

codify put into law

The CEQ was required by law to provide the president with an annual report on the state of the nation's environment. The report would establish the status and condition of the natural environment, the current and foreseeable environmental trends, the adequacy of natural resources for fulfilling the nation's needs, a review of other relevant programs and activities of government and nongovernment organizations, and a program for remedying existing environmental deficiencies. Throughout the 1970s, CEQ's annual report to the president was a treasure trove of information for citizens interested in environmental issues. Since then, CEQ has generally been underfunded, and as a consequence, its annual reports have shrunk in size and are not issued in a timely fashion.

Finally, CEQ acts as a referee in disputes between federal agencies implementing various aspects of NEPA. Although the statute assigns other general, environmentally related tasks to CEQ, the three noted above are the most important and most visible.

CEQ has had a checkered existence. Though active and visible through 1980, President Ronald Regan saw little need for it and sought to eliminate the CEQ. Failing in this endeavor, the President cut CEQ's funding by over 80 percent and failed to appoint any members to the council until the latter years of his presidency. President Bill Clinton prepared legislation that would eliminate CEQ, and transfer its functions to a new **cabinet**-level Department of the Environment. That legislation failed too. In 1995 the president rejuvenated the CEQ. Its greatest visibility in the Clinton years evolved when its chair, Kathleen McGinty, became the Executive Director of the President's Council on Sustainable Development (PCSD). The PCSD developed, and even began the implementation of, a broad plan for leading the country toward a more environmentally sustainable lifestyle. Though no activity on the part of CEQ may be currently apparent, President George W. Bush appointed a CEQ chairperson in 2001. SEE ALSO ENVIRONMENTAL IMPACT STATEMENT; NATIONAL ENVIRONMENTAL POLICY ACT (NEPA).

cabinet in government: collective name for the heads of federal departments that report directly to the president

Internet Resource

Executive Office of the President, Council on Environmental Quality. Available from http://www.whitehouse.gov/ceq.

James P. Karp

Progressive Movement

The Progressive Era, a term used to describe the period between approximately 1890 and 1920, witnessed an explosion of reform efforts in America. A great number of people, for a variety of reasons, participated in a vast number of diverse reforms, including women's suffrage, political reform, and prohibition. Progressive reformers initiated these changes in reaction to the increased level of, and problems associated with, urbanization and industrialization in late-nineteenth-century America. Taking advantage of new technological developments in transportation, communication, and organization, industry grew tremendously and immigrants flooded into unprepared cities for new jobs. With no government oversight or regulations, numerous problems erupted: Housing became overcrowded, dilapidated, and disease-ridden; industries failed to protect their employees financially, physically, or health-wise; and pollution became rampant.

Environmental activities formed part of progressive reformers' efforts. These environmental reformers generally viewed the environmental problems of the city in two different ways. The conservation and preservation activists, led by Gifford Pinchot and John Muir, respectively, pressed for the improvement and protection of "nature" outside the city.

They worked to set aside land either as undeveloped wilderness for its aesthetic values, or to maintain resources like forests for future use by humans.

Others interested in environmental problems, however, pressed for solutions within urban areas rather than outside of them. Jacob Riis, a muckraking journalist, published photographs of slum housing and their immigrant residents. His work outraged many and produced some reforms in living conditions. Upton Sinclair, perhaps one of the most famous muckrakers of the Progressive Era, published *The Jungle* in 1906, a startling, thinly fictionalized exposé of the meat-packing industry. Filled with stories of vile, unsanitary, and dangerous conditions for workers, the book led to legislative action in the form of the Meat Inspection Act and the Pure Food and Drug Act. In addition, reformers strived to improve working conditions in factories, resulting in factory inspection laws and child-labor laws.

Women also played a pivotal role in the antipollution movement of the Progressive Era. Alice Hamilton increased public awareness of toxic chemicals and their health effects. The Settlement House movement, led by women like Jane Addams, worked to better city services and conditions within immigrant neighborhoods. Smoke pollution also greatly concerned women at this time. Reacting to their increased laundry load in filthy conditions, as well as concerns about their husbands' and children's health, women dramatically altered the general public's conceptions of smoke. Up to this time, many had conceived of smoke as either a disinfectant or the necessary cost of progress. Women educated their fellow citizens on the health dangers of smoke, and their activism led to smoke-pollution-control laws in every major city in the United States by 1912. Men took control of this issue within legislative circles, stressing technology as a way to reduce smoke or burn the coal more efficiently.

Although progressive reformers generally raised awareness of environmental problems and changed public perceptions of pollution, their activism, in fact, remained quite limited. Reformers of this time generally accepted the beliefs of capitalism and industry. This caused them to limit their search for solutions to technological means, such as finding cleaner methods of burning coal, rather than examining consumption patterns of energy or other products. SEE ALSO ACTIVISM; ADDAMS, JANE; ENVIRONMENTAL MOVEMENT; HAMILTON, ALICE; INDUSTRY; LEAD; OCCUPATIONAL SAFETY AND HEALTH ADMINISTRATION (OSHA); POINT SOURCE; POLITICS; SETTLEMENT HOUSE MOVEMENT; SOLID WASTE; WATER POLLUTION; WORKERS HEALTH BUREAU.

Bibliography

Hoy, Suellen. (1995). *Chasing Dirt: The American Pursuit of Cleanliness.* New York: Oxford University Press.

Melosi, Martin, ed. (1980). *Pollution and Reform in American Cities, 1870–1930.* Austin, TX: University of Texas Press.

Stradling, David. (1999). *Smokestacks and Progressives: Environmentalists, Engineers, and Air Quality in America, 1881–1951.* Baltimore, MD: Johns Hopkins University Press.

Internet Resource

Library of Congress, Memory Gallery C. "The Progressive Era." Available from http://www.loc.gov/exhibits/treasures/tr11c.html#prog.

Elizabeth D. Blum

Property Rights Movement

The property rights movement has had a significant impact on the nation's environmental policies since 1980. The groups identified with the movement commonly oppose federal regulation or intrusion on land that is privately held, especially in cases where federal involvement is in the form of environmental laws that limit the owner's full or partial use of the land. The movement began with the Sagebrush Rebellion of the mid-1970s, when legislators from states in western United States sought the transfer of federal public lands to state control.

Researchers have identified numerous groups and organizations that fall under the general classification of the environmental opposition, one of which is the property rights movement. These groups commonly oppose federal regulation or intrusion related to land that is privately held, especially environmental laws that limit the owner's full or partial use of the land. This segment of activists is distinct from the wise use movement, which grew out of the Sagebrush Rebellion of the mid-1970s. Wise use advocates support an antigovernment regulatory agenda related to the use of public land and resources, where the property rights movement is based on the use of privately held land.

The property rights movement first surfaced in the early 1990s with local grassroots organizations made up of individuals seeking to develop their own property, usually by building a home, clearing out trees or brush, or draining a **wetland**. Many of the landowners had been unaware of federal regulations and permits that could thwart their efforts, such as provisions of the Clean Water Act or the Endangered Species Act. After being prohibited from developing their properties by the federal government, they often joined other frustrated property owners, usually in their area or neighborhood, who were similarly prohibited from doing what they wanted with their land. The "members" of the movement rarely joined a specific, formal organization; more commonly, they shared grievances against the government based on their individual disputes. They would, however, rely upon an organization for legal advice and updates on land regulations that would affect them.

wetland an area that is saturated by surface or ground water with vegetation adapted for life under those soil conditions, as swamps, bogs, fens, marshes, and estuaries

The property rights movement has been most active in regions in the eastern and southern United States, where title to land is often in a family's name for many generations. Historically, there has been an assumption that the right to control the land belongs to the **titleholder**, regardless of changes in the law or public policy. Many activists are farmers, ranchers, or rural or beachfront property owners who are unaware of the ecological value of their land until they decide to develop it. This has led to a national debate over competing land-related interests—the rights of the property owner to use the land versus the government's interest in controlling pollution, protecting wildlife and their habitat, and managing ecosystems or even other landowner's property.

titleholder the person or entity holding the legal title or deed to a property

Property rights stem from English common law and the **Magna Carta**, although there has been an evolution in legal interpretation of those rights since the 1920s. Most of the recent litigation has dealt with the concept of federalism, and more specifically, the Fifth Amendment to the U.S. Constitution. One of the clauses in the amendment refers to "takings"—a requirement that the government cannot take privately owned land for public use without compensating the owner for the value of the land. University of

Magna Carta English charter giving landowners rights under the king's authority

Chicago law professor Richard Epstein created an intellectual basis for the property rights movement in 1985 in *Takings: Private Property and the Power of Eminent Domain*. This book placed the takings clause in the context of wilderness designations, endangered species, and wetlands protection.

The takings issue has often resulted in a private property owner seeking compensation from the government by filing a suit before the U.S. Court of Federal Claims or the U.S. Supreme Court. Since 1987, the courts have frequently ruled that federal regulations like the Clean Water Act that deny the owner the economically viable use of the land must pay the owner for the loss of the use of the land. The government further expanded the rights of property owners with an executive order by President Ronald Reagan and with regulations that called for government agencies to evaluate the risk of unanticipated takings. The 1988 policy calls for the federal government to budget funds for a **takings impact analysis** that property owners feel protects their constitutional rights, although the law continues to evolve over these issues as movement activists continue to press for what they believe are their constitutional right to compensation. SEE ALSO ACTIVISM; ECONOMICS; POLITICS; WISE USE MOVEMENT

takings impact analysis analysis of the impacts due to government restriction on land use

Bibliography

Epstein, Richard. (1985). *Takings: Private Property and the Power of Eminent Domain*. Cambridge, MA: Harvard University Press.

Wise, Charles R. (1992). "The Changing Doctrine of Regulatory Taking and the Executive Branch." *Administrative Law Review* 44 (Spring):404.

Yandle, Bruce, ed. (1995). *Land Rights: The 1990s' Property Rights Rebellion*. Lanham, MD: Rowman and Littlefield.

Internet Resource

Meltz, Robert. (1995). "The Property Rights Issue." CRS Reports for Congress. Available from http://cnie.org/NLE.

Jacqueline Vaughn Switzer

Public Interest Research Groups (PIRGs)

Early in his career as a consumer advocate, Ralph Nader struck on an idea for a new type of organization. "How about a law office that worked for the public's interest—not that of corporations or just individuals?" he thought.

Out of this concept evolved the Public Interest Research Group (PIRG). It began its genesis with a staff of twelve lawyers and a physician, each bringing his or her expertise in a different field to the effort.

"It was like a law office, but for public interest," Nader said in *Ralph Nader: Battling for Democracy*, an authorized biography written by Kevin Graham. "We broke open a lot of new areas for several years. For instance, we were the first to bring action to create nonsmoking sections on public transportation. We presented the idea that nonsmokers had prior rights to those of smokers, which was unheard of back then."

In PIRG's early days, Donald Ross and Jim Welch—two of its original members—focused on organizing students on college campuses across the nation. With Nader's help, they created a student-led movement that still exists today. In its efforts, PIRG spread the notion that young people could make a difference in government and corporate America.

Nader's appearance at the University of Oregon in the fall of 1970 helped launch the idea of student activism and provided a successful example for other campuses to follow. Soon, all seven schools in the state college system approved the establishment of the Oregon Student Public Interest Research Group, known as OSPIRG. Students in other states then followed Oregon's lead.

Each student PIRG was financed and run by students, but guided by a small professional staff of attorneys, scientists, organizers, and other workers. The PIRGs distinguished themselves from many other movements at the time by actually participating in government processes, not by simply protesting against them. They became important players within the framework of the existing system and quickly discovered they could affect the outcome of government decisions.

For example, in Massachusetts, the Massachusetts Student Public Interest Research Group (MASSPIRG) placed an initiative on the state's ballot in 1986 aimed at reducing the use of toxic chemicals. Voters approved the measure by the largest margin of any initiative in the state's history. In 2002 a national set of laws and system of regulations are in place to deal with this same issue.

Other PIRGs tackled issues such as recycling, pollution, and public health and safety. The groups also provided training for thousands of students—training that continues today, producing wave after wave of students working to solve numerous environmental and other societal problems. PIRGs currently exist in twenty-four states, and seventeen more operate in Canada. Each is independent in operation, yet all share similar agendas and goals. SEE ALSO NADER, RALPH.

Bibliography

Isaac, Katherine, and Nader, Ralph. (1995). *Ralph Nadar Presents Practicing Democracy: A Guide to Student Action.* New York: St. Martin's Press.

Internet Resource

State PIRGs. Available from www.pirg.org.

Kevin Graham

Public Participation

Public participation is the general term for diverse formal processes by which public concerns, needs, and values are incorporated in governmental decisions. Public participation involves the use of techniques such as public meetings and hearings, advisory committees, interactive workshops, interviews, questionnaires, focus groups, and other methods to identify public concerns and preferences and address them during decision making. It does not include nonformal means of public involvement ranging from lobbying to letter campaigns and protests.

Most recent federal laws authorizing or establishing federal programs, including the latest environmental laws, contain requirements that agencies consult with the public during the design and implementation of the program. If money is given to the states, then these public participation requirements are also passed on to the states.

Who Is the "Public" in Public Participation?

Public participation does not mean taking a vote. The agencies offer the opportunity to participate, and people choose whether or not to participate. Because participation is self-selecting, most people who participate are those who have a "stake" in the issue (hence the term *stakeholders*). They may be affected economically, they may already use or want to use a resource (i.e., land or water), they may live in close proximity to a proposed project (and could be impacted by dust, noise, or traffic), or they may have a legal mandate that would be influenced by a project (i.e., a local government or regulatory agency). Often, the stake that people have in a decision primarily involves political philosophies or values. Because public participation typically involves only those who have a stake in the decision, some agencies have begun to use the term *stakeholder involvement* instead of the term *public participation*.

The public that is involved in public participation processes changes from issue to issue. People who are deeply concerned about environmental issues may not be involved in education issues, or decisions about welfare programs. People who live near a project may be very concerned about that project, but have little interest in similar issues elsewhere.

Despite these limitations, if some level of agreement (or at least acceptance) of a decision can be reached among the people who care most deeply, the agency has a stronger political mandate to act. Implementation of the decision is far less likely to be delayed by lawsuits or continued political opposition. One of the problems with public participation is that it is sometimes easier to inspire the participation of a small group of people who would be affected negatively, whereas it is difficult to motivate the participation of a much larger group who might benefit from a project, but not so greatly that it inspires them to participate.

What Does "Participation" Mean?

Some people use the term *public participation* when what they really mean is providing information to the public. Every good public participation program involves disseminating complete and objective information to the public, so people can participate on an informed basis. But public information alone is one-way communication. Public participation requires two-way communication.

Sometimes, the term public participation is used to describe a process whereby the public has a formal opportunity to comment on a proposed action, just before an agency announces its decision. The agency may or may not change any part of its decision in response to public input.

Another form of public participation occurs when an agency decides to fulfill only its minimum legal requirements. A number of agencies follow such an approach. For example, if an agency is making a decision that requires the preparation of an Environmental Impact Statement (EIS), it must meet three basic requirements:

1. It must conduct a "scoping process," typically a public meeting, to discuss the scope of the study.

2. Hold a public hearing after a draft version of the EIS has been distributed to the public.

3. Announce a public comment period (usually thirty days) with the agency providing a response to each comment in a final environmental report.

But if an issue is controversial, procedural public participation alone rarely resolves the issue. It is true that people may have been "heard," but when an agency goes ahead with a decision, that decision may still remain sufficiently controversial that it will never be implemented.

That is why, beginning in the 1980s, some agencies began to move beyond minimal procedural requirements to public participation that is characterized by genuine consultation or collaboration between the agency and the public, in an effort to address as many of the public's concerns as possible. The agency still makes the decision, but typically it does so by endorsing a solution that addresses as many of the public's concerns as the agency can, within the confines of its legal authorities, regulations, and budget. If there is still a minority who oppose the action, there has been enough interaction that the minority understands the decision-making process was fair and open, and why the decision was reached.

There are reasons why it is important that agencies retain this final decision-making authority. First, "the public" that participates on a particular issue consists of those people who choose to participate, not the entire public. A project that may have many undesirable impacts on a local community may benefit the country as a whole, and government agencies have to consider the broader state or national interest as well. Agencies operate within legal mandates, that is, authorities and budgets that constrain their options. Sometimes, this means they cannot implement options promoted by an advocacy group that fall outside their legal authority and would require the agency to violate legal requirements or far exceed their budget. Finally, the issue may be controversial not because of what the agency wants, but because the public itself is bitterly divided over the issue.

Sometimes, even this consultative approach to public participation will not result in sufficient agreement that an agency is able to make a decision which will ever be implemented. Opponents may throw up legal or political barriers that block any action. In recent years, agencies have experimented with a number of techniques, including mediation, arbitration, negotiated rule making, and interest-based negotiation to resolve such issues. These techniques are sometimes referred to as dispute resolution or alternative dispute resolution techniques, because they are frequently used as an alternative to litigation. Each of these techniques has proven to have value in special circumstances.

Why Should the Public Be Involved in "Technical" Decisions?

Many decisions that agencies think are "technical" involve choices between more than one thing society thinks is good. For example, when a regulatory agency sets a standard, such as air pollution standards, these standards can only be achieved by installing very expensive equipment, and sometimes only by shutting down existing factories, putting people out of work. The agency finds itself having to decide which is more important: clean air or jobs. If an agency creates a regulation prohibiting smoking in public buildings, it is making a choice about the relative importance of public health compared with freedom of choice. Both health and freedom of choice are good; the question is what weight or importance should be given to one over the other in a particular circumstance.

Many government agencies must make these kinds of value choices, and legislation alone does not provide sufficient guidance to determine what choice should be made in a particular decision. Although these decisions need to be informed with technical information, there is nothing about technical training that makes experts more qualified than the public in deciding which values are most important for society. Agencies need to consult with the public on these important value choices.

The Evolution of Public Participation

From the 1930s onward, the size of the U.S. federal government grew very rapidly, and government became involved in making many decisions that affected people's lives. As government grew, decisions previously made in a political process were increasingly delegated to technical experts. Over time, many people began to feel that impersonal bureaucrats were making decisions which controlled their lives.

After the Depression and World War II, there was broad general agreement in the United States that economic development should be the primary objective of domestic national policy. Leaving decisions to the experts worked well so long as this broad social consensus existed.

But in the 1960s, that consensus began to dissipate. The civil rights movement challenged the existing system of segregation, and when the public saw nightly images on television of African-Americans being brutalized by the police during nonviolent marches or demonstrations, and the aftermath of church bombings and other racially motivated violence, the social consensus began to change. Riots in Watts, a low-income area in Los Angeles, and the riots that spread throughout major cities following the assassination of the Rev. Martin Luther King Jr., caused many government officials to believe that the country was in serious trouble and the old ways needed to be reconsidered.

It was during this same time of turmoil and reflection that the environmental movement also began to grow. Many environmentalists questioned the belief that economic development should always be the primary goal. The environmental movement learned much from the civil rights movement and adopted a form of grassroots activism that challenged the existing political system. Natural resource agencies always seen as "the good guys" found themselves under increasing attack.

The controversy over the Vietnam War also challenged basic beliefs about America and its role in the world. The ensuing Watergate scandal during President Richard M. Nixon's first term, and other revelations concerning political corruption and dishonesty, further engendered public mistrust of government. The "leave it to the experts" mentality was effectively challenged on all fronts.

In response, Congress passed a series of laws designed to provide greater openness in governmental decision making, and a dialogue with the public before decisions are made. The key laws are shown in the table. In addition,

KEY LAWS PERTAINING TO PUBLIC PARTICIPATION

Year	Law	Significance
1946	Administrative Procedures Act	Established minimum standards for participation in agency rule making, including public notice, opportunity for group representation during trial-like hearings (adjudications), maintenance of a public record during such hearings, and holding public hearings (at the agency's discretion) on other matters.
1964	Economic Opportunity Act ("War on Poverty")	Required "maximum feasible participation" of the poor in decisions about community action programs. Agencies were obligated to encourage involvement of "target" populations.
1966	Demonstration Cities and Metropolitan Development Act ("Model Cities")	Required widespread participation among those affected by its program grants.
1966	Freedom of Information Act	Provided public access to most documents of government agencies.
1969	National Environmental Policy Act	Established the Council on Environmental Quality (CEQ), required intergovernmental consultation, and provided funding for citizen groups. CEQ implementing regulations (10 CFR 1500 through 1508 and 10 CFR 1021) established many of the public participation procedures that are the minimum standards for public participation in environmental decision-making.
1972	Federal Advisory Committee Act	Established procedures that must be followed by federal agencies when creating and working with citizen advisory groups.
1972	Federal Water Pollution Control Act of 1972	Stated that public participation was also required by states implementing programs under the law. Similar language was used in many subsequent laws affect environment, transportation, and social services.
1977	Government in the Sunshine Act	Required many government agencies, particularly regulatory agencies and advisory committees, to open most of their meetings to the public.
1986	Emergency Planning and Community Right to Know Act	Ensured that the public was informed about pollutant emissions from factories, energy facilities, and industrial operations (including privately owned enterprises) in their community.
1996	Administrative Dispute Resolution Act	Encouraged the use of alternative dispute resolution (ADR) techniques, and required agencies to designate an ADR officer and provide training in ADR.
1996	Executive Order 12988 – Civil Justice Reform	Encouraged and authorized the use of alternative dispute resolution (ADR) techniques in resolution of civil claims against federal agencies.
1998	Environmental Policy and Conflict Resolution Act	Created the U.S. Institute for Environmental Conflict Resolution, a new federal agency to support conflict prevention and resolution when a federal agency is involved.

SOURCE: James L. Creighton.

many agencies have issued policies and regulations concerning public participation that have created additional requirements.

Public participation has now developed sufficiently that many agencies require their planners and decision makers to attend public participation training. Public participation has also become a professional specialty. The International Association for Public Participation (IAP2) was established in 1992 and as of 2002 has approximately one thousand members. Some of these members define themselves as public participation practitioners, and provide these services on a full-time basis within agencies, or as consultants. Many of IAP2's members, however, are planners, engineers, or program managers who see public participation as an important tool for being effective in those professions.

In 1998 Congress created a new federal agency to promote and support processes to prevent and resolve conflict when a federal agency is involved. The U.S. Institute for Environmental Conflict Resolution is housed within the Udall Foundation in Tucson, Arizona.

Issues Facing the Field of Public Participation

Like any new field, the public participation field faces many challenges. Environmental activists and business leaders tend to be both white and middle class. Racial and ethnic minorities are underrepresented in many public participation processes. Language and cultural differences may account for some of the underrepresentation. But other barriers include a general fear of government agencies (who were sometimes sources of outright oppression in immigrants' countries of origin) and the belief that participation will not necessarily change the outcome.

The Internet provides a powerful new tool for participation, although all its potential uses are still being discovered. In the near future, agencies will not only use the Internet to provide information to the public, but almost all information repositories (places where copies of government documents are stored and made available to the public) are likely to become "virtual." Increasingly, the Internet is being explored as a tool for gathering public input and information. There is considerable concern, however, about a "digital divide," as the number of people with access to the Internet in the African-American, Latin, and Native American communities, and the poor in general, is considerably lower than in the public at large. People fear that the heavy use of the Internet by agencies will mean that minorities and the poor will not have the same access to the decision-making process as those people who are connected digitally. Activist groups, on the other hand, have embraced the Internet enthusiastically, and use it extensively for organizing and communication with other groups across the country.

Some developers and business have filed so-called strategic litigation against public participation (SLAPP) suits against citizen activists whose involvement in the decision-making processes may have caused delays or blocked issuance of building or environmental permits. Often, SLAPP suits have little basis in the law, but activists must hire lawyers to defend themselves in such actions, frequently at great personal expense. Many private individuals are unable to afford this, even if they would win ultimately, whereas large companies usually have the resources to hire attorneys and keep the process going as a threat against future participation. Several state courts have rejected SLAPP suits summarily, and this may begin to curtail their use. SEE ALSO ACTIVISM; AGENCIES, REGULATORY; ARBITRATION; CITIZEN SUITS; CONSENSUS BUILDING; ENVIRONMENTAL IMPACT STATEMENT; ENVIRONMENTAL JUSTICE; GOVERNMENT; MEDIATION; NATIONAL ENVIRONMENTAL POLICY ACT (NEPA); NONGOVERNMENTAL ORGANIZATIONS; POLITICS; PUBLIC POLICY DECISION MAKING; REGULATORY NEGOTIATION; RIGHT TO KNOW; WARREN COUNTY, NORTH CAROLINA.

Bibliography

Advisory Commission on Intergovernmental Relations. (1979). *Citizen Participation in the American Federal System*. Washington, D.C..

Carpenter, Susan L., and Kennedy, W.J.D. (2001). *Managing Public Disputes*. New York: John Wiley & Sons.

Creighton, James L. (1981). *Involving Citizens in Community Decision Making*. Washington, D.C.: National Civic League.

Gray, Barbara. (1989). *Collaborating*. San Francisco, CA: Jossey-Bass Publishers.

Herrman, Margaret S., ed. (1994). *Resolving Conflict*. Washington, D.C.: International City/County Management Association.

Langton, Stuart, ed. (1978). *Citizen Participation in America*. Lexington, MA: Lexington Books.

Susskind, Lawrence; McKearnan, Sarah; and Thomas-Larmer, Jennifer, eds. (1999). *The Consensus Building Handbook*. Thousand Oaks, CA: Sage Publications.

Thomas, John Clayton. (1995). *Public Participation in Public Decisions*. San Francisco, CA: Jossey-Bass Publishers.

Internet Resources

Creighton, James L. (1999). *How to Design a Public Participation Program*. Washington, D.C.: U.S. Department of Energy. Available from http://www.e.doe/ftlink/public.

Creighton, James L.; Delli Priscoli, Jerome; and Dunning, C. Mark, *et al.*, eds. (1998). *Public Involvement and Dispute Resolution*, Vols. 1 and 2. Alexandria, VA: Institute for Water Resources. Available from http://www.iwr.usace.army.mil/iwr/products.

International Association for Public Participation (IAP2) Web site. Available from http://www.iap2.org.

U.S. Institute for Environmental Conflict Resolution Web site. Available from http://www.ecr.gov.

James L. Creighton

Public Policy Decision Making

Public policy decision making refers to actions taken within governmental settings to formulate, adopt, implement, evaluate, or change environmental policies. These decisions may occur at any level of government.

The Scope of Environmental Policy

At the most general level, environmental policies reflect society's collective decision to pursue certain environmental goals and objectives and to use particular means to achieve them. Public sector decision making incorporates a diversity of perspectives on environmental problems, from those of industry to the views of activist environmental organizations. Ultimately, policies reflect the inevitable compromises over which environmental goals to pursue and how best to achieve them.

Private decision making by corporations and individuals also affects society's ability to respond to environmental challenges. Indeed, critics of governmental performance look to the private sector for initiatives. Yet, as a nation, the United States relies heavily on public decision making because only governments possess the necessary financial resources or have the requisite legal authority or political legitimacy.

Environmental policy is complex. Beyond the laws, regulations, and court rulings on the subject, it is strongly affected by agency officials who are charged with implementing and enforcing environmental law. Their decisions, in turn, are influenced by a range of political and economic forces, including the policy beliefs of elected officials, the health of the economy, anticipated costs and benefits of laws and regulations, federal–state relations, public opinion, media coverage of environmental issues, and efforts by corporations, environmental groups, and scientists to influence public policy.

The environmental quality standards that are set in laws and regulations reflect the uncertain and changing base of environmental science, as well as policy judgments concerning the extent of risk from air or water pollution or toxic chemicals that is acceptable to society. How clean is clean enough? A significantly safer or cleaner environment may be harder to achieve with existing technologies. Moreover, the effort may be both more costly and more controversial. Confronting tradeoffs among competing social values lies at the heart of environmental policy decision making.

Environmental policy covers a wide range of issues and has had a pervasive and growing impact on modern human affairs. It also goes well beyond federal and state actions on air and water pollution or control of hazardous waste and toxic chemicals. Increasingly, these actions are linked to decision making in many related areas that also affect environmental quality and human health. These include such disparate concerns as energy use, transportation, population growth, and agriculture and food production. Scientists and scholars use the concepts of **sustainability** and **sustainable development** to link these varied human influences on the natural environment. Reports from the 1992 Earth Summit and the President's Council on Sustainable Development firmly endorsed this more comprehensive and integrated view of environmental challenges.

At an even more fundamental level, environmental policy concerns the protection of vital global ecological, chemical, and geophysical systems that scientists increasingly believe to be put at risk through certain human activities. Climate change and loss of biological diversity are examples of such threats. Thus, environmental policy decision making addresses both long-term and global as well as short-term and local risks to health and the environment. For all these reasons, it has become one of the most important functions of government in both industrialized and developing nations.

Evolution of U.S. Environmental Policies

The fundamental framework for U.S. environmental policies, especially those dealing with pollution control, was established during the 1970s with the adoption of the Clean Air Act, Clean Water Act, Safe Drinking Water Act, Resource Conservation and Recovery Act (the major hazardous waste law), and Comprehensive Environmental Response, Compensation, and Liability Act (Superfund), among others. With later amendments, these statutes mandated a public policy system in which the federal government, usually the U.S. Environmental Protection Agency (EPA), set national environmental quality standards. Together with states, the EPA enforced those standards through direct regulation, or what critics call a "command-and-control" system. These same critics fault the pollution control system for its high costs and inefficiencies, a focus on remedial rather than preventive actions, and its complex, cumbersome, and adversarial rule-making and enforcement processes. Those who defend the prevailing approach cite evidence of its effectiveness and maintain that the decision-making processes on which it depends are essential to ensure fair treatment of all stakeholders. Public opinion has generally supported strong environmental protection activities, and environmental organizations have been reluctant to endorse many of the policy changes favored by industry and political conservatives.

Throughout the 1980s and 1990s and into the twenty-first century, new approaches to pollution control have been proposed, debated, and in some cases adopted. For instance, the federal government and states have experimented with market-based incentives such as the use of "green taxes" and marketable pollution allowances or permits, most notably in the acid rain

sustainable able to be practiced for many generations without loss of productivity or degradation of the environment

sustainable development economic development that does not rely on degrading the environment

control program established by the Clean Air Act amendments of 1990. In addition, industry often has advanced the idea of voluntary pollution prevention initiatives, including the use of new environmental management systems and disclosure of pollution information to the public. These are seen as supplements or alternatives to regulation.

The federal government has also encouraged such changes. For example, during the Clinton administration, the EPA attempted to improve pollution control through the use of more flexible and collaborative decision-making arrangements under the banner of "reinventing" regulation to make it "cleaner, cheaper, and smarter." These efforts continued under President George W. Bush, with some in his administration describing them as representing a "new era" in environmental protection. In both cases, emphasis was placed on improving federal–state relations. The fifty states handled most routine implementations of major federal pollution control statutes, although there was wide variation in the ability and commitment of individual states to assume these duties.

The Challenge of Environmental Policy Reform

Despite criticism of existing environmental policies and doubts about the capacity of the EPA and states to achieve the objectives outlined in these policies, reform has proved to be difficult. Studies continue to find fault with conventional pollution control policies and urge the adoption of new approaches (e.g., reports issued by the National Academy of Public Administration). However, conflicting political pressures on members of Congress have led more often to political stalemate than to constructive reform of existing statutes. These policies continue to result in substantial improvements in the nation's air and water quality, and thus in public and environmental health. Nonetheless, environmentalists and the business community usually are in substantial disagreement over most reform proposals, from greater reliance on benefit-cost analysis to increased dependence on the states for environmental enforcement.

The general verdict among both scholars and practitioners is that reform of U.S. environmental policy remains a much desired yet elusive goal. Environmentalists fear that such reform will come at the price of weakened existing laws and regulations. Industry representatives are equally adamant about the imperative to reduce what they believe to be excessively high costs for compliance. Compromise typically is difficult, particularly because few studies can point clearly to the absolute consequences of adopting proposed reforms—that is, whether reforms will improve the regulatory system as anticipated. Thus, policy change is seen as something of a gamble that many defenders of strong environmental protection are unwilling to take.

Despite these important constraints, one encouraging development in efforts to improve U.S. environmental policies can be found in the hundreds of initiatives taken at the state and local levels to reconcile environmental protection and economic development under the rubric of sustainability. Removed from the intense ideological battles in Congress, environmentalists, industry representatives, state and local officials, and concerned citizens have pioneered new collaborative arrangements that offer much promise for the future. These range from actions to promote "smart growth" land use practices, to efforts to improve air quality through better urban design and transportation initiatives, to collaborative efforts to clean up local rivers and bays and restore damaged habitat. SEE ALSO ACTIVISM; GOVERNMENT; NATIONAL ENVIRONMENTAL POLICY ACT (NEPA); NONGOVERNMENTAL ORGANIZATIONS (NGOs); PUBLIC PARTICIPATION.

Bibliography

Davies, J. Clarence, and Mazurek, Jan. (1998). *Pollution Control in the United States: Evaluating the System.* Washington, D.C.: Resources for the Future.

Kraft, Michael E. (2001). *Environmental Policy and Politics,* 2nd edition. New York: Addison Wesley Longman.

Mazmanian, Daniel A., and Kraft, Michael E., eds. (1999). *Toward Sustainable Communities: Transition and Transformations in Environmental Policy.* Cambridge, MA: MIT Press.

Portney, Paul R., and Stavins, Robert N., eds. (2000). *Public Policies for Environmental Protection,* 2nd edition. Washington, D.C.: Resources for the Future.

Rosenbaum, Walter A. (2002). *Environmental Politics and Policy,* 5th edition. Washington, D.C.: CQ Press.

Sexton, Ken; Marcus, Alfred A.; Easter, K. William; and Burkhardt, Timothy D., eds. (1999). *Better Environmental Decisions: Strategies for Governments, Businesses, and Communities.* Washington, D.C.: Island Press.

Sitarz, Daniel, ed. (1998). *Sustainable America: America's Environment, Economy and Society in the 21st Century.* Carbondale, IL: Earth Press.

United Nations. (1993). *Agenda 21: The United Nations Programme of Action from Rio.* New York: United Nations.

Vig, Norman J., and Kraft, Michael E., eds. (2000). *Environmental Policy,* 4th edition. Washington, D.C.: CQ Press.

Internet Resources

Center for American Politics and Public Policy Web site. Available from http://depts.washington.edu/ampol.

National Academy of Public Administration. (2000). *Environmental Government: Transforming Environmental Protection for the 21st Century.* Washington, D.C. Available from http://www.napawash.org.

Michael E. Kraft

Racism, Environmental *See Environmental Racism*

Radioactive Fallout

The term radioactive fallout, or just fallout, refers to the debris and radioactive materials that settle out of the air after the detonation of a nuclear weapon or after a nuclear accident that produces a cloud of airborne material, or plume. Detonation of a nuclear weapon results in the immediate propagation of a shock wave and intense heat. As the superheated fireball rises, a vacuum is formed that draws in scorched building material, soil, and other materials from the epicenter of the blast. In addition, **radionuclides** produced in the nuclear chain reaction leading to the explosion and any weapon material not consumed in that reaction will also be a part of the subsequent plume. Any similar thermal process, such as the intense fire during the Chernobyl I reactor accident in 1986, will introduce radioactive and other materials into the atmosphere, as well.

The direction and distance the fallout travels depends largely on weather conditions. Wind speed, wind direction, atmospheric stability, and the amount of rain all factor into the extent and timing of the fallout and subsequent

radionuclide radioactive particles, human-made or natural, with a distinct atomic weight number; can have a long life as soil or water pollution contamination. The amount of radioactive contamination depends on the initial amount of radioactive material contained, for instance, in a nuclear weapon. In the case of a reactor fire or steam explosion, the damage to the reactor, the amount of material at risk, and the length of time until the event is under control are all factors.

Exposure to radioactive materials, either while still in the plume, or after the fact as contamination, is the basis for potential health concerns. While alpha-emitting fallout material is not the external hazard that beta-, gamma- and x-ray-emitting materials can be, all of these materials are a potential internal hazard concern when the contamination spreads to sources of groundwater and surface water, livestock, crops, and other foodstuffs. Fallout effects can be long lasting, contaminating an area for hundreds or even thousands of years. Fallout also enters the food chain. Cows eating contaminated grass produce contaminated milk, which can pose a widespread health risk.

As with anything to do with radiation, it is the amount of absorbed energy, or the radiation absorbed dose, that matters. Remaining indoors, with doors and windows shut and air conditioning systems turned off until the plume has passed, can reduce exposure to fallout. Traveling out of the path of an incoming plume, if this can be predicted accurately, may also help avoid or reduce exposure to the fallout. Certain foodstuffs, especially water and milk, may have to be brought in from unaffected areas. Time will be one of the best countermeasures should such an event occur. The "seven–ten" rule for nuclear detonations states that for every seven-fold increase in time after a weapon detonation, there will be a concomitant tenfold decrease in the amount of dose afforded by the fallout. SEE ALSO CANCER; DISASTERS: NUCLEAR ACCIDENTS; NUCLEAR ENERGY; TERRORISM; WAR.

Ian Scott Hamilton

Radioactive Waste

Radioactive waste (or nuclear waste) is a material deemed no longer useful that has been contaminated by or contains radionuclides. Radionuclides are unstable atoms of an element that decay, or disintegrate spontaneously, emitting energy in the form of radiation. Radioactive waste has been created by humans as a by-product of various endeavors since the discovery of radioactivity in 1896 by Antoine Henri Becquerel. Since World War II, radioactive waste has been created by military weapons production and testing; mining; electrical power generation; medical diagnosis and treatment; consumer product development, manufacturing, and treatment; biological and chemical research; and other industrial uses.

There are approximately five thousand natural and artificial radionuclides that have been identified, each with a different half-life. A half-life is a measure of time required for an amount of radioactive material to decrease by one-half of its initial amount. Half-life values for each known radionuclide are unique. The half-life of a radionuclide can vary from fractions of a second to millions of years. Some examples of radionuclides with a range of different half-lives include sodium-26 (half-life of 1.07 seconds), hydrogen-3 (half-life of 12.3 years), carbon-14 (half-life of 5,730 years), and uranium-238 (half-life of 4.47 billion years). The decay process of a radionuclide is the

Workers at a nuclear power plant standing near a storage pond filled with spent fuel. (©Tim Wright/Corbis. Reproduced by permission.)

mechanism by which it spontaneously releases its excess energy. Typical mechanisms for radioactive decay are alpha, beta, and gamma emission. Alpha decay is a process that is usually associated with heavy atoms, such as uranium-238 and thorium-234, where excess energy is given off with the ejection of two neutrons and two protons from the nucleus. Beta decay involves the ejection of a beta particle, which is the same as an electron, from the nucleus of an excited atom. A common example of a beta-emitter found in radioactive waste is strontium-90. After an alpha or beta decay, the nucleus of an atom is often in an excited state and still has excess energy. Rather than releasing this energy by alpha or beta decay, energy is lost by gamma emission—a pulse of electromagnetic radiation from the nucleus of an atom.

Everything on Earth is exposed to radiation. However, exposure to radiation at levels greater than natural background radiation can be hazardous.

Exposure to certain high levels of radiation, such as that from high-level radioactive waste, can even cause death. Radiation exposure can also cause cancer, birth defects, and other abnormalities, depending on the time of exposure, amount of radiation, and the decay mechanism. High-level radioactive waste from nuclear reactors can be hazardous for thousands of years. Radioactive waste can be categorized by its source or point of origin. Because of this, the governments of many nations have developed waste classification systems to regulate the management of radioactive waste within their borders. The proper treatment, storage, and disposal of radioactive waste are prescribed based on the waste classification system defined in a nation's laws, rules, and regulations. The table outlines common categories of radioactive waste.

Radioactive Waste Description

Radioactive waste can vary greatly in its physical and chemical form. It can be a solid, liquid, gas, or even something in between, such as sludge. Any given radioactive waste can be primarily water, soil, paper, plastic, metal, ash, glass, ceramic, or a mixture of many different physical forms. The chemical form of radioactive waste can vary as well. Radioactive waste can contain radionuclides of very light elements, such as radioactive hydrogen (tritium), or of very heavy elements, such as uranium. Radioactive waste is classified as high, intermediate, or low level. Depending on the radionuclides contained in it, a waste can remain radioactive from seconds to minutes, or even for millions of years.

Radioactive Waste Management

Radioactive waste management includes the possession, transportation, handling, storage, and ultimate disposal of waste. The safe management of radioactive waste is necessary to protect public health. If handled improperly, potential exposures of humans to high-level radioactive waste can be dangerous, even deadly. Some radioactive wastes such as certain types of **transuranic waste** can cause biological effects in humans only if the radionuclides contained in the waste are directly inhaled or ingested. Most low-level radioactive wastes can be handled by humans without any measurable biological effects. Nevertheless, good handling practices of all radioactive materials and waste should be the goal to provide optimum protection to humans and the environment. There have been historic practices associated with the use of radioactive material where workers were unaware of potential risks. The radium watch dial painters of the 1920s illustrate the health effects that can be associated with improper handling practices. The painters experienced high occurrences of cancer of the larynx and tongue due to ingestion of radium.

transuranic waste waste containing one or more radioactive elements heavier than uranium, created in nuclear power plants or processing facilities

The transportation of radioactive waste can occur via roadway, aircraft, ship/barge, and rail. The classification and physical size of radioactive waste dictate the method of transport, the packaging required, and the labeling necessary to allow for the shipment of a specific waste. There are international transportation requirements for radioactive waste, as well as more specific regulations in individual countries.

Radioactive Waste Disposal

Various methods to manage and dispose of radioactive waste have been considered. Proposed management and disposal methods have included the

COMMON CATEGORIES OF RADIOACTIVE WASTE

Waste Category	Description of Waste Category	Common Sources of Waste	Common Radionuclides in Waste and Their Half-Life (y=years)
High-Level Radioactive Waste	Highly radioactive material that is deemed a waste that requires special precautions by humans, including remote handling and use of shielding; also includes spent fuel and waste resulting of used fuel	Partially used fuel from nuclear power reactors; liquid waste from the reprocessing of spent fuel taking place outside the United States	strontium-90 half-life: 29.78 y cesium-137 half-life: 30.07 y
Transuranic Waste	Material that is deemed a waste that contains radionuclides with an atomic number greater than that of uranium (92)	Weapons-production waste included mixed transuranic waste	plutonium-238 half-life: 87.7 y americium-241 half-life: 432.7 y
Mixed Waste	Material that is deemed a waste that contains both radionuclides and a characteristic or listed hazardous waste	Weapons-production waste and some research wastes	plutonium-239 half-life: 24,100 y plutonium-241 half-life: 14.4 y
Naturally Occurring Radioactive Material (NORM) Waste	Material that is deemed a waste that contains radionuclides that are present on Earth without any human interaction	Scale buildup on pipe walls that carry petroleum products	radium-226 half-life: 1,599 y radium-228 half-life: 5.76 y
Uranium or Thorium Mill Tailings Waste	The tailings material created as a by-product by the extraction of uranium or thorium from natural ore formations	Production exclusively at the site of milling for rare earth extraction	radium-226 half-life: 1,599 y thorium-230 half-life: 75,400 y
Low-Level Radioactive Waste (LLRW)(and Intermediate Waste outside U.S.)	Material that is deemed a waste that generally has been contaminated by or contains short-lived radionuclides or longer-lived radionuclides in relatively low concentrations. Low-level radioactive waste is further segregated into classes (see below)	Industrial trash from nuclear power plants; medical, research, and academic trash such as paper, plastic, and glass	hydrogen-3 half-life:12.32 y cobalt-60 half-life: 5.27 y
Class A:	Lowest level of LLRW, generally decays in 100 y		
Class B:	Moderate level of LLRW, generally decay in 300 y		
Class C:	Special controls required for this high level of LLRW, including shielding/barriers that must isolate for 500 y		
Greater than Class C:	Exceed the Class C limits and cannot be disposed in LLRW facilities; must be disposed with high-level radioactive waste		
Exempt Material or Very Low Activity Waste	Material that is deemed a waste that contains trace concentrations of short half-life radionuclides that are considered below regulatory concern	Various medical procedures	iodine-131 half-life: 8.027 days

following scenarios: shallow land burial; engineered disposal vaults; vertical shafts drilled into granite, salt, basalt, or volcanic rock; disposal cavities mined into specific rock formations such as salt; deeper-earth disposal into the submantle layer; above-ground isolation in engineered, concrete structures; recycling and reuse of waste material; radionuclide transmutation into nonradioactive material; ocean and seabed disposal; ice-sheet disposal; isolation disposal on a remote island; and even disposal in space.

Most of the civilian high-level radioactive waste throughout the world is currently being stored at nuclear power reactor sites. The spent nuclear fuel generated from the 103 operating civilian power reactors in the United

States is currently being stored on-site at the point of generation. In Europe, prior to on-site storage, spent fuel is first sent to either the Sellafield site in the United Kingdom or the La Hague site in France to be reprocessed in order to recover usable fuel. No reprocessing of commercial spent fuel is being conducted in the United States. In the United States, spent fuel and other high-level radioactive waste awaits the construction of a central, permanent repository. It is currently stored in spent fuel pools or, in some cases, in dry casks. Spent fuel pools are water-filled, lead-lined chambers that are adjacent to reactors on civilian power reactor sites. Dry-cask storage has become necessary in some cases where the on-site spent fuel pools have reached capacity. The Office of Civilian Radioactive Waste Management at the U.S. Department of Energy (DOE) is charged with developing this federal repository. Amid local opposition, Yucca Mountain, Nevada, is presently under study to evaluate its suitability as a central repository for all U.S. high-level radioactive waste. The Yucca Mountain site has been officially designated by President George W. Bush and Congress for full-scale studies. There has been further emphasis placed on the security of spent fuel, and in general on nuclear reactor sites following the September 11, 2001, terrorist attacks. Nuclear reactor sites that store spent fuel have been identified as possible terrorist targets and, therefore, have been subject to heightened security and debate over potential vulnerabilities. France, Germany, the United Kingdom, and Japan also have plans to develop centralized repositories for high-level radioactive waste at various times in the future.

Transuranic waste generated by the DOE has an operational final repository. The Waste Isolation Pilot Project located near Carlsbad, New Mexico, accepts transuranic waste and mixed transuranic waste (i.e., transuranic waste that also has a hazardous waste component) from federal facilities throughout the United States. This facility is comprised of disposal cavities mined into a salt formation some 2,150 feet underground.

The disposal method used in the 1960s and 1970s for low-level radioactive waste was shallow land burial in earthen trenches. The infiltration of water into these trenches resulted in the migration or movement of certain radionuclides into surrounding soil and groundwater. To respond to such problems, engineered disposal units have been developed to replace shallow land burial, utilizing enhanced cover systems to reduce the potential for water infiltration. The trial-and-error nature of early radioactive waste disposal sites has rendered new facility development a slow and cautious process.

Historical Perspective

The first commercial site for the disposal of low-level radioactive waste was opened in Beatty, Nevada, in 1962. Within the next ten years, five more sites opened in the United States: in Washington, Illinois, South Carolina, New York, and Kentucky. Private companies operated these sites on land leased from state governments. Prior to 1979, the DOE routinely used commercial sites for the disposal of federal waste.

Migration problems at commercial disposal sites in the United States were first discovered in the late 1960s. Four of the six commercial low-level radioactive waste disposal sites in the United States closed. Three of the four sites that closed developed leaks due to erosion by surface water, subsidence on tops of trenches, or buried waste immersed in water. Several of these

Superfund the fund established to pay for the cleanup of contaminated sites whose owners are bankrupt or cannot be identified

locations became federal **Superfund** sites due to radionuclides migrating beyond the disposal trenches, complicated by the presence of hazardous waste within the same facilities.

The historical problems experienced with commercial radioactive waste disposal in the United States resulted in the development of new regulatory requirements for site selection, construction parameters, operating practices, and waste-acceptance criteria at future disposal sites. A new U.S. disposal regulation, Title 10, Code of Federal Regulations, Part 61, "Licensing Requirements for Land Disposal of Radioactive Wastes" was introduced in 1982. This regulation outlines the requirements necessary to ensure public health, safety, and the long-term protection of the environment. Since the development of this new regulation in the United States, only one site, in Clive, Utah, has been licensed and opened for disposal of low-level radioactive waste.

Summary

Radioactive waste is being generated in the United States and throughout the world as a result of research, mining, electricity production, nuclear weapons production, and medical uses. There are many possible beneficial activities due to the use of radioactive material. Laws, rules, and regulations are made on a global scale to help ensure the safe handling of radioactive waste to protect human and environmental health. However, the question of the safe final deposition of all radioactive waste generated worldwide is still problematic. SEE ALSO CLEANUP; ENERGY, NUCLEAR; SUPERFUND; WASTE, TRANSPORTATION OF; YUCCA MOUNTAIN.

Bibliography

League of Women Voters Education Fund. (1993). *The Nuclear Waste Primer.* New York: Lyons & Burford, Publishers.

Murray, Raymond L. (1994). *Understanding Radioactive Waste*, 4th edition. Columbus, OH: Battelle Press.

Parrington, Josef R.; Knox, Harold D.; Breneman, Susan L.; Baum, Edward M.; and Feiner, Frank. (1996). *Nuclides and Isotopes*, 15th edition. San Jose, CA: General Electric Company.

Internet Resources

International Atomic Energy Agency. "World Atom." Available from http://www.iaea.or.at/worldatom.

U.S. Department of Energy, Office of Civilian Radioactive Waste Management. "The Yucca Mountain Project." Available from http://www.ymp.gov.

U.S. Nuclear Regulatory Commission. "Radioactive Waste." Available from http://www.nrc.gov/waste.html.

Waste Link Directory. "Guide to Radioactive Waste." Available from http://www.radwaste.org/general.htm.

Susan M. Jablonski

isotope a variation of an element that has the same atomic number of protons but a different weight because of the number of neutrons; various isotopes of the same element may have different radioactive behaviors, some are highly unstable

Radon

Radon is an odorless, colorless, radioactive, though chemically unreactive gas. It has an atomic number of eighty-six, which corresponds to the number of protons found in the nucleus of any **isotope** of radon. There are more than thirty known isotopes of radon, and each one emits some combination

PATHWAY FOR RADON ENTERING THE HOME

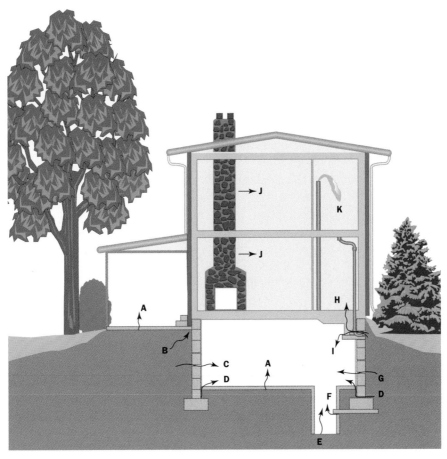

A. Cracks in concrete slabs
B. Spaces behind brick veneer walls that rest on uncapped hollow block foundation
C. Pores and cracks in concrete blocks
D. Floor–wall joints
E. Exposed soil, as in a sump
F. Weeping (drain) tile, if drained to open sump
G. Mortar joints
H. Loose fitting pipe penetrations
I. Open tops of block walls
J. Building materials such as some rocks
K. Water (from some wells)

SOURCE: Adapted from Texas A & M University.

of **alpha**, **beta**, and **gamma radiation** when undergoing radioactive transformation, commonly referred to as "decay."

Radon gas is ubiquitous in the natural environment. This is because the precursors to radon, such as the aforementioned radium isotopes, and others such as radium, thorium, and uranium isotopes, are present in some rock formations. Radon is also found in the man-made environment because many of the materials, consumer products, and foodstuffs of everyday life come from the naturally radioactive environment.

Radon is one of the few examples in nature of a gaseous element that results from the decay of a solid element and then decays into another solid element. This increases its potentially harmful effect in humans. For example,

alpha radiation fast-moving particle composed of two protons and two neutrons (a helium nucleus), emitted by radioactive decay

beta radiation high-energy electron, emitted by radioactive decay

gamma radiation very high-energy light with a wavelength shorter than x rays

radon-222, the most common isotope of radon, is a product of the alpha decay of radium-226 atoms, found in rocks. Radon-222 atoms subsequently produce polonium-218 in a similar alpha-decay process, and it is this solid substance that can lodge in human tissue.

Solid-state radionuclides remain where created by decay processes unless they are redistributed by dissolving in groundwater or by becoming airborne. Given the chemically inert nature of radon, there are no known compounds that include this element. Thus isotopes of radon may diffuse away from their place of origin and usually end up dissolved in ground water or mixed with air above the soil and rocks that bear their solid precursors.

People's exposure to radon primarily occurs when radon seeps out of air spaces above soil or rocks and into surrounding indoor or outdoor air, such as the basements of houses built over radium-bearing rocks. It is not exposure to radon gas that actually may lead to harm, but exposure to the decay products of radon, specifically the ones with short **half-lives** that emit alpha radiation. Radon-222 offspring, like polonium-218 and polonium-214, become attached to dust particles that may be breathed in by people exposed to the gas and become lodged in the respiratory tract. Decay of the radon progeny while in the lungs is the means by which the radiation dose is delivered to the lungs. This dose, which is the energy of alpha particles absorbed by cells that line the lungs, is what gives rise to the potential for lung cancer associated with exposure to radon.

half-life the time required for a pollutant to lose one-half of its original concentration; for example, the biochemical half-life of DDT in the environment is fifteen years

Radon has been labeled by the Environmental Protection Agency as the second-leading cause of lung cancer in humans (after tobacco smoke), based on mathematical risk estimates derived from many published studies of exposure of subsurface uranium miners to highly elevated levels of the gas, primarily radon-222. Many radiation health scientists have challenged such findings because of the vast difference in exposure levels between homes and buildings on the one hand, and subsurface mines on the other. However, a variety of action levels and exposure limits for radon gas exposure have been recommended or set into law for the protection of the public. The Surgeon General and the EPA recommend that radon levels of four picocuries or more inside homes be reduced. The EPA states that radon levels less than four picocuries still pose a risk, especially for smokers.

Methods to both detect and mitigate indoor radon exposure have been devised as well. Detection and measurement methods usually make use of a device to collect radon gas atoms or the offspring particles. The simplest real-time method would be a "grab sample," in which air is drawn into an evacuated flask that is then taken back to a laboratory for analysis. The most popular short-term measurement device is the activated charcoal canister, a small container of steam-treated charcoal that is opened and left at the sampling location for a prescribed time. Radon is adsorbed by the charcoal, and the decay products of the radon are analyzed after the canister has been resealed and retrieved. The simplest **mitigation** methods include sealing cracks and penetrations through foundations, as well as diverting the radon away from the slab or out of the ground, with vacuum or ventilation systems. SEE ALSO CANCER; HEALTH, HUMAN; RISK.

mitigation measures taken to reduce adverse impacts

Bibliography

National Research Council. (1999). *Health Effects of Exposure to Radon—BEIR VI.* Washington, D.C.: National Academy Press.

National Research Council. (1999). *Risk Assessment of Radon in Drinking Water.* Washington, D.C.: National Academy Press.

Other Resources

U.S. Environmental Protection Agency. "Indoor Air Quality: Radon." Available from http://www.epa.gov/radon.

Ian Scott Hamilton

Recycling

Recycling is any process that involves the recovery and reuse of materials that were once considered trash. Recycling can be as simple as reusing something—such as a coat or computer—by passing it on for someone else to use. Or, it can be as involved as reprocessing materials in metals, plastics, paper, or glass to make new products.

An Old Idea Is Rediscovered

There is nothing new about recycling. People have found ways to reuse pottery, gold, silver, and bronze for thousands of years. Old swords were melted and reshaped to use as plows. Gold and silver jewelry were melted down and reshaped into other forms. As recently as one hundred years ago, traveling peddlers in the United States and Europe collected rags, bones, and scrap metal waste from household garbage and sold them to manufacturers to make into new products.

During the early twentieth century, Americans relied less and less on recycling. By the 1950s the United States was labeled a "throw-away economy" because Americans were consuming increasing amounts of goods that ended up in garbage landfills.

Recycling was revived in many Western countries back in the 1960s and 1970s as the public became interested in conservation and looked for ways to reduce damage to the environment. In the United States, the first Earth Day in 1970 is often viewed as the official beginning of the modern recycling movement. On that day, hundreds of new recycling centers opened across the country.

The recycling movement caught on in many other Western countries during the next thirty years. Today, Germany recycles 30 percent of all of its trash. Japan recycles over 50 percent of its trash, half of all wastepaper and glass bottles, and more than 60 percent of its drink and food cans.

At the start of the twenty-first century, the United States recycling efforts are behind many European nations. Americans generate twice the amount of trash as Germans, but recycle less. According to the Environmental Protection Agency, the United States recycled 28 percent of its waste in 2002. States vary widely in their recycling programs. Minnesota is the nation's leader in recycling with a rate of recycling 45 percent of all domestic waste. Montana and Wyoming are at the bottom of the list, recycling less than 5 percent.

Why Recycle?

Recycling is one of the easiest steps anyone can take to reduce the impact of humans on the environment. On average, each American produces

Aluminum cans in recycling bin at Portsmouth Recycling Center. (©Ian Harwood; Ecoscene/CORBIS. Reproduced by permission.)

approximately 3.5 pounds of garbage per day. That is 1,500 pounds per person each year—or 90,000 pounds in a lifetime. Without recycling, all this trash ends up in landfills.

In the 1970s many people believed that recycling's greatest benefit was the reduction of the number of landfills because this would reduce the pollution associated with landfills and preserve the land. More recently, researchers have found multiple benefits to recycling.

1. **Recycling saves natural resources.** Recycling reduces the demand for new materials from the environment. For example, by recycling paper, fewer trees are needed to produce new paper.

2. **Recycling saves habitats such as rain forests.** By reducing the demand for new materials (such as metals that must be mined and refined) from the environment, more land and habitats can be preserved and/or conserved.

3. **Recycling saves energy and reduces emissions.** In most cases, it takes less energy to make new products from recycled materials than from virgin raw materials. For example, it takes 95 percent less energy to produce aluminum products from recycled aluminum than from the raw materials of bauxite ore. In general, recycling of materials also produces less pollution than processing raw materials.

4. **Recycling can be economical.** Recycling is often less expensive than the combined cost of processing new materials and managing waste disposal.

5. **Recycling reduces the need for new landfills and incinerators.** Landfills and incinerators can emit hazards to the environment. When landfills leak, hazardous solvents can contaminate underlying groundwater—water that may be used for agriculture or as drinking water. Landfills and incinerators also emit pollution into the air.

6. **Recycling reduces the improper disposal of trash, such as littering.**

Internal and External Recycling

Most people associate recycling with items such as newspapers, magazines, plastics, aluminum, and glass. The recovery, reprocessing, and reuse of materials from used items is called external recycling and requires public participation.

A second type of recycling, internal recycling, is the reuse of waste materials from manufacturing and does not involve the general public. For example, the manufacture/production of copper items results in wasted copper pieces; with internal recycling, these pieces are melted down and recast. Although internal recovery is possible in many industries, it is most common in the metal industry.

Because industrial waste accounts for 98 percent of all waste in the United States, many critics of recycling advocate that more attention should be paid to internal recycling than external recycling.

How External Recycling Works

External recycling involves three basic steps:

1. **Recovery.** Recovery is the collection of used items that can be recycled. Many cities have drop-off centers or special curbside pickup programs

to collect recyclables. Recovery may include sorting and separation of collected materials.

2. **Reprocessing.** Reprocessing is the conversion of used items into reusable products. For example, glass is melted down and molded into new bottles or paper is reprocessed into new paper. There are three kinds of reprocessing: primary, secondary, and tertiary:

   - Primary recycling is the reprocessing of materials into the same type of product, such the recycling of used glass bottles into new glass bottles.

   - Secondary recycling is the reprocessing of materials into different but similar products, such as processing corrugated cardboard boxes into cereal boxes.

   - Tertiary recycling is the reprocessing of a material into a product that cannot be recycled again—for example, when mixed office paper is reprocessed into bathroom tissue.

3. **Marketing and sale of new items.** One of the most challenging parts of recycling is creating markets for recycled items. Recycling programs depend on their ability to advertise and sell recycled items at competitive prices. Recycling does not accomplish its goals if recycled items are not used.

What Things Are Recycled?

There are four groups of materials that are commonly recycled today.

1. **Standard recyclables.** The most commonly recycled materials are aluminum, glass, paper products, steel, and plastics.

2. **Hazardous wastes.** Hazardous wastes include items such as antifreeze, motor oil, paint, and batteries. Many cities have special centers to recycle hazardous wastes.

3. **Newer products.** Some recycling centers have systems to reprocess newer products such as compact and floppy disks.

4. **Used automobiles and parts.**

Aluminum. Aluminum cans are the most widely recycled metal. In 1999 roughly two-thirds of all aluminum cans produced in the United States were recycled. However, not all forms of aluminum are recycled. For example, aluminum foil can be recycled, but not all recycling centers are set up to process it.

Paper. Paper recycling is one of this country's most successful recycling programs. By weight, more paper is recycled each year than all other materials combined. The success of this program is in part due to the successful marketing and sale of recycled paper. Recycled paper is widely used today. Unfortunately, paper can only be recycled a limited number of times, because the paper fibers become too short to continue reprocessing after awhile.

Newspaper. Every part of a newspaper can be recycled—including the newspaper and inserts. Newspaper recycling has been profitable for decades.

Steel. Steel cans can be recycled many times. Recycled steel is used for many products such as tin cans.

RECYCLING RATES OF SELECTED MATERIALS, 1994

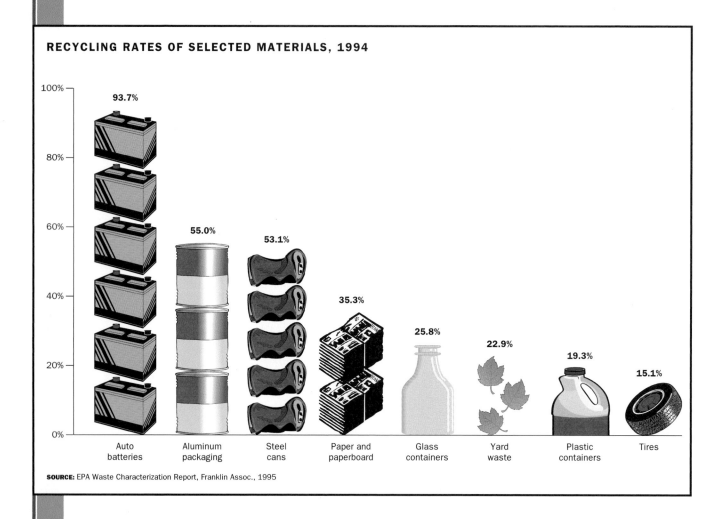

SOURCE: EPA Waste Characterization Report, Franklin Assoc., 1995

Plastics. Plastics are not biodegradable, so the best choice is to recycle them. But plastics are a challenge for recycling centers. There are so many different kinds of plastics that they are difficult for recycling centers to reprocess; in fact, many plastics cannot be recycled. Those plastics that can be recycled can only be recycled a few times. Today, most plastic containers are marked on the bottom with a number in a triangle. Each number indicates a different kind of plastic. This information allows recycling center staff to identify plastic containers that can or cannot be recycled. Containers marked one or two are the most commonly accepted plastics for recycling.

Hazardous wastes. Hazardous wastes include toxic materials such as paints, solvents, motor oil, antifreeze, herbicides, and batteries. If these materials end up in landfills, the risk exists that they may leak into underlying groundwater which people use for drinking. If incinerated, these materials end up in the air. Many recycling centers have special programs for handling hazardous wastes.

Batteries. Batteries contain many toxic ingredients, such as lead and cadmium, which can cause serious environmental damage if they are buried in landfills. Many recycling centers direct customers to special dealers who accept used batteries.

Computers. Used computers are a challenge for recycling, because they need to be completely disassembled. Recently, a number of companies have

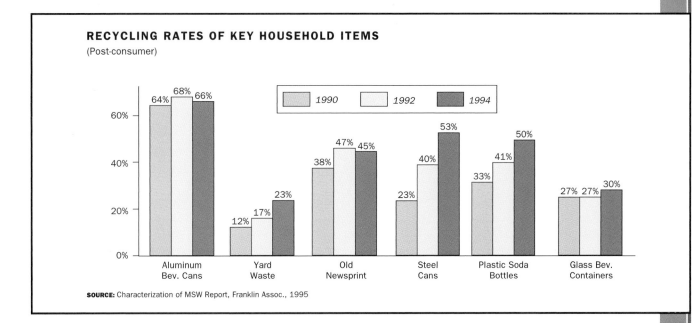

RECYCLING RATES OF KEY HOUSEHOLD ITEMS
(Post-consumer)

Legend: 1990 | 1992 | 1994

Item	1990	1992	1994
Aluminum Bev. Cans	64%	68%	66%
Yard Waste	12%	17%	23%
Old Newsprint	38%	47%	45%
Steel Cans	23%	40%	53%
Plastic Soda Bottles	33%	41%	50%
Glass Bev. Containers	27%	27%	30%

SOURCE: Characterization of MSW Report, Franklin Assoc., 1995

started exploring ways to do this efficiently and cost effectively. Recycling of computers is becoming increasingly important as the number of used computers continues to grow. One computer manufacturer, Dell, is now offering to take back old computers for reuse or recycling.

Automobile Recycling. For years, the economic incentives of recycling parts from cars, trucks and other motor vehicles has made automobile recycling a big business. In the United States, each year, more than eleven million vehicles are sent to the junkyard because they have been damaged in accidents or have reached the end of their life. About three-quarters of the scrapped vehicles are recycled or their parts are resold. Every part from the doors and windows to engines and transmissions are sold; other recyclable metal parts are magnetically separated from other materials. The rest are shredded and buried in landfills.

In the future, a smaller percentage of automobile parts will be recyclable as cars are built with more nonmetal, nonrecyclable materials, unless the automobile makers give serious attention to designing new cars that can be recycled. New cars are being built with more and more high-tech gear and hundreds of different materials that cannot be recovered.

Countries in the European Union have been exploring ways to encourage automobile manufacturers to take greater responsibility for the recycling of "end of life" automobiles. Several countries have already implemented "end of product responsibility" programs. For example, in the Netherlands, car manufacturers are liable to pay a recycling fee when they market a vehicle. The fee is then used to cover possible recycling costs.

Composting—Recycling Organic Materials

Composting is a method of recycling organic materials, such as certain food waste and yard clippings, directly into the soil. Although there are many ways to make composts, the basic idea is to mix yard clippings and food waste into a pile with soil and let it decompose; worms, insects, and other organisms

The Netherlands recycled more than three quarters (77%)of the approximately 65 million tons of garbage it generated in 2000. Public pressure to reduce dioxin emissions from incineration plants and pollution from land-fills led to landfill taxes begin-ning in 1995 and a landfill ban on combustible waste in 1997. In addition, government-owned incineration plants were operated below full capacity at the same time as incentives to expand the recyclables market and encour-age end-of-life producer respon-sibility were initiated. Mandatory separation of different types of industrial wastes, with recycling of construction and demolition waste within a government financed infrastructure, and municipal curbside pickups of organic waste for composting, along with separated household recyclables, has decreased land-filling from 35 percent in 1985 to 9 percent in 2000.

help break it down. Once the material in a compost has broken down, the degraded material can be tilled into the soil and applied as nutrient-rich mulch or material for plants.

Composting offers an opportunity to provide a rich source of nutrients for gardens and to reduce the amount of waste taking up space in landfills. Food and yard wastes currently make up about 30 percent of all wastes going into landfills. The airtight design of landfills slows down the decomposition of organic materials because they need oxygen to decompose. One commu-nity that has taken composting seriously is Halifax, Nova Scotia. Roughly 30 to 50 percent of their waste is organic matter. In 1997 the Nova Scotia Department of Environment passed a law banning the disposal of food, leaf and yard waste from landfills. Through heightened use of composting and other programs, between 1989 and 2000, Nova Scotia's per capita waste pro-duction dropped from 720 kg to 356 kg. SEE ALSO COMPOSTING; PLASTICS; POLLUTION PREVENTION; REUSE; SOLID WASTE; WASTE REDUCTION.

Bibliography

Ackerman, Frank. (1997). *Why Do We Recycle?* Washington, D.C.: Island Press.

Cothran, Helen, ed. (2003). *Garbage and Recycling: Opposing Viewpoints.* Chicago: Greenhaven Press.

The Earthworks Group. (1989). *50 Simple Things You Can Do to Save the Earth.* Berke-ley, CA: The Earthworks Press.

The Earthworks Group. (1990). *The Recyclers Handbook: Simple Things You Can Do.* Berkeley, CA: The Earthworks Press.

League of Women's Voters. (1993). *The Garbage Primer: A Handbook for Citizens.* New York: Lyons and Burford Publishers.

Mc Donough, William, and Braungart, Michael. (2002). *Cradle to Cradle: Remaking the Way We Make Things.* New York: Northpoint Press.

Nova Scotia Department of the Environment. (2001). *Status Report 2001 of Solid Waste-Resource Management in Nova Scotia.* Halifax, NS: Nova Scotia Department of the Environment.

Thompson, Claudia G. (1992). *Recycled Papers: The Essential Guide.* Cambridge, MA: The MIT Press.

Internet Resources

U.S. Environmental Protection Agency. "Municipal Solid Waste." Available from http://www.epa.gov/epaoswer/non-hw/muncpl/recycle.htm.

Global Recycling Network Web site. Available from http://grn.com.

Recycling Today Web site. Available from http://www.recyclingtoday.com.

Corliss Karasov

Regulatory Negotiation

Regulatory negotiation (also called negotiated rule making, policy dialogue, shared decision making, or "reg-neg") is a consensus-building process in which representatives of affected parties and sectors of the public (termed "stakeholders") work together with government officials to develop policies or regulations. Issues subjected to regulatory negotiation include car-emission levels, risk from lead exposure, and contamination cleanup levels. These com-plex interest-based processes utilize impartial process facilitators—often peo-ple who are experienced mediators. Those interests participating in the process are expected to abide by any resulting agreement and implement its terms. This agreement-seeking process usually occurs only after a thorough conflict assessment has been conducted, and is generally undertaken with the assistance of a skilled neutral mediator or facilitator. SEE ALSO ARBITRATION; CONSENSUS BUILDING; ENFORCEMENT; LITIGATION; MEDIATION; PUBLIC POLICY DECISION MAKING.

Bibliography

Cormick, Gerald; Dale, Norman; Emond, Paul; Sigurdson, Glenn; and Stuart, B. (1996). *Building Consensus for a Sustainable Future: Putting Principles into Practice.* Ottawa, ON: National Round Table on the Environment and Economy.

Susskind, L., and Cruikshank, J. (1987). *Breaking the Impasse: Consensual Approaches to Resolving Public Disputes.* New York: Basic Books.

Internet Resource

U.S. Institute for Environmental Conflict Resolution Web site. Available from http://www.ecr.gov.

Susan L. Senecah

Remediation   *See Abatement; Cleanup*

Renewable Energy

Renewable energy is energy that is **regenerative** or, for all practical purposes, virtually inexhaustible. It includes solar energy, wind energy, hydropower, biomass (derived from plants), geothermal energy (heat from the earth), and ocean energy. Renewable energy resources can supply energy for heating and cooling buildings, electricity generation, heat for industrial processes, and fuels for transportation. The increased use of renewable energy could reduce the burning of fossil fuels (coal, petroleum, and natural gas), eliminating associated air-pollution and carbon dioxide emissions, and contributing to national energy independence and economic and political security.

regenerative able to be regenerated or created anew

Historical and Current Use

Before the 1900s, the world as a whole used wood (including wood converted to charcoal) for heat in homes and industry, vegetation for feeding draft animals, water mills for grinding grain and milling lumber, and wind for marine transportation and grain milling and water pumping. By the 1920s, however, coal and petroleum had largely replaced these energy sources in industrialized countries, although wood for home heating and hydroelectric power generation remained in wide use. At the end of the twentieth century, nearly 90 percent of commercial energy supply was from fossil fuels.

Renewable energy, however, makes important contributions to world energy supplies. Hydroelectric power is a major source of electrical energy in many countries, including Brazil, Canada, China, Egypt, Norway, and Russia. In developing countries many people do not have access to or cannot afford electricity or petroleum fuels and use biomass for their primary energy needs. For example, most rural people in Africa use wood, scrub, grass, and even animal dung for cooking fuel. Small-scale renewable energy technologies are often the only practical means of supplying electricity in rural areas of these countries. The table indicates the relative consumption of energy sources in the United States.

anaerobic a life or process that occurs in, or is not destroyed by, the absence of oxygen

Major Types of Renewable Energy Sources

Biomass.   Biomass includes wood, agricultural crops and residues, municipal refuse, wood and paper products, manufacturing process waste, and human and livestock manure. It can be used to heat homes and buildings, produce electricity, and as a source of vehicle fuel. Wood and paper manufacturers and sugar mills use biomass residues for process heat and electricity production. There are power plants that burn wood, agricultural residues, and household trash to produce electricity. Biogas (composed of methane, carbon dioxide, and other gases) produced by decomposing biomass in **anaerobic** conditions is captured from landfills, municipal sewage treatment plants, and livestock waste management operations. This gas can be used for heat or to generate electricity.

Ethanol is used as a transportation fuel in the United States, Brazil, and a few other countries. Nearly all the fuel ethanol in the United States is made from corn, although it can also be produced from other sources, including wastepaper. There is a small but growing consumption of "biodiesel" made from grain oils and animal fats.

Geothermal systems.   Geothermal energy (heat from the earth) created deep beneath the earth's surface is tapped to produce electricity in twenty-two countries, some of which include the United States, Iceland, Italy, Kenya, and the Philippines. Geothermal hot springs can also heat buildings, greenhouses, fish farms, and bathing pools.

Hydropower.   Hydropower, produced from flowing water passing through hydroelectric **turbines**, is the leading renewable energy source, contributing to approximately 9 percent of the electricity generated in the United States. Most hydropower is produced at large dams, although there are many small systems operating around the world, such as the small hydropower plant in Namche Bazar, Nepal, which provides power for the tourist and market town near Mt. Everest. The production of hydroelectricity from year to year varies with precipitation.

Ocean energy.   The world's oceans are a vast and practically untapped source of energy. There are a few operating wave and tidal power plants around the world, and several experimental ocean thermal energy conversion (OTEC) plants have also been built. A small wave power plant in Norway captures water from waves in a dam and lets the water out through a turbine. A 240-megawatt tidal power facility on the Rance River in France produces electricity as tidal flows move back and forth through turbines located at the mouth of the river. In Hawaii, a small OTEC plan was built which uses the temperature of warm surface water to evaporate cold seawater in a vacuum to produce steam that turns a turbine and generator.

Solar energy systems.   The simplest uses of solar energy are for drying crops, and heating buildings and water. Solar-heated homes and solar water heaters can be found in nearly every country around the world. Crops can be simply laid in the sun to dry, or more sophisticated collectors can be used to heat air to dry food stored on drying racks. Solar water heaters use collectors

An acid rain monitor, monitoring in a high elevation in forest. (U.S. EPA.)

turbine machine that uses a moving fluid (liquid or gas) to gas to turn a rotor, creating mechanical energy

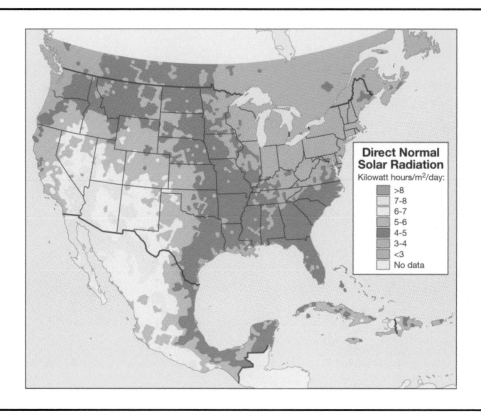

Direct Normal Solar Radiation

Kilowatt hours/m²/day:
- >8
- 7-8
- 6-7
- 5-6
- 4-5
- 3-4
- <3
- No data

to heat water that is stored in a tank for later use. Homes can be heated by using a masonry floor to absorb sunshine coming through windows, or by using solar collectors to heat a large tank of water than can be distributed for heating at night.

Concentrated sunlight can be used to produce high-temperature heat and electricity. Nine concentrating solar parabolic trough power plants, with a combined generation capacity of 354 megawatts, are located in the Mojave Desert in California. (A megawatt is 1 million watts, or 1,000 kilowatts.) The U.S. Department of Energy built and tested a ten-megawatt solar thermal central receiver power plant near Barstow, California, which operated successfully for about seven years. Another type of concentrating solar thermal power system is a parabolic dish. Systems with a capacity of up to twenty-five kilowatts have been developed.

Photovoltaic (PV) systems are based on solar electric cells, which convert sunlight directly to electricity. They can be used to power hand calculators or in large systems on buildings. Many PV systems are installed in remote areas where power lines are expensive or unfeasible, although the number of systems connected to electricity transmission systems is increasing, and range in size from 1 to several kilowatts on houses, to systems over one hundred kilowatts on large buildings. PV systems are very suitable for use in developing countries where people have no electricity from electric power lines.

Wind energy systems. Water-pumping and grain-milling windmills have evolved into electric power turbines. There are now tens of thousands of wind turbines operating around the world. They range in size from tiny turbines on the back of sailboats to very large units that can produce as much as

U.S. ENERGY CONSUMPTION AND ELECTRICITY GENERATION, 1999

Energy Source	(Quads*)	(%Total)	(Bill. kWh**)	(%Total)
Total	96	85	3,641	
Coal/Coal Coke	22	23	1,891	52
Petroleum	38	39	116	3
Natural Gas	22	23	546	15
Nuclear	8	8	674	19
Renewables (Total)	7.2	7.5	419	12
Hydro	3.5	3.6	339	9.4
Biomass/Biofuels	3.2	3.3	58	1.6
Geothermal	0.4	0.4	17	0.46
Solar	0.07	0.07	0.85	0.02
Wind	0.05	0.05	4.5	0.12

*A quad is quadrillion British Thermal Units (BTUs), and is the equivalent of about 180 million barrels of crude oil.
**Bill. kWh = a billion kilowatt-hours; One kilowatt-hour (kWh) is the equivalent of running a 100-watt lightbulb for 10 hours.
Note: values are rounded.

SOURCE: Energy Information Administration, U.S. Department of Energy.

2 to 3 megawatts of electricity, with 100-foot (30-meter) blades. They can be installed on land and in shallow water in coastal areas.

The Future for Renewable Energy

Renewable energy has many advantages that will help to maintain and expand its place in world energy supply:

- Renewable energy resources are enormous—hundreds of times beyond the needs of world energy consumption in 2000.

- Advances in technologies are reducing manufacturing costs and increasing system efficiencies, thereby reducing the cost of energy from renewable resources.

- Negative environmental and health impacts of renewable energy use are much fewer than those of fossil fuels and nuclear power.

- Many renewable energy technologies can produce energy at the point of use, allowing homeowners, businesses, and industry to produce their own power.

- There is strong support for renewable energy from people around the world.

- Many governments have programs that support renewable energy use to limit the emission of greenhouse gases and thereby reduce the threat of global warming.

As fossil fuels such as oil and natural gas become scarce, they will become more expensive. Some experts believe that demand for oil will exceed production capability within the next twenty years.

Using energy conservatively and efficiently, no matter how it is produced or where it comes from, is the most economical way to consume energy. Simply turning off lights and computers when they are not in use can save an individual household or business money and reduce the environmental impact associated with producing electricity.

Bibliography

U.S. Energy Information Administration. (2001). *Annual Energy Review 2000*. Washington, D.C.: U.S. Department of Energy.

U.S. Energy Information Administration. (2001). *International Energy Annual 1999*. Washington, D.C.: U.S. Department of Energy.

U.S. Energy Information Administration. (2001). *Renewable Energy Annual 2000, with Data for 1999*. Washington, D.C.: U.S. Department of Energy.

Internet Resource

Renewable Energy World. London: James & James Science Publishers. Available from http://www.jxj.com.

U.S. Department of Energy, Office of Energy Efficiency and Renewable Energy. Available from http://www.eren.doe.gov.

Paul Philip Hesse

Resource Conservation and Recovery Act

The Resource Conservation and Recovery Act (RCRA) of 1976 is a federal law aimed at protecting human health and the environment by safely managing and reducing hazardous and solid nonhazardous waste. It gives the U.S. Environmental Protection Agency (EPA) the task of controlling hazardous waste, through safety regulations, permits, and inspections, from its creation to disposal or from "cradle to grave." RCRA also aims to conserve energy and natural resources by giving states or regions the job of developing programs for nonhazardous waste, such as recycling and waste reduction programs. RCRA is an amendment to the 1965 Solid Waste Disposal Act. It became effective in 1980 but does not apply to sites abandoned before this date, which are addressed by the Comprehensive Environmental Response, Compensation, and Liability Act (CERCLA).

The 1984 Hazardous and Solid Waste Amendments (HSWA) to RCRA, sometimes called the "land ban," were a response to concern about hazardous wastes leaking into groundwater. HSWA states that only treated hazardous wastes may be disposed of on or beneath the ground, unless it can be guaranteed that they will not leak out. It also imposes safety requirements on landfills and other land-based hazardous waste disposal facilities. These include leakproof liners and systems to monitor and capture leachate. One consequence of the costly treatment requirements for land disposal of hazardous waste has been a reduction in the amount of hazardous waste; manufacturers have been motivated to substitute nonhazardous materials. HASW also regulates the three to five million underground storage tanks (USTs) containing petroleum and hazardous products, as distinct from waste. In September 1988 the EPA gave tank owners and operators ten years within which to replace, upgrade, or close existing USTs. Regular inspections are required to help prevent leaks.

In 2002, in one of the largest-ever hazardous waste settlements, Mobil Oil Corporation agreed to pay $11.2 million for the alleged mismanagement of benzene-contaminated waste in Staten Island, New York. Despite this and other successes, many facilities holding hazardous waste permits have not been inspected between 2000 and 2002, as required by RCRA, according to information made public by the EPA. SEE ALSO UNDERGROUND STORAGE TANK.

Internet Resource

U.S. Environmental Protection Agency. "Enforcement and Compliance History." Available from http://www.epa.gov.

U.S. Environmental Protection Agency. "RCRA." Available from http://www.epa.gov.

Patricia Hemminger

Reuse

The reuse of products, materials, and parts can have significant environmental and economic benefits. Waste is not just created when consumers throw items away. Waste is generated throughout the life cycle of a product, from extraction of raw materials, to transportation to processing and manufacturing facilities, to manufacture and use. Reusing items or making them with less material decrease waste dramatically. Ultimately, less material will need to be recycled or sent to landfills or waste-combustion facilities.

Used goods are widely available to industries, businesses, institutions, and individuals. There are secondhand markets for entire industrial production facilities, such as breweries and chemical production plants, as well as for industrial, construction, and medical equipment. Used goods for individuals include cars, clothes, books, furniture, household items, sports equipment, and musical instruments. Sources of used goods include on-line auctions and markets, secondhand merchandise stores, classified advertisements, estate sales, auctions, rummage sales, yard sales, salvage yards, materials exchanges, trash salvaging or "dumpster diving."

Amount of Reuse

In the United States, several secondhand markets are $100 billion dollar industries, and several more fall in the $1 to $10 billion range. Each year 40 million used cars are sold in the United States, nearly three times the number of new cars purchased. Overall, secondhand markets are almost as large as consumer recycling in terms of the amount of material processed (approximately fifty million tons of paper and ten million tons of glass are recycled annually in the United States), and the economic value of secondhand markets is far greater than those for recycling.

A considerable percentage of secondhand goods are exported from the United States, especially clothing; automobiles; and industrial, construction, and medical equipment. In a number of countries, including the Czech Republic, Nigeria, Uganda, and Zimbabwe, imports of used clothing compete strongly with the domestic production of new clothes.

Theory of Reuse

Reuse can reduce the pollution and resource use associated with manufacturing a new item, and can delay or eliminate disposal of the item. In order to experience the greatest environmental benefits, reuse of an item needs to replace, at least partially, the purchase and production of a new item. In some situations, reuse may not incur any real benefits. For example, if a car owner sells or gives a car to someone who would not otherwise possess a car, and then buys a new car to replace the old one, the result is that there are now two operating cars rather than one. In other situations, the reuse of an item may

---

IRISH PLASTIC BAG TAX

In March 2002, stores across Ireland began to charge an extra fifteen euro-cents for each plastic shopping bag, formerly given away to hold purchased merchandise. Before the "green" tax was implemented, Ireland's 3.9 million people used about 1.2 billion plastic bags each year. After just five months, this number was cut by ninety percent and 3.5 million euros had been raised by the tax program to be spent on environmental projects. Shoppers now bring sturdy reusable shopping bags along and enjoy the beautiful Irish countryside without the eyesore of plastic bags caught on hedgerows and blown into gutters.

A woman donating clothes to the Salvation Army at a deposit drop-off site. (M. Stone, U.S. EPA. Reproduced by permission.)

have zero effect on the production or purchase of new items. For example, if someone buys a "white elephant" at a rummage sale (perhaps a necklace or a used compact disc), that purchase will not in any way prevent or replace the purchase of a new item. However, even if reuse has no tangible environmental benefits, it can have economic and social welfare benefits. If the car example above is reconsidered, for instance, two people, not just one, now own a useful vehicle. In the compact disc example, the buyer acquires another disc for his or her pleasure, and the seller earns some perhaps much needed cash.

Reuse can replace the production and purchase of new items, especially when the first owner does not sell in order to be able to buy a new item. Examples of this sort include clothes and furniture, which are typically given away or sold at low prices by the first owners, and which second-hand buyers often buy instead of new items.

Role of Government and Industry

The U.S. government is one of the largest purveyors of used goods in the United States; it regularly sells surplus items through sealed bids, auctions, silent auctions, and fixed-price sales. On the other hand, government regulations largely prevent the purchase of used items by the U.S. government and require the labeling of products containing used parts in a way that may discourage the use of used parts by industry.

There are both incentives and disincentives for reuse by industry. Reuse, remanufacturing, repair, and refurbishment of products and parts can be economically beneficial for industry. For example, used copiers are often remanufactured and refurbished. A number of companies now sell modular, reusable carpet. On the other hand, firms in some cases have an incentive to discourage reuse of their products, in order to maintain and increase production of new goods.

Reuse by the Individual

Individuals can maximize the environmental and economic benefits of their own reuse efforts by carefully contemplating their reuse strategies, by developing the ability to make repairs, and by learning about local sources of used goods and replacement parts. The environmental and economic benefits of reuse typically increase as the size and cost of the item increase. For example, new furniture is both resource-intensive and expensive. Repair, repainting, and reupholstering of used furniture can replace the purchase of new furniture. The regular repair of shoes can considerably extend their life. Used clothing, ranging from designer clothes at consignment stores to basic items at rummage sales, is widely available. Used books, sports equipment, and musical instruments are also available at local stores and on-line. Used building materials (doors, windows, hardware, etc.) are increasingly available at salvage yards such as Urban Ore in Berkeley, California.

Reuse can have significant environmental and economic benefits by replacing the purchase of a new item. Secondhand items range from large industrial facilities and equipment to cars, sports equipment, clothes, and toys for individuals. Businesses can benefit from secondhand markets both by buying secondhand equipment and by selling surplus equipment for reuse. Individuals can make a valuable contribution to the environment and their own finances by learning to make repairs, by wisely shopping for secondhand goods, and by selling or donating their unwanted goods so that others may use them. SEE ALSO RECYCLING; WASTE; WASTE REDUCTION.

Bibliography

Dacyczyn, Amy. (1998). *The Complete Tightwad Gazette: Promoting Thrift as a Viable Alternative Lifestyle*. New York: Villard Books.

Goldbeck, Nikki and David. (1995). *Choose to Reuse*. Woodstock, NY: Ceres Press.

Internet Resources

Reuse Development Organization (ReDO). Available from http://www.redo.org.

U.S. Environmental Protection Agency, Office of Solid Waste. "Source Reduction and Reuse." Available from http://www.epa.gov/epaoswer.

Valerie M. Thomas

Right to Know

An industrial democracy requires well-informed citizens. The use of public information as a means of reducing harm from pollution evolved throughout

By reclaiming parts from eleven million vehicles each year, automotive salvage yards in North America save both raw materials and millions of barrels of oil that would otherwise be used to manufacture new replacement parts. Municipal collection programs for latex paints have provided considerable savings in hazardous waste disposal fees while providing usable paints to nonprofit organizations. Many reuse activities such as thrift shops and rummage sales benefit charities and provide low-cost or free goods to those in need.

—Source: Reuse Development Organization (ReDo). Available from http://www.redo.org.

the twentieth century. The Pure Food and Drug Act of 1906 and the Insecticide Act of 1910 established mandatory content labeling for all products. Consumers remained the primary recipients of such information until pressure from unions and public interest groups led to the enactment of the Hazardous Communication Standard Regulations in 1983, administered by the Occupational Safety and Health Administration (OSHA). These regulations required all private employers using hazardous substances to label containers in the workplace, to train employees in safe practices, and to provide readily available, action-oriented material safety data sheets (MSDS) for each controlled substance. Each MSDS explains health risks from exposure and provides step-by-step procedures for accident response.

Nine months after the 1984 disaster in Bhopal, India, an accidental release at a pesticide factory in West Virginia injured 150 people. It became readily apparent that communities near industrial sites were both ignorant about substances used in factories and poorly prepared for emergency response. In the 1986 reauthorization of the Superfund Act (SARA), the U.S. Congress added Title III, the Emergency Planning and Community Right to Know Act (EPCRA). EPCRA requires states to establish local emergency response planning committees that include elected officials and representatives from emergency agencies, industry, the mass media, and the public. Companies using regulated substances must provide an inventory of materials to the local committee along with the corresponding MSDSs. A separate section of EPCRA requires facilities to annually provide states with a Toxic Release Report. These reports specify the quantity of toxic material released or disposed of and where it ends up (landfill, underground injection, air, water, and recycling). States forward these reports to the Environmental Protection Agency (EPA) for analysis and public distribution in what is known as the annual Toxic Release Inventory (TRI), available on the Internet.

Citizens now have access to information regarding potentially harmful substances in the products they buy, those used in their workplace, and those released in their community. These data are used in making individual decisions about place of employment and residence, as well as schools. Since expenditures for operating each emergency planning committee are a local decision, citizens and community groups play a critical role in generating adequate support. Releases reported in TRI, except for emergency spills, are regulated by discharge permits. TRI can help reveal when the permits are being violated or government enforcement is lax. This may be the basis for a citizen lawsuit. The negative publicity associated with media coverage, coupled with the pressure exerted by well-organized and persistent community groups, has led many companies to reduce emissions well below legally permissible limits.

Numerous examples demonstrate the power of a well-informed public to create change. A Michigan-based nongovernment organization (NGO), the Ecology Center, has acted in conjunction with the Great Lakes Auto Pollution Alliance to work with industry to implement major reductions in **toluene** air releases, the source of noxious community odors. Combined with U.S. Census data on the population, TRI now plays a central role in "environmental justice" analyses of the distribution of pollution in low-income and minority communities. Since 1993, businesses in Canada have also been required to report similar releases. These data are available on the Internet as the Canadian National Pollution Release Inventory (NPRI). They play a critical role in

toluene carbon-containing chemical used in fuel and as a solvent

local and regional environmental initiatives. For example, a Montréal NGO used NPRI to compare discharges from local refineries. It found one facility with double the benzene emissions of a similar facility. Public pressure led the refineries to voluntarily pledge a reduction in emissions.

The right to know (RTK) laws have led to significant increases in worker safety, the emergency preparedness of communities, and some major voluntary reductions in facility emissions. Citizens can obtain company inventory and MSDS information from their local emergency planning committee. Via the Internet they can access TRI and MSDS information from the EPA. Many environmental organizations also provide on the Internet TRI or NPRI information combined with additional analysis tools such as geographic information system maps. Users should be cautioned, however, about the limits of the data. Not all chemicals are regulated, and uses below specified quantities are exempt. Some facilities fail to meet the self-reporting requirements. A complete hazard assessment involves the analysis of releases, pathways (such as an air plume), human exposures, and dose–response relationship by population type. Most RTK information only includes estimates of annual source releases. SEE ALSO ACTIVISM; COMPREHENSIVE ENVIRONMENTAL RESPONSE, COMPENSATION, AND LIABILITY ACT (CERCLA); ENVIRONMENTAL JUSTICE; GOVERNMENT; INFORMATION, ACCESS TO; NONGOVERNMENTAL ORGANIZATIONS (NGOS); OCCUPATIONAL SAFETY AND HEALTH ADMINISTRATION (OSHA); PUBLIC PARTICIPATION.

Bibliography

Emergency Planning and Community Right-to-Know Act (EPCRA). SARA Title III, 42 U.S.C. 11001 *et seq.*

Hadden, Susan. (1989). *A Citizen's Right to Know.* Boulder, CO: Westview Press.

Internet Resources

Environment Canada. "National Pollutant Release Inventory." Available from http://www.ec.gc.ca/pdb.

U.S. Environmental Protection Agency. "Toxic Release Inventory: Community Right to Know." Available from http://www.epa.gov/tri.

Wolf, Sidney. (1996). "Fear and Loathing about the Public Right to Know: The Surprising Success of the Emergency Planning and Community Right to Know Act." *Journal of Land Use and Environmental Law* 11(2):218–319. Also available from http://www.law.fsu/edu/journals.

John P. Felleman

Risk

Risk is the potential for harm. Although the concept of risk—and some of the same analytic tools—are also used in finance and actuarial science, as well as to describe threats from natural events, this discussion focuses on risks to human health and the environment from toxic pollution.

risk=f(hazard, exposure)

The magnitude and severity of risk are a function of the types of harm (i.e., the hazards, or what can go wrong) and the extent and likelihood of exposure. If the elements of hazard and exposure are not both in play, there is no biophysical risk to health or the environment. However, the *perception* of risk can be as damaging, with potential for destroying trust and sapping resources and

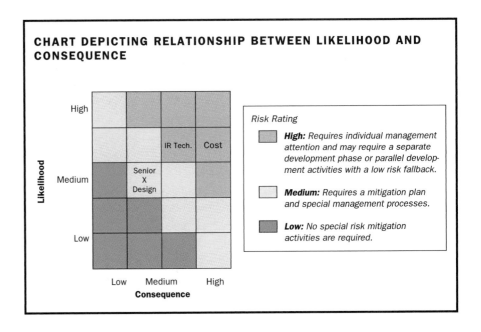

CHART DEPICTING RELATIONSHIP BETWEEN LIKELIHOOD AND CONSEQUENCE

emotional energy. Maintaining an appropriate balance between the level of social concern about a threat on the one hand and the extent of its social impact or risk on the other hand is an ongoing challenge for risk communicators, an engaged citizenry, and policymakers.

Hazards

Hazards to human health include cancers, asthma, skin rashes, infectious diseases, eye and lung irritation, developmental problems, and broken bones. Population hazards also include habitat destruction, resource degradation, threats to public health from contamination of drinking water, bacterial resistance to antibiotics, famine, and such macroconcerns as global climate change. Of greatest consequence are hazardous effects that are irreversible or long lasting, or which seriously compromise the length or quality of lives in current and future generations. Hazards that will affect future generations, or groups spatially removed from the root of the problem, may go unidentified or be discounted in formulating an assessment of risk.

The Dose–Response Concept

The toxicity or severity of a hazard can be described by a *dose–effect* (also called *dose–response*) relationship. This concept is conveyed graphically by plotting dosage (amount or concentration of a toxin) against population.

Data for describing dose–response relationships are gathered from tests in which groups of organisms are exposed to a toxin at a range of doses. Typically, as the dose increases, the toxic effect of concern is produced in more of the population.

The dosage at which the specified effect is measured is called the *effective dose* (ED). The percentage of the population affected is indicated by a subscript. So for example, ED_{10} refers to the dose at which 10 percent of the population would be affected by the toxin. When the measured effect is mortality, the term *lethal dose* (LD) or *lethal concentration* (LC) is used. An LD_{50} is the dose at which 50 percent of a population is killed.

CANCER CLUSTERS

When a number of people in a neighborhood or workplace develop the same disease within a short period of time, it may signal a *disease cluster*. A disease cluster is defined by having more cases of an illness within a particular geographic area and time period than would be statistically expected for a population with the same characteristics. Disease clusters can result from exposures to hazardous materials in the local environment or from similar lifestyle risk factors (i.e., people who live or work together may have similar eating, exercising, or smoking habits).

Disease clusters can provide clues to the cause—or risk factors—associated with a disease. They are easier to identify when a number of people show the same symptoms soon after exposure, such as when nausea follows soon after eating spoiled food in a restaurant. With a longer lag time or small number of sick people,

or with symptoms that are dissimilar or not obvious, real disease clusters may not be noticed. Conversely, clusters may be suspected due to misperception of a higher-than-average incidence of cases, or when different diseases are perceived to be the same or to have stemmed from the same cause.

Cancer clusters are particularly difficult to prove because (1) there are more than one hundred types of cancer, each with different associated risk factors; (2) there is a typically long lag time between exposure to environmental risk factors and noticeable development of the cancer; and (3) the location of the suspected cluster may be different than where a diseased person lived, worked, or went to school at the time they were exposed. Cancer clusters are more likely to be identified if a large number of individuals are diagnosed with a rare cancer or one that is rare for their age group.

At the same dose, chemicals that are more hazardous affect a greater proportion of the population than do chemicals that are less hazardous. Thus chemicals that are less hazardous have a higher ED_{50} or LD_{50} than do those that are more hazardous.

While human beings are the population of ultimate interest in dose–response studies of human health hazards, rodents are typically used as surrogates in lab tests of the effects of the toxic materials. The process of extrapolating results from rodents (or other indicator organisms) to people introduces layers of uncertainty because of physiological, developmental, and size differences between the species. Hazardous effects on plants and animals are also studied using the same conceptual methods, both because of the intrinsic value of these species and also, in some cases, because they are indicators of indirect effects on the human population.

Population Variability

As dose–response relationships show, populations are not equally susceptible to toxic hazards. Differences among individuals are due to gender, age, inherited genetic makeup, and the wear and tear and immunities that develop during the course of life. For example some people have inherited the genes that enable them to detoxify certain pesticide poisons. These people do not get sick from exposure at levels that make other people ill. Current research is linking biomarkers for genetic risk factors to disease outcomes. As it becomes clearer why people are differently vulnerable (or resistant), it also becomes more apparent that the same risk-based standards may not be applicable across populations. For example dietary iron is a risk factor for heart disease among middle-aged men at concentrations considered beneficial to women of reproductive age.

DOSE–EFFECT RELATIONSHIP

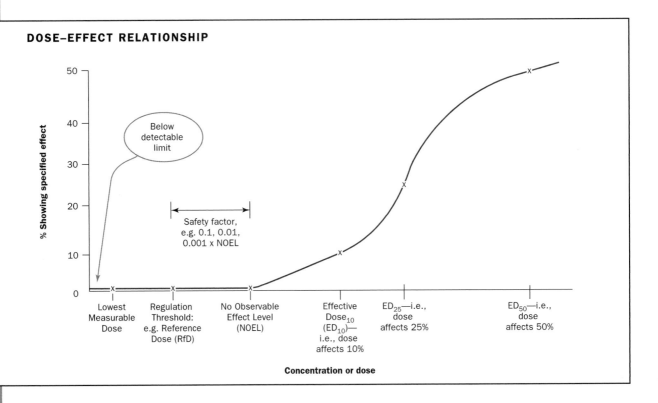

Vulnerability to hazards also changes during our lifetimes, with greater sensitivity to many toxins during fetal development, the rapidly developing stages of early childhood, and puberty—although negative effects may not be manifest until much later in life. For these reasons, among others, exposures are not easily tied to disease outcomes (see sidebar). Just try to imagine, for example, how you or your parents would struggle to respond accurately to a survey asking which pesticides you were exposed to in early childhood, and in what quantities! Some toxins and infections are particularly hazardous to those with weakened immune systems and defenses, such as the elderly and those whose systems are compromised due to other diseases or by interactive effects with medical treatments or other chemical pollutants.

Exposure

Individuals are not at risk from the consequences of a hazard if they are not exposed to it. The critical factor linking exposure to risk is the quantity of toxin that is bioavailable to vulnerable organs or processes. However, **bioavailability** is difficult to measure directly, so various measurement endpoints are used as surrogates for exposure. For pesticides, these have included sales and use data, application dosages, residues on food, and fate-and-transport data (i.e., what happens to a pesticide after application, where it goes, and how fast it degrades). Exposures are sometimes estimated from mathematical or simulation models that extrapolate from data collected by empirical studies (e.g., the amount of pesticide reaching skin or clothing, tracked indoors on shoes, or leached through soil into groundwater).

Estimates of exposure can vary widely, depending on the method for collecting data, the surrogate indicator used, and whether the assumed range of possible exposures is limited to permitted quantities or also includes accidental or purposeful exposures at much higher levels.

bioavailability degree of ability to be absorbed and ready to interact in organism metabolism

Risk-Based Decisions

Whereas risk assessments are a product of the quality and choice of input data and the assumptions incorporated into the assessment model, the usefulness and relevance of a risk characterization depend largely on how the risk problem is perceived and formulated. A well-formulated problem must engage the perspectives of multiple "publics" and be integrated with decision-management options. Perceived options for risk management are constrained by societal values that determine what are considered acceptable risks and by the resources invested for risk mitigation (i.e., for preventing or remediating the problem).

The National Research Council framework for using risk to "inform decisions in a democratic society" (1996) iteratively builds from a multifaceted formulation of the problem incorporating all aspects of risk analysis. As defined by the Society for Risk Analysis, the premier professional organization in the field, these components include risk assessment—or the quantification and description of hazards and exposure, risk characterization, risk communication, risk management, and policy relating to risk. Criteria for a successful risk-based decision process are listed in the following table.

The assessment and regulation of sewage-sludge disposal provides a good illustration of the potential and foibles of risk-based decision making, underscoring the importance of a participatory and iterative analytical and deliberative process in setting risk standards and developing protective environmental policies.

Risks from Sewage Sludge: A Cross-Country Comparison

Sewage sludge is the semisolid or concentrated liquid residue generated during the treatment of wastewater. In addition to biodegradable organic material, sludges can contain pathogens (disease organisms) and industrial pollutants (such as heavy metals) that can be damaging to human health. Among the means for disposing of sludges—by incineration, landfilling, or spreading across farmland and other open space—only land application has the benefit of returning the fertilizing nutrients in sludge to the soil.

However, land application also has associated risks, including the long-term effects of increasing the concentration of nondegradable contaminants in the soil. These elements can be taken up into food plants, ingested by children who put soiled hands into their mouths, eroded into surface waters, or leached into groundwater.

The benefits and risks of sludge disposal accrue to different groups: The advantages of cheap disposal are reaped by those generating waste. The benefits from fertilizing nutrients are reaped by farmers and other land managers. Risks accrue to those who may ingest the toxins through the media of food, soil, water, or air, now and especially in the future, when toxins will have accumulated to higher levels.

In 1993 the U.S. Environmental Protection Agency established standards for land-application of sludge, setting limits for permitted quantities of nine pollutants—arsenic, cadmium, copper, lead, mercury, molybdenum, nickel, selenium, and zinc—on the basis of a risk assessment. A *maximum*

The "X" factor is a major stumbling block in communicating risk. Health standards are expressed in terms of 1×10^{-4} or 1×10^{-6}. This is a shorthand way of expressing the *increased* number of deaths that exposure to the contaminant of concern is likely to cause over a given period of time. A 1×10^{-4} risk is a 1 in 10,000 (4 zeroes) risk; a 1×10^{-6} risk is a 1 in 1,000,000 risk. Since risk is dose (level of exposure) times time (length of exposure), a 30-year 1×10^{-6} health standard for cancer risk is the level of exposure that would be expected to cause one additional case of cancer in a population of one million people exposed at that level for 30 years.

<table>
<tr><td colspan="2">CRITERIA FOR SUCCESSFUL RISK-BASED DECISION PROCESSES</td></tr>
<tr><td>Criterion</td><td>Measurement Procedure</td></tr>
<tr><td>Getting the science right</td><td>Ask risk analytic experts who represent the spectrum of interested and affected parties to judge the technical adequacy of the risk-analytic effort</td></tr>
<tr><td>Getting the right science</td><td>Ask representatives of the interested and affected parties how well their concerns were addressed by the scientific work that informed the decision</td></tr>
<tr><td>Getting the right participation</td><td>Ask public officials and representatives of the interested and affected parties if there were other parties that should have been involved</td></tr>
<tr><td>Getting the participation right</td><td>Ask representatives of the parties whether they were adequately consulted during the process; if there were specific points when they could have contributed but did not have the opportunity</td></tr>
<tr><td>Developing accurate, balanced, and informative synthesis</td><td>Ask representatives of the parties how well they understand the bases for the decision; whether they perceived any bias in information coming from the responsible organization</td></tr>
</table>

SOURCE: National Research Council (1996). *Understanding Risk: Informing Decisions in a Democratic Society.* Washington D.C., National Academy Press.

concentration load (MCL) per unit quantity of sludge was derived from an assessment of how much of each element a person could be exposed to in their lifetime without causing unacceptable harm. To calculate the MCL, assumptions were made about the body size of this person and what they would eat in the course of a lifetime (and therefore how much of each pollutant would be consumed). The model person used for the calculations was a young adult male who did not eat many vegetables (the food group that accumulates the heavy metals). Some therefore argue that this risk assessment is not sufficiently protective of children and of people who eat many vegetables or would otherwise be exposed to greater contaminant levels.

Several European countries (as well as Canadian provinces) have established more conservative standards, permitting only much lower contaminant levels in sludges that will be recycled through land application. The policy objective of these standards is to prevent the concentration of contaminants from accumulating above the level in soils where sludge has not been applied (i.e., above background levels). However, this approach lacks risk-based criteria, since background levels of contaminants vary greatly with the type of soil and how they have been used over time. (European soils have been the site of industrial and agricultural activities for centuries.)

It is entirely possible that comparably protective standards could have emerged in the United States from a risk-based policy that was more appropriately sensitive to vulnerable subpopulations, that incorporated protective buffers to compensate for current scientific uncertainties about the hazards of these elements, and that assumed higher levels of possible exposure through food, soil, and airborne particles.

The summary lesson to be taken from this comparison is that no matter what framework or assumptions are used—whether it be risk analysis or some other—decisions regarding health and the environmental protection are based on an intermixed combination of social values and science, neither of which is objective nor without uncertainty. While the view and measure of "risk" are not the same for all, the concept of "risk" remains meaningful and useful; "risk reduction" is a critical objective across all policy arenas; and the framework and tools of risk analysis offer a structured approach for evaluating, prioritizing, and acting on environmental and health issues.

Bibliography

Harrison, E.Z.; McBride, M.B.; and Bouldin, D.R. (1999.) "Land Application of Sewage Sludges: An Appraisal of the US Regulations." *International Journal of Environment and Pollution* 11(1):1–36.

National Research Council. (1996). *Understanding Risk: Informing Decisions in a Democratic Society.* Washington, D.C.: National Academy Press.

Internet Resources

Centers for Disease Control and Prevention. "Cancer Clusters." Available from http://www.cdc.gov/nceh/clusters.

Cornell University, Environmental Risk Analysis Program. "Links to Risk Analysis Resources & Organizations." Available from http://environmentalrisk.cornell.edu/ERAP/RiskLinks.cfm.

National Cancer Institute. "Cancer Clusters, Cancer Facts." Available from http://cis.nci.nih.gov/fact/3_58.htm.

Society for Risk Analysis. "Risk Glossary." Available from http://www.sra.org/glossary.htm.

Lois Levitan

Rivers and Harbors Appropriations Act

The modern form of the Rivers and Harbors Act was enacted in 1890, and amended by the Rivers and Harbors Appropriation Act of 1899, also known as the Refuse Act. It was amended again several times during the twentieth century. In general, the act prohibits the dumping of refuse into navigable waters or the creation of any navigational obstruction, and it regulates the construction of wharves, piers, jetties, bulkheads, and similar structures in ports, rivers, canals, or other areas used for navigation.

Although the Clean Water Act now predominates in the regulation of surface water pollution, the Rivers and Harbors Act remains valid law. It provides useful supplemental jurisdiction for addressing certain kinds of water pollution, and especially for dredge and fill activities. As with the Clean Water Act, discharges of refuse or fill material, or construction activities in waterways, require a permit. The permitting agency is the Army Corps of Engineers rather than the Environmental Protection Agency, reflecting the essentially navigational concerns of this legislation.

The Rivers and Harbors Appropriations Act imposes civil and criminal penalties. Criminal convictions discourage activities that either directly or indirectly seek to evade permitting requirements. Statutory shortcomings include the absence of a state role and the act's inapplicability to municipal discharges. In addition, earlier case law restricted its application to actual interference with navigation, rather than construing the act as widely applicable to activities in navigable waters. More recent case law, however, has broadly reinterpreted the act's purpose and specifically the term "obstruction."

As a result of this trend, the U.S. Army Corps' inclusion of environmental considerations, such as the effect of a structure on vegetal habitat and the impacts of resulting shadows, in reaching its permitting decisions has been upheld. Nonetheless, recent law has also upheld, against environmental challenges, Army Corps' environmental assessments and environmental impact statements that minimized or rejected claims of adverse impacts, even when the Army Corps differed with the EPA on practical alternatives. As a consequence, the Rivers and Harbors Appropriations Act is useful for environmental challenges, especially in view of its criminal penalties, but challengers should not assume that the statute will always be successful in a legal setting in achieving environmental goals. Its usefulness as a means of reducing or eliminating pollution is restricted. SEE ALSO LAWS AND REGULATIONS, UNITED STATES.

Bibliography

Weinberg, Philip, and Reilly, Kevin A. (1998). *Understanding Environmental Law.* New York: Matthew Bender & Co.

Internet Resource

Hudson Watch Web site. "The Nation's Original Environmental Statute." Available from http://www.hudsonwatch.net/fyi.html.

Kevin Anthony Reilly

Sagebrush Rebellion   *See Wise Use Movement*

Science

Scientists collect samples of air, water, soil, plants, and tissue to detect and monitor pollution. Pollutants are most often extracted from samples, then isolated by a technique called **chromatography** and analyzed by appropriate detection methods. Many pollutants are identified by their spectral fingerprints, unique patterns of absorbed or emitted radiation in the ultraviolet (UV), visible, or infrared (IR) region of the **electromagnetic spectrum**. **Biomonitoring** and technologies including satellite observation, sidescan sonar, and **bioluminescent** reporter chips are also used for pollution monitoring. In the United States, the U.S. Environmental Protection Agency (EPA) approves the methods for monitoring regulated pollutants such as pesticide residues and those in air and drinking water.

Sampling and Extraction

Air can be actively or passively sampled. Actively sampled air is pumped through a filter or chemical solution. For example, airborne lead, mostly originating from metals processing plants, is collected on filters by active sampling and then analyzed spectroscopically. Air that is not pumped but allowed to flow or diffuse naturally is passively sampled. Nitrogen oxides, resulting from vehicle emissions and combustion, can be monitored in passive sampling tubes by their reaction with triethanolamine to form nitrates. The tubes are taken to a laboratory and the amount of nitrate analyzed.

Liquid or solid extraction removes a mix of pollutants from samples. In liquid extraction, samples are shaken with a solvent that dissolves the pollutants. Solid extraction involves the **adherence** or **absorption** of pollutants to a solid that is then heated to release a mix of vaporized pollutants which are subsequently analyzed.

chromatography means of resolving a chemical mixture into its components by passing it through a system that retards each component to a varying degree

electromagnetic spectrum the range of wavelengths of light energy, including visible light, infrared, ultraviolet, and radio waves

biomonitoring the use of living organisms to test the suitability of effluents for discharge into receiving waters and to test the quality of such waters downstream from the discharge; analysis of blood, urine, tissues, etc. to measure chemical exposure in humans

bioluminescence release of light by an organism, usually a bacterium

adherence substances: sticking to; regulation: abiding by

absorption the uptake of water, other fluids, or dissolved chemicals by a cell or an organism (as tree roots absorb dissolved nutrients in soil)

SELECTED INSTRUMENTAL DETECTION METHODS

Chemical	Method
Anions in water (e.g., nitrate, phosphate, sulfate, bromide, fluoride, chloride)	Ion exchange chromatography/conductivity detector
Criteria pollutants sulfur dioxide, ozone, nitrogen oxides	Ultraviolet absorption spectroscopy
Dioxins and furans	High-resolution gas chromatography/high-resolution mass spectrometry
Greenhouse gases carbon dioxide, methane and nitrous oxide	Infrared absorption spectroscopy
Herbicides diquat and paraquat in drinking water	High-performance liquid chromatography/ultraviolet spectroscopy
Chlorinated disinfection by-products, haloacetic acids	Gas chromatography/electron capture detector or mass spectrometry
Hydrocarbons in vehicle emissions	Infrared absorption spectroscopy
Metals	Inductively coupled plasma–atomic emission spectrometry or mass spectrometry or graphite furnace atomic absorption spectrometry for trace amounts (e.g. arsenic and lead)
Mercury	Cold vapor atomic absorption spectrometry
Organophosphate pesticides (e.g. malathion, parathion)	Gas chromatography/nitrogen/phosphorus detector
PCBs, chlorinated pesticides (e.g. DDT, lindane) and herbicides in water	Gas chromatography/electron capture detector or mass spectrometry
Phthalates in water or biological samples	Gas chromatography/electron capture or photoionization detector or mass spectrometry
Toxic gases such as hydrogen sulfide, ammonia, styrene, hydrogen fluoride	Ultraviolet or infrared absorption spectroscopy
Volatile organic compounds (VOCs) in water	Gas chromatography/photoionization and electrolytic conductivity detectors in series
Volatile organic compounds in air	Fourier transform infrared spectroscopy

Chromatography

Chromatography is the method most often used in environmental chemistry to separate individual pollutants from mixtures. The mixture to be analyzed is added to a liquid or gas, depending on whether liquid or gas chromatography is employed. The liquid or gas, called the mobile phase, is then forced through a stationary phase, often a column packed with solid material that can be coated with a liquid. The stationary and mobile phases are chosen so that the pollutants in the mixture will have different solubilities in each of them. The greater the affinity of a pollutant for the stationary phase, the longer it will take to move through the column. This difference in the migration rate causes pollutants to separate.

A chromatogram is a graph of intensity peaks that are responses to a detection method, indicating the presence of a pollutant, plotted against time.

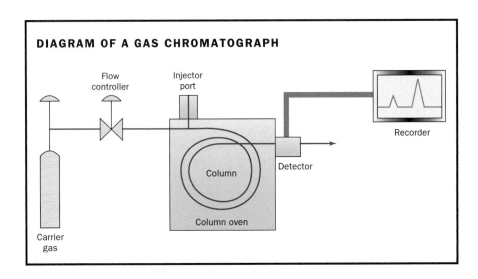

Individual pollutants are identified by comparing their chromatogram to one for the suspected compounds under the same conditions. The pollutant concentration is determined from the height of the peaks and area under them.

Different kinds of chromatography work best for different pollutants. Gas chromatography separates organic chemicals that vaporize easily (VOCs). Benzene and ethylbenzene are VOCs in vehicle exhaust and are monitored in drinking water. Many pesticides, polychlorinated biphenyls (PCBs), and dioxin are separated by gas chromatography. Less volatile substances such as the herbicide diquat are isolated by high-performance liquid chromatography (HPLC). Ion exchange chromatography separates inorganic ions such as nitrates that can pollute water when excess fertilizer or leaking septic tanks wash into it.

Detectors

Chromatographic methods are routinely automated. A detector that responds to the pollutants' physical or chemical properties analyzes the gas or liquid leaving the column. Detectors can be specific for individual pollutants or classes of pollutants, or nonspecific.

Nonspecific Detectors. Flame ionization, thermal conductivity, and mass spectrometry are common nonspecific detection methods that detect all molecules containing carbon and hydrogen. In mass spectrometry, molecules of a gas are energized in a variety of ways, such as bombardment with electrons or rapid heating, causing them to gain or lose electrons. Because they have different masses and charges, the resulting ions are separated when they pass through magnetic and electric fields. The size and distribution of peaks for ions with different mass-to-charge ratios, known as the mass spectrum, identify the gas and determine its concentration. Gas chromatography coupled with high-resolution mass spectrometry definitively identifies PCBs and is the most accurate way to determine their concentration. Portable gas chromatograph/mass spectrometers can measure VOCs in soil and water to parts per billion (ppb).

Specific Detectors. Methods that detect classes of pollutants include nitrogen/phosphorous detectors for organophosphate pesticides, thermionic ionization detectors that detect molecules containing NO_2, nitro groups, such as dinitrotoluene and electron capture. Electron capture is particularly sensitive to compounds, such as organohalide pesticides that contain the halogen atoms, chlorine, bromine, or fluorine. These atoms strongly attract electrons. The electron capture detector emits electrons that are captured by the halogens atom. The reduction in electric current corresponds to the concentration of pollutant. Chlorinated disinfection by-products, haloacetic acid, and phthalates in drinking water can be separated by gas chromatography and measured by electron capture. Sulfur hexafluoride, an ozone-depleting gas, can be measured to parts per trillion (ppt) by electron capture. Spectroscopic detection methods including IR, UV, and atomic absorption and emission spectroscopy are unique for specific compounds.

Spectroscopic Detection. The electromagnetic spectrum encompasses all forms of electromagnetic radiation from the most energetic cosmic and gamma rays to the least energetic radio waves. The part of the spectrum that is particularly useful in identifying and measuring pollutants consists of radiation that interacts with the atoms and molecules that make up life on Earth. This includes radiation in the UV, visible, and IR regions.

Atomic Spectra. Atoms of different elements may be thought of as having different arrangements of electrons around the nucleus in increasing energy levels. When metals such as lead, copper, and cadmium are vaporized at high temperatures, some electrons jump to higher energy levels. When the electrons drop to their original levels, the metal atoms emit radiation in a range of wavelengths from IR to UV, including visible light. The colors in fireworks result from such emissions. The wavelengths emitted constitute a unique "fingerprint" for each element and their intensity reflects the metal concentration. Inductively coupled plasma emission spectra (ICP–AES), in which a high-temperature gas or plasma excites metal atoms, are used to identify and quantify heavy metal contamination.

The same spectral fingerprint is obtained from the wavelengths of light that each element absorbs. Trace amounts of certain metals such as mercury and arsenic are more accurately measured from their absorption, rather than their emission spectra.

UV and IR Spectra. Many pollutants can be identified by their UV and IR spectra because all molecules that absorb strongly at specific wavelengths exhibit spectral fingerprints. Pollutants separated by liquid chromatography are often detected by spectroscopy. Gases such as those from vehicle emissions, landfills, industrial manufacturing plants, electric power plants, and hazardous incineration smokestacks can be monitored by spectroscopic methods. Gas and chemical leaks may also be monitored by spectroscopy.

UV Absorption Spectra. Toxic gases such as hydrogen sulfide, ammonia, and styrene can be monitored by their UV absorption spectra. **Open path monitors** emit UV radiation from a source, such as a bulb containing excited xenon gas, across the area to be monitored. Detectors record the absorbed wavelengths to produce a spectral fingerprint for each gas. Ammonia is often used as a coolant for turbine generators in power plants. It can be monitored for worker safety by its UV spectrum. The EPA has established National Ambient Air Quality standards for the six criteria pollutants: carbon monoxide, lead, nitrogen dioxide, ozone, particulate matter, and sulfur dioxide.

open path monitor detection device that employs a beam of light passing through an open space

DIAGRAM OF THE ELECTROMAGNETIC SPECTRUM

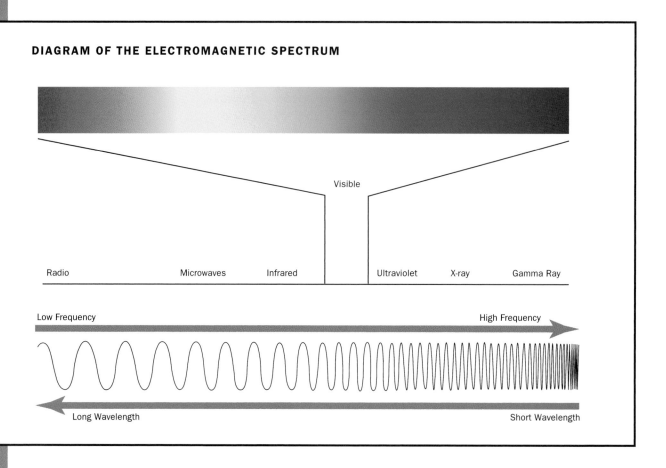

Satellite instruments monitoring stratospheric ozone generally measure the decrease in intensity in UV solar radiation due to ozone absorption. The total ozone mapping spectrometer on the Earth probe satellite (TOMS/EP) scans back and forth beneath the satellite to detect six individual frequencies of UV light that are scattered by air molecules back through the stratosphere. The more ozone in the stratosphere, the more "backscattered" UV radiation will be absorbed compared to UV radiation directly from the sun.

Some IR open path monitors use a tunable diode laser source in the near IR. The laser emits the specific frequency at which a monitored gas absorbs, so there is no interference from other gases or particles such as rain or snow. Such lasers are widely employed in the telecommunications industry. Pollutants that absorb at specific wavelengths in this range include hydrogen fluoride, an extremely toxic gas used in the aluminum smelting and petroleum industries. Hydrogen fluoride can be monitored to one part per million (ppm) for worker safety by this method.

The greenhouse gases carbon dioxide, nitrous oxide, and methane may also be monitored by IR spectroscopy. Currently, emissions of carbon dioxide from power plants are not generally measured directly but are estimated. However, the amount of carbon dioxide in the atmosphere over Mauna Loa has been measured continuously by IR spectroscopy since 1958. The Mauna Loa Observatory is located on the earth's largest active volcano on the island of Hawaii. It is relatively remote from human activity and changes in carbon dioxide concentration above it are considered a reliable indicator of the trend of carbon dioxide concentration in the troposphere. Data from Mauna

Loa show a 17.4 percent increase in carbon dioxide concentration from 315.98 parts per million (ppm) by volume of dry air in 1959 to 370.9 ppm in 2001.

Remote sensors for vehicle emissions contain units that detect and measure carbon monoxide, carbon dioxide, and hydrocarbons by their IR spectra. Because IR absorption bands from water and other gases found in car exhaust interfere with the IR spectrum of NO_x, the sensor also contains a unit that measures NO_x from their UV absorption spectra.

Fourier transform IR spectroscopy (FTIR) analyzes the **absorption spectrum** of a gas mixture to detect as many as twenty gases simultaneously. The technique involves analyzing the spectra mathematically and then comparing the observed fingerprints with calibrated reference spectra stored on the hard drive of the computer to be used for analysis. Reference spectra for more than one hundred compounds are stored, including most of the VOCs considered hazardous by the EPA. Instruments that use UV Fourier transform analysis are now available. The instruments are generally installed at one location, but are portable and can be battery operated for short-term surveys. Multiple gas-monitoring systems are used in a variety of industries, including oil and gas, petrochemical, pulp and paper, food and beverage, public utility, municipal waste, and heavy industrial manufacturing.

absorption spectrum "fingerprint" of a compound generated when it absorbs characteristic light frequencies

Biomonitoring

Biomonitoring is the study of plants, vertebrate, and invertebrate species to detect and monitor pollution. Moss and lichens absorb heavy metals, mainly from air, and have been analyzed by scientists studying air pollution.

Water pollution can be studied by recording changes in the number and type of species present and in specific biochemical or genetic changes in individual organisms. Blue mussels accumulate metals in certain tissues over time and are monitored in the United States and international waters for changes in pollution levels. The index of biotic integrity (IBI), first developed by James Karr in 1981 to assess the health of small warmwater streams, uses fish sampling data to give a quantitative measure of pollution. Twelve indicators of stream health, appropriate to the geographical area, including the total number of fish, the diversity of species, and food chain interactions, are numerically rated with a maximum of five points each. An IBI close to sixty corresponds to a healthy stream, whereas a rating between twenty and twelve implies a considerable pollution. Versions of the IBI with appropriate indicators are used to assess rivers and streams in France, Canada, and different regions of the United States.

Bioluminescent Reporter Technology

In bioluminescent reporter technology, bacteria that break down pollutants are genetically modified to emit blue green light during the degradation process. The bacteria are embedded in a **polymer** porous to water and combined with a light sensor integrated with a silicon computer chip. The sensor measures the intensity of the glow to determine the amount of pollution, and that information is transmitted to a central computer.

Bioluminescent reporter technology is still being studied by researchers, but is currently employed in some wastewater treatment plants in the United

polymer a natural or synthetic chemical structure where two or more like molecules are joined to form a more complex molecular structure (e.g., polyethylene)

When New Jersey inventors John Mooney and Carl Keith invented the three-way catalytic converter in 1974, the *Wall Street Journal* called it a $20 million mistake. Industry estimates today credit the catalytic converter with preventing fifty million tons of carbon monoxide and fifty million tons each of hydrocarbons and nitrogen oxides from polluting the air worldwide. In addition, the use of catalytic converters required that lead be removed from gasoline.

epidemiological epidemiology: study of the incidence and spread of disease in a population

Kingdom. Incoming wastewater is monitored for chemicals that inhibit the bacterial activity necessary for efficient wastewater treatment. The incoming water is automatically sampled and mixed with freeze-dried luminescent bacteria from the treatment plant. A reduction in light intensity compared to a control with pure water indicates the chemical inhibition of wastewater microorganisms. This technology is also being used to identify petroleum pollutants, such as napthelene.

Sidescan Sonar

Sidescan sonar instruments bounce sound off surfaces both vertically and at an angle to produce images of sea and riverbeds. Because PCBs tend to stick preferentially to organic matter, there is a greater possibility of finding them in small-grain aquatic sediments, since these contain more organic material. The EPA has analyzed sound reflection patterns from sidescan sonar data to identify areas of small grain size and selectively sample for PCBs in the Hudson River, New York. Sidescan sonars are also used to detect sea grass, an indicator of marine health, and sewage or oil leaks from underwater pipelines.

Regulations

Once a potentially harmful pollutant is measured in trace amounts, then regulators, such as the EPA, have to decide on a safe limit. Risk analysis is the method used to set limits on harmful pollutants in the United States. Risk is calculated based on laboratory tests, sometimes on animals, and **epidemiological** studies that relate human health to exposure.

Risk analysis is conducted for individual pollutants, but people can be exposed to multiple pollutants simultaneously, such as pesticides, heavy metals, dioxins, and PCBs. Even though a person's exposure to individual chemicals may fall within regulated limits, the pollutants may interact to cause as yet unknown adverse health effects. It is known, for instance, that exposure to both asbestos and tobacco smoke geometrically increases the risk of cancer. Because there are so many potentially harmful chemicals in the environment scientists cannot predict all their possible interactions and consequent health effects on the body. SEE ALSO AIR POLLUTION; ARSENIC; DIOXIN; GREENHOUSE GASES; HEAVY METALS; LEAD; MERCURY; OZONE; PCBs (POLYCHLORINATED BIPHENYLS); PESTICIDES; RISK; VEHICULAR POLLUTION; VOCs (VOLATILE ORGANIC COMPOUNDS); WATER TREATMENT.

Bibliography

Csuros, Maria. (1997). *Environmental Sampling and Analysis Lab Manual.* Boca Raton, FL: Lewis Publishers.

Manahan, Staley E. (2001). *Fundamentals of Environmental Chemistry*, 2nd edition. Boca Raton, FL: Lewis Publishers.

Schnelle, Kard B., Jr., and Brown, Charles A. (2002). *Air Pollution Control Technology Handbook.* Boca Raton, FL: CRC Press.

Internet Resources

Carbon Dioxide Information Analysis Center Web site. "Atmospheric Carbon Dioxide Record from Mauna Loa." Available from http://cdiac.esd.ornl.gov/trends/co2/sio-mlo.htm.

Goddard Space Flight Center Web site. "Ozone Measurements, TOMS on Earth Probe Satellite." Available from http://toms.gsfc.nasa.gov/eptoms.

University of Tennessee. Center for Environmental Biotechnology Web site. "Bioreporter Research Projects." Available from http://www.ceb.utk.edu.

U.S. Environmental Protection Agency, Office of Water. "Approved Methods for Inorganic Chemicals and Other Parameters." Available from http://www.epa.gov/safewater.

U.S. Environmental Protection Agency, Technology Transfer Network Emissions Measurement Center. "CFR Promulgated Test Methods." Available from http://www.epa.gov/ttn.

U.S. Geological Survey. National Environmental Methods Index Web site. Available from http://www.nemi.gov.

Patricia Hemminger

Scrubbers

Scrubbers are air-pollution-control devices that remove harmful gases and particulates from the smokestacks of incinerators, chemical manufacturing facilities, and electric power plants before they enter the atmosphere. There are different types of scrubbers, including wet and dry, regenerative and non-regenerative. Regenerative scrubbers recycle the material that extracts the pollutants.

The nonregenerative wet scrubber is most commonly used to capture sulfur dioxide emitted from coal and oil burning power plants. It works by spraying limestone and water slurry into the flue gases. Sulfur dioxide reacts with limestone to form gypsum or calcium sulfate. The gypsum sludge is disposed of in landfills or recycled in saleable byproducts such as wallboard, concrete, and fertilizer. Regenerative scrubbers can also be used; one reacts sodium sulfite with sulfur dioxide to form sodium bisulfite, from which sodium sulfite is recovered by adding alkali. The released sulfur is trapped in water to produce sulfuric acid, which is sold to offset the cost of installing the scrubber.

Particulates can be removed using venturi and centrifugal or condensation scrubbers. Flue gas enters through the top of the cone-shaped venturi scrubber and water, injected horizontally, forms droplets that absorb dust and other particles. The resulting slurry discharges from the bottom of the unit or can be separated from the clean gas by centrifugation or spinning at high speed. Copper oxide regenerable scrubbers that absorb sulfur and simultaneously convert nitrogen oxides to nitrogen are being researched.

In 1971 the EPA set a maximum limit on sulfur dioxide in air. To help meet this limit, revisions to the Clean Air Act in 1977 required all new power plants to install scrubbers to remove sulfur dioxide. Most spray tower scrubbers remove at least 90 percent of sulfur dioxide, according to the EPA. In 1990 further revisions to the Clean Air Act under the Acid Rain Program allotted allowable amounts of sulfur dioxide emissions to electric utilities, which could trade allowances to meet their quotas. Sulfur dioxide emissions from power plants in 2001 were 33 percent lower than in 1990 and 5 percent lower than in 2000 according to the EPA. SEE ALSO AIR POLLUTION; CLEAN AIR ACT.

Bibliography

Schnelle, Karl B. Jr., and Brown, Charles A. (2002). *Air Pollution Control Technology Handbook*. Boca Raton, FL: CRC Press.

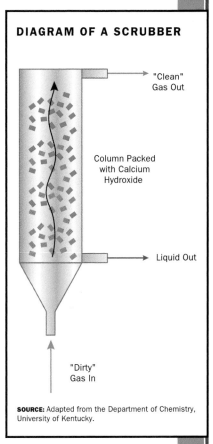

DIAGRAM OF A SCRUBBER

"Clean" Gas Out

Column Packed with Calcium Hydroxide

Liquid Out

"Dirty" Gas In

SOURCE: Adapted from the Department of Chemistry, University of Kentucky.

Internet Resources

EPA Air Pollution Technology Fact Sheets. "Condensation Scrubbers" and "Spray-Chamber/Spray-Tower Wet Scrubber." Available from http://www.epa.gov/ttn.

Illinois Clean Coal Institute Annual Report (2002). Available from http://www.icci.org.

Patricia Hemminger

La Secretaría del Medio Ambiente y Recursos Naturales
See Mexican Secretariat for Natural Resources

Sedimentation

substrate surface on which an organism, i.e. mold, grows

Sediments in the aquatic ecosystem are analogous to soil in the terrestrial ecosystem as they are the source of **substrate** nutrients, and micro- and macroflora and -fauna that are the basis of support to living aquatic resources. Sediments are the key catalysts of environmental food cycles and the dynamics of water quality. Aquatic sediments are derived from and composed of natural physical, chemical, and biological components generally related to their watersheds.

bed load transport movement of sediments that remain at the bottom of a moving water body

hydrodynamic condition related to flow of water

Sediments range in particle distribution from micron-sized clay particles through silt, sand, gravel, rock, and boulders. Sediments originate from **bed load transport**, beach and bank erosion, and land runoff. They are naturally sorted by size through prevalent **hydrodynamic conditions**. In general, fast-moving water will contain coarse-grained sediments and quiescent water will contain fine-grained sediments. Mineralogical characteristics of sediments vary widely and reflect watershed characteristics. Organic material in sediments is derived from the decomposed tissues of plants and animals, from aquatic and terrestrial sources, and from various point and nonpoint wastewater discharges. The content of organic matter increases in concentration as the size of sediment mineral particles decreases. Dissolved chemicals in the overlying and sediment **pore waters** are a product of inorganic and organic sedimentary materials, as well as runoff and ground water that range from fresh to marine in salinity. This sediment/water environment varies significantly over space and time and its characteristics are driven by complex **biogeochemical interaction** between the inorganic, living, and nonliving organic components. The sediment biotic community includes micro-, meso-, and macrofauna and -flora that are interdependent of each other and their **host** sediment's biogeochemical characteristics.

pore waters water present in the pores or cavities in sediments, soil, and rock

biogeochemical interaction interactions between living and nonliving components of the biosphere

host in genetics, the organism, typically a bacterium, into which a gene from another organism is transplanted; in medicine, it is an animal infected or parasitized by another organism

Sedimentation is the direct result of the loss (erosion) of sediments from other aquatic areas or land-based areas. Sedimentation can be detrimental or beneficial to aquatic environments. Moreover, sediment impoverishment (erosion or lack of replenishment) in an area can be as bad as too much sedimentation. Sedimentation in one area is linked to erosion or impoverishment in another area and is a natural process of all water bodies (i.e., lakes, rivers, estuaries, coastal zones, and even the deep ocean). As an example, detrimental effects can be related to the burial of bottom-dwelling organisms and beneficial effects can be related to the building of new substrates for the development of marshes. These natural physical processes will continue whether or not they are influenced by the activities of humankind.

Human activities, however, have significantly enhanced sedimentation as well as sediment loss. Sedimentation activities can be land-based (i.e., agriculture, forestry, construction, urbanization, recreation) and water-based (i.e., dams, navigation, port activities, drag fishing, channelization, water diversions, wetlands loss, other large-scale **hydrological** modifications). Sediment impoverishment or loss is generally due to retention behind dams, bank or beach protection activities, water diversions, and many of the aquatic activities cited here. Morphological changes (physical changes over a large area) to large aquatic systems can also result in major changes in natural sediment erosion and sedimentation patterns. As an example, the change in the size and shape of a water body will result in new water flow patterns leading to erosion or sediment removal from sensitive areas.

hydraulic related to fluid flow

The environmental impacts of sedimentation include the following: loss of important or sensitive aquatic habitat, decrease in fishery resources, loss of recreation attributes, loss of coral reef communities, human health concerns, changes in fish migration, increases in erosion, loss of wetlands, nutrient balance changes, circulation changes, increases in **turbidity**, loss of submerged vegetation, and coastline alteration.

turbid containing suspended particles

Abatement or control of sedimentation can be successful if implemented on a broad land area or watershed scale and is directly related to improvement in land-use practices. Agriculture and forestry (logging) improvements where soil loss is minimized are not only technically feasible: They can be carried out at a moderate cost and with net benefits. The U.S. Department of Agriculture has a wide range of training and implementation programs for these types of activities. The United Nations Environmental Programme also has global programs, their Regional Seas activities, to guide countries in the management of land-based activities negatively impacting the coastal zone. Improved land-use practices are the primary measures to control sediment sources: terracing, **low tillage**, modified cropping, reduced agricultural intensity (e.g., no-till buffer zones), and wetlands construction as sediment interceptors. Forestry practices such as clear-cutting to the water's edge without replacement tree planting must be seriously curtailed because base soil in exposed areas will erode and import sediment to sensitive aqueous areas. Wetlands that separate upland areas from aquatic areas serve as natural filters for the runoff from the adjacent land. Wetlands thus serve to trap soil particles and associated agricultural contaminants. The construction of natural buffer zones and wetlands replenishment adjacent to logging areas are effective techniques. Watershed construction activities such as port expansion, water diversions, channel deepening, and new channel construction must undergo a complete environmental assessment, coupled with predictive sediment resuspension and transport modeling, so alternative courses of action and activities to minimize the negative impacts of sedimentation may be chosen.

low tillage reduced level of plowing

Sediment impoverishment is equally important in coastal areas, such as coastal Louisiana where twenty-five to thirty square miles of wetlands are being lost each year. This loss primarily results from the Mississippi River levee system halting the annual natural replenishment of sediments that rebuilds the marsh system. Engineered water diversion can replace sediment in the natural system to decrease losses due to dams, levees, jetties, and other structures built to control the flow of water and thus sediments. Proper placement of sediments from navigation dredging can also be a useful abatement technique.

sediment impoverishment loss of sediment

Sediments are absolutely necessary for aquatic plant and animal life. Managed properly, sediments are a resource; improper sediment management results in the destruction of aquatic habitat that would have otherwise depended on their presence. The United Nations Group of Experts on the Scientific Aspects of Marine Environmental Protection recently recognized that on a global basis, changes in sediment flows are one of the five most serious problems affecting the quality and uses of the marine and coastal environment. SEE ALSO DISASTERS: ENVIRONMENTAL MINING ACCIDENTS; DREDGING; PARTICULATES; WATER POLLUTION.

Bibliography

Huber, M.E., *et al.* (1999). "Oceans at Risk." *Marine Pollution Bulletin* 38 (6):435–438.

Fischetti, Mark. (2001). "Drowning New Orleans." *Scientific American* 285 (4):76–85.

Internet Resource

Joint Group of Experts on the Scientific Aspects of Marine Environment Protection. (2001). "Sea of Troubles." GESAMP Study No. 70. Geneva: United Nations Environmental Programme. Also available from http://gesamp.imo.org/no70.

USDA-ARS National Sedimentation Laboratory. Available from http://www.sedlab.olemiss.edu.

Robert M. Engler

Septage   *See Wastewater Treatment*

Settlement House Movement

As more women gained access to a college education in the late nineteenth century, many hoped to use their skills and talents for more than homemaking and child rearing. Jane Addams, born in 1860 to a Quaker miller in Illinois, was one of these women who hoped to improve the life of others and society at large. After completing her education, Addams took a trip to Europe, where social activism in the slums of London had a dramatic effect on her. She returned to Chicago to found her own version of London's "settlement houses" in 1889. The British settlement houses, which inspired Addams, were residences located within destitute neighborhoods with programs designed to improve living conditions. Addams's Hull House, located in an immigrant area of the city with appalling living conditions, provided numerous women with the opportunity to serve the poor neighborhood and reform conditions there. Environmental reforms became an important component of their work, but settlement houses also organized kindergartens for immigrant children; provided classes on ethnic culture and art; and gave immigrants a place to meet, visit, bathe, and see health professionals.

Addams incorporated a large number of environmental reforms in her agenda for Hull House. One of the most notable included her efforts to address the unhealthy piles of garbage in immigrant neighborhoods because of a lack of municipal attention. The mayor of Chicago eventually appointed Addams garbage inspector for her area, a job she took very seriously. Addams supervised garbage collectors and took violators of garbage regulations to court. Although Addams and her cohorts often initiated reforms, the immigrants played an active role too, assisting in information gathering and its communication to their neighbors. Alice Hamilton, also a resident of Hull House, worked extensively on occupational health and safety issues, demonstrating the dangers of lead and other toxic substances.

The Settlement House Movement, begun by Addams and a part of national Progressive Era reform movements, spread quickly to other industrial urban areas. Lillian Wald established Henry House in New York. Initially hoping to focus on the delivery of modern health care, Wald quickly became outraged over immigrant living conditions and shifted her focus to improving city services, establishing parks for children, and educating immigrants about sanitation issues.

Although the most famous settlement house workers were middle- and upper-class white women, African-American women also participated in the movement throughout the United States. They focused on issues similar to those of white women, but had to cope with the additional problems of racism, segregation, disfranchisement, and discrimination facing black communities in general. They worked tirelessly to educate other African-Americans about sanitation and health issues and to improve neighborhoods by pressing for garbage pickup and better city services like sewers and lighting.

Although settlement houses failed to eliminate the worst aspects of poverty among new immigrants, they provided some measure of relief and hope to their neighborhoods. Nonetheless, historians have found that settlement house workers held a very condescending attitude toward immigrant populations, one that dismissed native cultures and sought to impose decidedly white middle-class values. Despite any such limitations, settlement house workers raised public awareness of pollution issues, especially in the areas of health, sanitation, and city services. They influenced politicians and forced them to consider issues of importance to immigrants. Finally and equally importantly, settlement house workers provided a legitimate venue for women to become active in city politics and other national issues, such as the burgeoning women's suffrage movement. SEE ALSO ACTIVISM; ADDAMS, JANE; ENVIRONMENTAL MOVEMENT; HAMILTON, ALICE; INDUSTRY; LEAD; OCCUPATIONAL SAFETY AND HEALTH ADMINISTRATION (OSHA); POLITICS; PROGRESSIVE MOVEMENT; SOLID WASTE; WORKERS HEALTH BUREAU.

Bibliography

Addams, Jane. (1911). *Twenty Years at Hull House, with Autobiographical Notes.* New York: Macmillan.

Lasch-Quinn, Elisabeth. (1993). *Black Neighbors: Race and the Limits of Reform in the American Settlement House Movement, 1890–1945.* Chapel Hill: University of North Carolina Press.

Levine, Daniel. (1971). *Jane Addams and the Liberal Tradition.* Madison: State Historical Society of Wisconsin.

Internet Resources

"Settlement Houses: New Ideas in Old Communities." Available from http://www.socialworker.com/sethouse.pdf.

United Neighborhood House Web site. Available from http://www.unhny.org.

Elizabeth D. Blum

Settling Ponds　*See Wastewater Treatment*

Sewage Sludge　*See Biosolids*

Sick Building Syndrome　*See Indoor Air Pollution*

Smart Growth

The "smart growth" movement arose in the 1990s to combat the perceived negative aspects of the dominant growth patterns of the time: rapidly spreading development that tended to draw people and resources away from existing neighborhoods and created new, look-alike communities where vehicle use was mandatory and walking was discouraged. Proponents of smart growth—a group that includes city planners, environmentalists, urban designers, neighborhood activists, and others—do not try to stop development, but instead work to make development improve life in existing cities and towns, rather than degrade it. They generally agree on several core principles:

infrastructure the basic facilities, services and installations needed for the functioning of a system, i.e., the various components of a water supply system

1. Revitalizing communities by directing public investment toward areas where the **infrastructure** to support development is already in place or planned.

2. Creating walkable neighborhoods by locating housing, shopping, schools, and offices in closer proximity to each other and providing sidewalks and attractive streetscapes.

3. Offering a choice in transportation modes, whether by foot, car, bike, bus, or train.

4. Involving citizens in deciding how and where their community should grow.

5. Fostering distinctive, attractive communities with a unique sense of place.

6. Providing housing for people of all income levels in close proximity to jobs and activities.

7. Preserving open space, farmland, natural beauty, and critical environmental areas.

8. Saving taxpayers the unnecessary cost of building the infrastructure required to support spread-out development.

SEE ALSO SPRAWL.

Internet Resource

Smart Growth Online. Available from http://www.smartgrowth.org.

David Goldberg

Smelting

Mined ores are processed to concentrate the minerals of interest. In the case of metal ores, these mineral concentrates usually need to be further processed to separate the metal from other elements in the ore minerals. Smelting is the process of separating the metal from impurities by heating the concentrate to a high temperature to cause the metal to melt. Smelting the concentrate produces a metal or a high-grade metallic mixture along with a solid waste product called slag.

The principal sources of pollution caused by smelting are contaminant-laden air emissions and process wastes such as wastewater and slag.

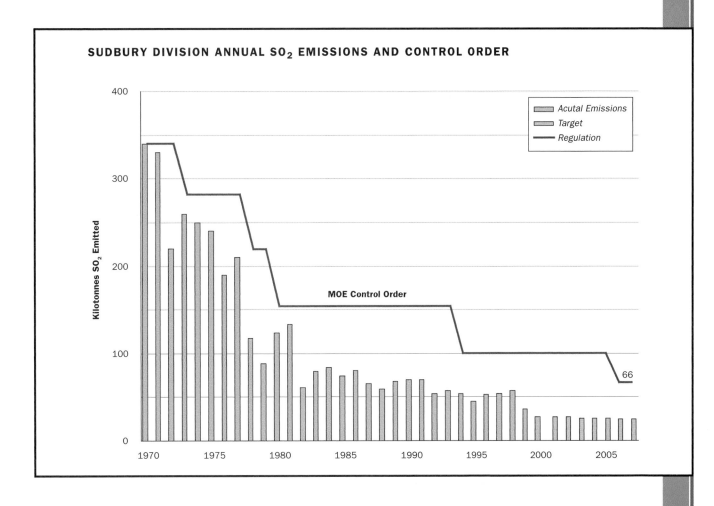

SUDBURY DIVISION ANNUAL SO₂ EMISSIONS AND CONTROL ORDER

One type of pollution attributed to air emissions is acid rain. The smelting of sulfide ores results in the emission of sulfur dioxide gas, which reacts chemically in the atmosphere to form a sulfuric acid mist. As this acid rain falls to the earth, it increases the acidity of soils, streams, and lakes, harming the health of vegetation and fish and wildlife populations.

In older smelters, air emissions contained elevated levels of various metals. Copper and selenium, for example, which can be released from copper smelters, are essential to organisms as trace elements, but they are toxic if they are overabundant. These metals can **contaminate** the soil in the vicinity of smelters, destroying much of the vegetation. In addition, **particulate** matter emitted from smelters may include oxides of such toxic metals as arsenic (cumulative poison), cadmium (heart disease), and mercury (nerve damage).

When compared to pollution caused by air emissions, process wastes and slag are of less concern. In modern smelters, much of the wastewater generated is returned to the process. If the economic value of the metal concentrate in slag is high enough, the slag may be returned to the process, thereby reducing the amount requiring permanent disposal.

New technologies are playing an important role in reducing or even preventing smelter pollution. Older smelters emitted most of the sulfur dioxide generated, and now almost all of it is captured prior to emission using new technologies, such as electrostatic precipitators, which capture dust particles and return them to the process. Raw material substitution or elimination,

contaminant any physical, chemical, biological, or radiological substance or matter that has an adverse effect on air, water, or soil

particulate fine liquid or solid particles such as dust, smoke, mist, fumes, or smog, found in air or emissions; they can also be very small solids suspended in water, gathered together by coagulation and flocculation

such as recycling lead batteries and aluminum cans, decreases the need to process ore, which reduces pollution.

Some of the major federal statutes and regulations that apply to smelting are the same as those that have applied to mining since the Clean Air Act (CAA) of 1970 became law. The CAA established nationally uniform standards that control particular hazardous air pollutants.

Sudbury, in Ontario, Canada, is one of the world's largest smelting complexes, with an international reputation as a highly polluted area that has been mined for more than one hundred years. The environmental impact was completely or partially denuded vegetation on over 46,000 hectares and 7,000 acid-damaged lakes. Smelting caused much of the ecological damage via acid rain and elevated levels of copper and nickel in the vicinity of the smelters. Efforts by government and industry since the 1970s have eliminated most of the sulfur dioxide emissions in the area, and there has been significant progress toward achieving sustainable ecosystems. SEE ALSO ACID RAIN; AIR POLLUTION; LEAD; MINING; SUPERFUND.

Bibliography

Gunn, John M. (1995). *Restoration and Recovery of an Industrial Region*. New York: Springer-Verlag.

Weiss, Norman L., ed. (1985). *SME Mineral Processing Handbook*. Kingsport, TN: Kingsport Press.

Internet Resources

U.S. Environmental Protection Agency. (1995). "EPA Office of Compliance Sector Notebook Project: Profile of the Nonferrous Metals Industry." U.S. EPA Document No. EPA/310-R-95-010. Available from http://es.epa.gov.

U.S. Geological Survey. (2001). "Mine and Mineral Processing Plant Locations—Supplemental Information for USGS Map I-2654." Available from http://pubs.usgs.gov.

U.S. Geological Survey. (2001). "USGS Tracks Acid Rain." Fact Sheet FS-183-95. Available from http://pubs.usgs.gov.

Michael J. McKinley

Smog

Originally, the term *smog* was coined to describe the mixture of smoke and fog that lowered visibility and led to respiratory problems in industrial cities. More recently, the term has come to mean any decrease in air quality whether associated with reduced visibility or a noticeable impact on human health. Smog occurs when emissions of gases and particles from industrial or transportation sources are trapped by the local meteorology so the concentrations rise and chemical reactions occur. It is common to distinguish between two types of smog: *London smog* and *Los Angeles smog*.

London, or sulphurous, smog was noted following the introduction of coal into cities. It is most prevalent in the fall or winter when cool conditions naturally produce a thick surface fog. This fog mixes with the smoke and gases from burning coal to produce a dark, thick, acrid sulphurous atmosphere. Normally, the unpolluted fog would disperse during the day and be reformed at night. However, the presence of smoke particles makes the fog so thick that sunlight cannot penetrate it and so only a major change in meteorology can disperse it. The smog has been shown to contribute to an

A thick cloud of smog covering Santiago, Chile. (AP/Wide World Photos. Reproduced by permission.)

increased death rate, primarily due to respiratory problems. The most notable example of this kind of smog occurred in London, from December 4 to 10, 1954, when some four thousand deaths in excess of normal averages resulted. A similar episode in Donora, Pennsylvania, in 1948 involved approximately twenty excess deaths. Most jurisdictions have instituted control measures to prevent this level of disaster from happening again. They have moved industries out of cities, demanded lower industrial emissions, and increased the heights of smokestacks so emissions are not trapped by local meteorology. These approaches have been largely successful, at least in controlling the most extreme events.

Los Angeles, or photochemical, smog first became apparent in the late 1940s in warm sunny cities that did not have significant coal-burning industries. It is a daytime phenomenon characterized by a white haze and contains oxidants, such as ozone, that cause eyes to water, breathing to become labored, and plants to be damaged. It results from the action of sunlight on the combination of **hydrocarbons** and nitrogen oxides (NO_x), known as precursor gases. These are emitted from combustion sources to produce a range of oxidized products and oxidants. These compounds have been shown to produce respiratory and cardiac problems in individuals sensitive to pollution, and the damage inflicted on crops can cause significant decreases in

hydrocarbon compounds of hydrogen and carbon

yield. In most cities, the automobile is the primary contributor of smog's precursor gases. As the name would suggest, the most notable example of this type of smog occurs in Los Angeles, California, but it has also been experienced in a large number of cities where the weather is dry, sunlight is plentiful, and there are many automobiles or petroleum industries (e.g., Houston, Athens, and Mexico City.)

The control of photochemical smog is more difficult than for sulphurous smog because the compounds responsible for human and crop impacts are not directly emitted, but produced by chemistry in the atmosphere. Thus, greater knowledge on the emissions of gases, their reactions in the atmosphere, and their lifetime is needed. Most jurisdictions continue to focus their control strategies on reducing ozone concentrations, although particle concentrations are receiving increasing attention. Because smog results from the sunlight-initiated chemistry of hydrocarbons and nitrous oxides, the most common approach to smog control is to decrease the emission of these compounds at their source. Lower volatility gasolines and systems to capture gasoline vapors are used to reduce hydrocarbon emissions while tailpipe controls (catalytic converters) reduce emissions of both hydrocarbons and nitrogen oxides. The emission control systems of the twenty-first century mean that a car typically emits 70 percent less nitrogen oxides and 80 to 90 percent less hydrocarbons than the uncontrolled cars of the 1960s. The expected improvement in air quality, as a result of increasing controls, is estimated by using computer models of the atmosphere and its chemistry. SEE ALSO AIR POLLUTION; ASTHMA; DONORA, PENNSYLVANIA; HEALTH, HUMAN; OZONE.

Bibliography

Brimblecombe, Peter. (1987). *The Big Smoke: A History of Air Pollution in London since Medieval Times.* London: Methuen.

Turco, Richard. (1997). *Earth under Siege.* Oxford: Oxford University Press.

Internet Resources

U.S. Environmental Protection Agency. "Air Quality Index: A Guide to Air Quality and Your Health." Available from http://www.epa.gov/airnow/aqibroch.

U.S. Environmental Protection Agency. "National Air Pollutant Emission Trends, 1900–1998." Available from http://www.epa.gov/ttn.

Donald R. Hastie

Snow, John

BRITISH ANESTHESIOLOGIST
(1813–1858)

In 1854, John Snow was a well-regarded London anesthesiologist, tending to Queen Victoria, among others. He was born in 1813 of humble stock, but through education and intellectual perseverance—he obtained his M.D. degree in 1844—was able to rise to a position of scientific prominence. Snow became interested in the emerging field of **epidemiology**, especially as it applied to cholera, a disease of unknown cause (attributed thirty years later by Dr. Robert Koch to *Vibrio cholerae*). Two population-based studies—both occurring in 1854—established Snow's reputation, and focused scientific attention away from the fallacious notion of airborne transmission towards the role of contaminated water in the spread of cholera.

epidemiology study of the incidence and spread of disease in a population

Snow's first study occurred after the government had mandated that water companies along the polluted Thames River should move their inlets upstream where the quality of water was better. One company moved its intake pipes in 1852 but still maintained the same local water distribution system. A second company kept its intakes in place (but finally moved in 1855), providing contaminated water to portions of the same area as the first company. When cholera next arrived in London in 1853 and 1854, Snow was able to compare cholera among households according to water source. The populations were very similar—consumers of the two water companies lived side by side. Using existing mortality data, Snow was able to measure the impact the two companies had on cholera, thereby linking water source and quality to the disease.

His second study took place in 1854 near his home in the Soho region of London. It followed what he described as: "The most terrible outbreak of cholera which ever occurred in this kingdom." With skillful assembling of data, analysis, and use of maps, he identified a single water pump on Broad Street as the likely source, suggesting that the pump water was contaminated with an unseen microbial agent. Water pumps, at that time, were hand-operated pumps with spigots—people pumped their water into buckets to be carried home. There was no "running water," as we know it, in people's homes. Snow recommended to local politicians that the pump handle be removed, which was done during the declining days of the outbreak. For this, he is remembered as a public health hero.

John Snow died in 1858 at age forty-five. During his short life, he became a pioneer in both anesthesiology and epidemiology, and clarified the role of water, rather than air, in cholera transmission.

Ralph R. Frerichs

Soil Pollution

Soil pollution comprises the pollution of soils with materials, mostly chemicals, that are out of place or are present at concentrations higher than normal which may have adverse effects on humans or other organisms. It is difficult to define soil pollution exactly because different opinions exist on how to characterize a pollutant; while some consider the use of pesticides acceptable if their effect does not exceed the intended result, others do not consider any use of pesticides or even chemical fertilizers acceptable. However, soil pollution is also caused by means other than the direct addition of xenobiotic (man-made) chemicals such as agricultural runoff waters, industrial waste materials, acidic precipitates, and radioactive fallout.

Both organic (those that contain carbon) and inorganic (those that don't) contaminants are important in soil. The most prominent chemical groups of organic contaminants are fuel hydrocarbons, polynuclear aromatic hydrocarbons (**PAHs**), polychlorinated biphenyls (**PCBs**), chlorinated aromatic compounds, detergents, and pesticides. Inorganic species include nitrates, phosphates, and heavy metals such as cadmium, chromium and lead; inorganic acids; and **radionuclides** (radioactive substances). Among the sources of these contaminants are agricultural runoffs, acidic precipitates, industrial waste materials, and radioactive fallout.

PAHs polyaromatic hydrocarbons; compounds of hydrogen and carbon containing multiple ring structures

PCBs polychlorinated biphenyls; two-ringed compounds of hydrogen, carbon, and chlorine

radionuclide radioactive particle, man-made or natural, with a distinct atomic weight number; can have a long life as soil or water pollutant

An area of Karabache, Russia, where soil has been poisoned by high concentrations of lead, arsenic, nickel, cobalt, and cadmium. (©Gyori Antoine/ Corbis Sygma. Reproduced by permission.)

denitrification the biological reduction of nitrate or nitrite to nitrogen gas, typically by bacteria in soil

Soil pollution can lead to water pollution if toxic chemicals leach into groundwater, or if contaminated runoff reaches streams, lakes, or oceans. Soil also naturally contributes to air pollution by releasing volatile compounds into the atmosphere. Nitrogen escapes through ammonia volatilization and **denitrification**. The decomposition of organic materials in soil can release sulfur dioxide and other sulfur compounds, causing acid rain. Heavy metals and other potentially toxic elements are the most serious soil pollutants in sewage. Sewage sludge contains heavy metals and, if applied repeatedly or in large amounts, the treated soil may accumulate heavy metals and consequently become unable to even support plant life.

In addition, chemicals that are not water soluble contaminate plants that grow on polluted soils, and they also tend to accumulate increasingly toward the top of the food chain. The banning of the pesticide DDT in the United States resulted from its tendency to become more and more concentrated as it moved from soil to worms or fish, and then to birds and their eggs. This occurred as creatures higher on the food chain ingested animals that were already contaminated with the pesticide from eating plants and other lower animals. Lake Michigan, as an example, has 2 parts per trillion (ppt) of DDT in the water, 14 parts per billion (ppb) in the bottom mud, 410 ppb in amphipods (tiny water fleas and similar creatures), 3 to 6 parts per million (ppm) in fish such as coho salmon and lake trout, and as much as 99 ppm in herring gulls at the top of the food chain.

The ever-increasing pollution of the environment has been one of the greatest concerns for science and the general public in the last fifty years. The rapid industrialization of agriculture, expansion of the chemical industry, and the need to generate cheap forms of energy has caused the continuous release of man-made organic chemicals into natural ecosystems. Consequently, the atmosphere, bodies of water, and many soil environments have become polluted by a large variety of toxic compounds. Many of these compounds at high concentrations or following prolonged exposure have the potential to produce adverse effects in humans and other organisms: These include the danger of acute toxicity, mutagenesis (genetic changes), carcinogenesis, and teratogenesis (birth defects) for humans and other organisms. Some of these man-made toxic compounds are also resistant to physical, chemical, or biological degradation and thus represent an environmental burden of considerable magnitude.

Numerous attempts are being made to decontaminate polluted soils, including an array of both *in situ* (on-site, in the soil) and off-site (removal of contaminated soil for treatment) techniques. None of these is ideal for remediating contaminated soils, and often, more than one of the techniques may be necessary to optimize the cleanup effort.

The most common decontamination method for polluted soils is to remove the soil and deposit it in landfills or to incinerate it. These methods, however, often exchange one problem for another: landfilling merely confines the polluted soil while doing little to decontaminate it, and incineration removes toxic organic chemicals from the soil, but subsequently releases them into the air, in the process causing air pollution.

For the removal and recovery of heavy metals various soil washing techniques have been developed including physical methods, such as attrition scrubbing and wet-screening, and chemical methods consisting of treatments with organic and inorganic acids, bases, salts and chelating agents. For example, chemicals used to extract radionuclides and toxic metals include hydrochloric, nitric, phosphoric and citric acids, sodium carbonate and sodium hydroxide and the chelating agents EDTA and DTPA. The problem with these methods, however, is again that they generate secondary waste products that may require additional hazardous waste treatments.

In contrast to the previously described methods, *in situ* methods are used directly at the contamination site. In this case, soil does not need to be excavated, and therefore the chance of causing further environmental harm is minimized. *In situ* biodegradation involves the enhancement of naturally occurring microorganisms by artificially stimulating their numbers and activity. The microorganisms then assist in degrading the soil contaminants. A number of environmental, chemical, and management factors affect the biodegradation of soil pollutants, including moisture content, pH, temperature, the microbial community that is present, and the availability of nutrients. Biodegradation is facilitated by aerobic soil conditions and soil pH in the neutral range (between pH 5.5 to 8.0), with an optimum reading occurring at approximately pH 7, and a temperature in the range of 20 to 30°C. These physical parameters can be influenced, thereby promoting the microorganisms' ability to degrade chemical contaminants. Of all the decontamination methods bioremediation appears to be the least damaging and most environmentally acceptable technique. SEE ALSO ABATEMENT; BIOREMEDIATION; CLEANUP; DDT (DICHLORODIPHENYL TRICHLOROETHANE); SCIENCE; SUPERFUND; TECHNOLOGY.

Bibliography

Adriano, D.C.; Bollag, J.-M.; Frankenberger, W.T.; and Sims, R.C., eds. (1999). *Bioremediation of Contaminated Soils*. Agronomy monograph 37. American Society of Agronomy.

Miller, R.W., and Gardiner, D.T. (1998). *Soils in Our Environment*, 8th edition. Upper Saddle River, NJ: Prentice Hall.

Pierzynski, G.M.; Sims, J.T.; and Vance, G.F. (2000). *Soils and Environmental Quality*, 2nd edition. Boca Raton, FL: CRC Press.

Internet Resources

Ministry of the Environment Web site. "Environmental Quality Standards for Soil Pollution." Available from http://www.env.go.jp/en/lar/regulation/sp.html.

U.S. Environmental Protection Agency Web site. "Soil and Groundwater Pollution Remediation Act." Available from http://www.epa.gov.tw/english/laws/soil.htm.

Brigitte Bollag and Jean-Marc Bollag

Solar Energy   *See Renewable Energy*

Solid Waste

The garbage that is managed by local governments is known as municipal solid waste (MSW). Specifically, MSW is waste generated by commercial and household sources that is collected and either recycled, incinerated, or disposed of in MSW landfills. The U.S. Environmental Protection Agency (EPA) separates MSW into several categories, including containers and packaging, yard wastes, durable goods, and nondurable goods. Examples of

---

PHYTOREMEDIATION

Plants can absorb, accumulate and in some cases break down pollutants such as heavy metals, pesticides, and explosives in soil and groundwater. Now the United States Department of Agriculture and the Department of Energy are conducting pilot studies to investigate whether plants can also remove radionuclides from soil. By adding soil amendments such as ammonium compounds, the pigweed plant, *Amaranthus retroflexus,* will absorb cesium-137 that contaminates soil at some DOE sites due to aboveground nuclear testing during the Cold War era.

A bulldozer moving on top of a large mound of garbage. (United States Environmental Protection Agency. Reproduced by permission.)

durable goods, which are designed to last longer than three years, include appliances, tires, batteries, and electronic equipment. Newspapers, clothing, disposable tableware, office paper, wood pallets, and diapers, which all have a lifetime of less than three years, are types of nondurable goods. MSW does not include domestic sewage and other municipal wastewater treatment sludges, demolition and construction debris, agricultural and mining residues, combustion ash, and wastes from industrial processes. These types of waste, known collectively as industrial solid waste, are largely excluded from hazardous waste regulation; programs addressing industrial solid waste are still in their infancy.

During the 1980s, solid waste management issues emerged in the United States due to the increasing amounts of solid waste generated, shrinking landfill capacity, rising disposal costs, and strong opposition to the siting of new solid waste facilities. This problem was illustrated by the much-publicized *Mobro* garbage barge, which traveled on a six-month odyssey before the garbage was finally disposed of in New York state, where it was originally generated.

With millions of households and businesses generating garbage in the United States, developing a national management program is challenging. Instead of federal regulations dictating how solid wastes should be managed, solid-waste programs are managed by states and municipalities on the local level according to individual community needs. With the exception of federally mandated landfill design and operating criteria to ensure the protection of groundwater and requirements for the federal purchase of products containing recovered materials, the EPA's role in implementing solid-waste management programs includes setting national goals, providing leadership and technical assistance, and developing educational materials.

MSW Stream

The generation of MSW has grown steadily over the past thirty years, from 88 million tons per year, or 2.7 pounds per person per day in 1960, to 229.9 million tons, or 4.62 pounds per person per day in 1999. The largest component of the MSW stream is paper and paperboard products (38.1%), with yard trimmings the second most predominant component (12.1%). The top of two pie charts on the next page breaks down this waste by material category. While the generation of waste has grown steadily, so too have its recycling and recovery. In 1960 about 7 percent of MSW was recycled, and in 1999 this figure had increased to 27.8 percent. How MSW is managed is shown in the bottom of two pie charts on the next page. Although the majority of solid waste is still sent to landfills, statistics indicate that there is a clear trend away from reliance on this method. Combustion of MSW and recovery through recycling are now a common practice in the United States.

MSW Management

In response to mounting solid waste problems, EPA published *The Solid Waste Dilemma: An Agenda for Action* in 1989, which presents goals and recommendations for action by the EPA, state and local governments, industry, and consumers to address the solid waste problems facing the United States. The EPA recommends an integrated, hierarchical approach to waste management using four components: source reduction, recycling, combustion, and landfills. This comprehensive approach addresses critical junctures in the manufacture, use, and disposal of products and materials to minimize wastefulness and maximize value. This strategy favors **source reduction** to decrease the volume and toxicity of waste and to increase the useful life of products. After source reduction, recycling, including composting, is the preferred waste management approach to divert waste from combustors and landfills. Combustion is used to reduce the volume of waste being disposed as well as to recover energy, whereas landfills are used for the final disposal of nonrecyclable and noncombustible material.

source reduction reducing the amount of materials entering the waste stream from a specific source by redesigning products or patterns of production or consumption (e.g., using returnable beverage containers); synonymous with waste reduction

The goal of the integrated management hierarchy is to use a combination of all these methods to handle the MSW stream safely and effectively with the least adverse impact on human health and the environment. The EPA encourages communities to develop community-specific assessments of potential source reduction, recycling, combustion, and landfill programs and to customize programs according to local needs, keeping in mind the strategies preferred in the national hierarchical structure. Because each community's waste profile (i.e., the amounts and types of waste generated), infrastructure, social and economic structure, and policies differ, decision makers at the local level are the most qualified to assess community needs and develop an appropriate solid waste management strategy.

Source Reduction

Source reduction, also known as waste prevention, is a front-end approach to addressing MSW problems by changing the way products are made and used.

(top) Breakdown of the 229.9 million tons of MSW generated in the United States in 1999 by material category. Generation amounts represent the percent of total generation by weight in millions of tons. (bottom) Demonstrates how 229.9 million tons of MSW generated in the United States in 1999 were managed: via combustion, recovery for recycling (including composting), and shipment to landfills. Described by the percent of total generation by weight. (Both based on statistics from EPA, Municipal Solid Waste in the United States: 1999 Facts and Figures.)

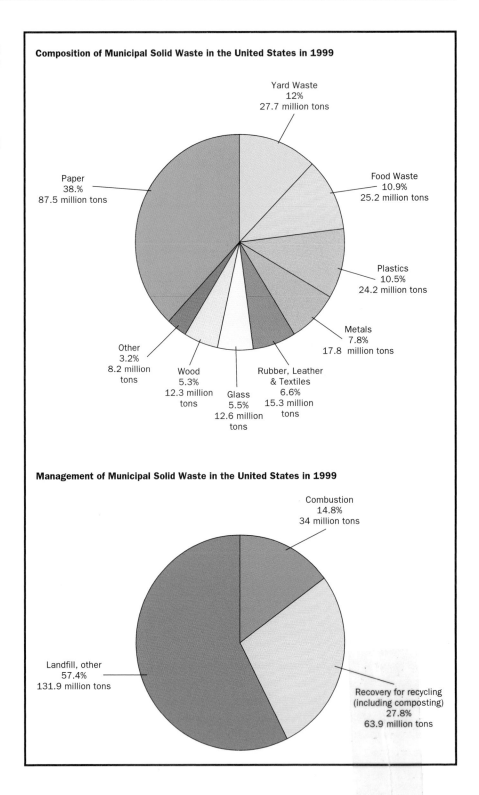

Composition of Municipal Solid Waste in the United States in 1999

Yard Waste
12%
27.7 million tons

Food Waste
10.9%
25.2 million tons

Plastics
10.5%
24.2 million tons

Metals
7.8%
17.8 million tons

Rubber, Leather
& Textiles
6.6%
15.3 million tons

Glass
5.5%
12.6 million tons

Wood
5.3%
12.3 million tons

Other
3.2%
8.2 million tons

Paper
38.%
87.5 million tons

Management of Municipal Solid Waste in the United States in 1999

Combustion
14.8%
34 million tons

Recovery for recycling
(including composting)
27.8%
63.9 million tons

Landfill, other
57.4%
131.9 million tons

It represents an attempt to move away from the traditional "end-of-the-pipe" waste management approach used in the past. Source reduction at the "beginning of the pipe" is defined as the design, manufacture, and use of products in a way that reduces the quantity and toxicity of waste produced when products reach the end of their useful lives. Waste-prevention activities include product reuse (e.g., reusable shopping bags), product material volume reduction (e.g., eliminating unnecessary product packaging), reduced toxicity of products (e.g., use of substitutes for lead, mercury, and other toxic substances), increased product lifetime (e.g., design of products with a longer useful life), and decreased consumption (e.g., changing consumer buying practices, bulk purchasing). In 1996 the EPA reported that 23 million tons of MSW had been source-reduced, approximately 11 percent of the 209.7 million tons of MSW generated that year. Businesses, households, and state and local governments all play an active role in implementing successful source reduction programs.

Recycling

Recycling refers to the separation and collection of wastes and their subsequent transformation or remanufacture into usable or marketable materials. Recycling, including composting, diverts potentially large volumes of material from landfills and combustors, and prevents the unnecessary waste of natural resources and raw materials. Other environmental benefits offered by recycling include a reduction in greenhouse gas emissions, energy conservation, and the preservation of biodiversity and habitats that would otherwise be exploited for virgin materials. In addition, recycling programs create new manufacturing jobs, boost the economy, and facilitate U.S. competitiveness in the global marketplace.

Like any other part of the integrated waste management hierarchy, recycling programs should be carefully designed and implemented to address the needs of the community, including attention to their cost-effectiveness. Recycling collection and separation programs vary in degree of implementation: Some may be simple drop-off programs, whereas others may involve comprehensive curbside collection and complex source separation at a recovery facility. Successful recycling, however, requires more than the separation and collection of postconsumer materials. Recycling programs must identify and develop markets for recovered material; only when the materials are reused is the recycling loop complete.

Although markets and uses for recovered materials are constantly expanding, reuse opportunities will vary by material. For example, recycling options for plastic are contingent on the type of resin used. Soft drink bottles are currently incorporated into products such as carpeting, household cleaner bottles, and fiberfill for coats and pillows, whereas polystyrene food containers and cups are being recycled into insulation, cafeteria food trays, and children's toys. Depending on their condition, tires can be used for artificial reefs, playground equipment, floor mats, and road construction materials. Recycled-content newspapers, stationery, corrugated containers, and toilet paper are some examples of how discarded paper is recycled.

Recycling activities also include centralized composting of yard and food wastes. Composting refers to the controlled decomposition of organic matter by microorganisms into a stable **humus** material that is used primarily on the land to improve soil quality. Many communities conduct large-scale centralized composting of yard waste in an effort to save landfill capacity. Individuals are also helping to reduce waste by composting yard waste in their backyards, and by not bagging grass clippings or other yard wastes—these activities are actually classified as source reduction. The composting of yard waste has seen tremendous growth in the past ten years. In 1980 the amount of yard waste recovered was negligible (less than 5,000 tons, or 0.05%). By

humus rich soil component derived from plant breakdown and bacterial action

1999 the amount of yard waste recovered had grown to 12.6 million tons, or 45.3 percent.

Combustion

combustion burning, or rapid oxidation, accompanied by release of energy in the form of heat and light

Burning has been a popular method of reducing the volume and odor of garbage for centuries. With the onset of the 1970s energy crisis and the Clean Air Act, a more sophisticated system of incineration was developed that could use waste as a fuel to produce energy. Modern combustion facilities no longer just destroy garbage, but instead are designed to recover energy that is used to produce steam and electricity. Developing a successful waste-to-energy system involves numerous decisions that will dictate whether such a project is effective in a given community. Over the past two decades communities have demonstrated an increased interest in **combustion** as a waste management option. Between 1980 and 1999, the combustion of solid waste increased 5.8 percent, with approximately 2.6 million tons of MSW burned in 1999. In addition to the benefits of energy recovery, combustion residues consume less landfill space; combustion ash amounts to approximately 25 percent (dry weight) of the MSW input. However, citizens often oppose the building of incinerators close to communities and farmland because of the perception of health risks due to emission of pollutants including mercury and **dioxin** that are toxic, persistent, and bioaccumulate.

Landfills

dioxin any of a family of compounds known chemically as dibenzo-p-dioxins. Concern about them arises from their potential toxicity as contaminants in commercial products; tests on laboratory animals indicate that it is one of the more anthropogenic (human-made) compounds

Even with the use of source reduction, recycling, and combustion, there will always be waste that ultimately must be disposed of in landfills. According to the EPA's *Municipal Solid Waste in the United States: 1999 Facts and Figures*, landfill disposal still remains the most widely used waste management method (accounting for approximately 57.4% of the total). Many communities now face difficulties siting new landfills largely because of increased citizen and local government concerns about the potential health risks and aesthetics of situating a landfill in their neighborhoods. The EPA issued new technical standards for MSW landfills in 1991. These addressed several aspects of landfill management, including location restrictions, design and operating criteria, and groundwater monitoring. Even with national landfill standards, decreasing landfill capacity and the difficulties associated with the construction of new landfills remain significant issues.

The EPA has explored several solutions to conserving landfill capacity, including the viability of engineering materials such as plastics to be less resistant to degradation or, in other words, biodegradable. Biodegradable materials can be broken down into simpler substances (e.g., elements and compounds) by bacteria or other natural decomposers. Paper and most organic wastes such as food and leaves are biodegradable. In contrast, non-biodegradable substances cannot be broken down in the environment by natural processes. In general, degradation in landfills occurs very slowly due to modern landfill design criteria, which minimize waste exposure to sunlight, air, and moisture. In fact, even biodegradable organic materials might take decades to decompose in a landfill; carrots and cabbage have been discovered in recognizable form after several years of burial. Studies indicate that biodegradable materials may help diminish risks to wildlife and aesthetic damage (i.e., discarded six-pack beverage rings and wrappers), but will not reduce the volume or toxicity of waste nor provide a solution to decreasing landfill capacity.

In continuing efforts to conserve landfill space and reduce waste toxicity, the EPA is currently investigating the potential benefits and drawbacks associated with the use of bioreactor landfills. Bioreactor landfills are designed to transform and more quickly stabilize the decomposable organic constituents of the waste stream through the controlled injection of liquid or air to enhance microbiological degradation processes. In other words, by controlling the moisture content, bioreactor landfills facilitate microbial decomposition of waste. Recent findings show that bioreactor landfills successfully expedite the degradation process (e.g., from decades to years), offer a 15 to 30 percent gain in landfill space, and may reduce postclosure care and **leachate** disposal costs. In addition, the bioreactor technology significantly increases landfill gas emissions, which are captured and often used beneficially for energy recovery. Due to their complexity, however, bioreactor landfills may be more costly, and concerns have been raised regarding increased odors, liner instability, and surface seeps. Working in conjunction with state and local governments and private companies, the EPA has initiated several research and pilot projects to examine the effectiveness of this innovative technology.

leachate water that collects contaminants as it trickles through wastes, pesticides, or fertilizers; leaching may occur in farming areas, feedlots, and landfills, and may result in hazardous substances entering surface water, ground water, or soil

International Solid Waste Management

Because solid waste is generated everywhere, addressing the environmentally safe management of solid waste is not limited to the United States. Management strategies vary by country and region, although most programs address waste issues with models consisting of some combination of source reduction, combustion, recycling, and landfills. For example, the European Environment Agency (EEA) offers solid-waste management guidance analogous to EPA's integrated hierarchy. Specifically, the Community Strategy on Waste recommends that the agency's eighteen-member countries make waste prevention their top priority, followed by materials recovery, energy recovery, and, finally, the safe disposal of waste.

The international Organisation for Economic Co-operation and Development (OECD) compiles worldwide data, including environmental statistics, for its thirty member countries. The bar graphs on the next page compare the total amount of municipal waste generated annually and the annual amount of municipal waste generated per capita, respectively, by the United States and other selected OECD member countries in 1997. Per capita waste generation rates vary significantly by country; factors contributing to such discrepancies may include individual lifestyle and national economic structure. Although individual national definitions may differ, for the purpose of analysis here, OECD regards municipal waste as waste collected by or on the order of municipalities, including that originating from households, commercial activities, office buildings, institutions such as school and government buildings, and small businesses.

The environmentally safe management of municipal solid waste may always be an issue, simply because societies will continue to generate trash due to increasing populations and the growing demands of modern society. Working together, federal, state, and local governments, industry, and citizens have made substantial progress in effectively responding to solid waste

(top) Comparison of annual amounts of MSW generated by the United States and other selected OECD countries in 1997. Generation amounts are in 1,000 tons. (bottom) Comparison of annual amounts of MSW generated per capita by the United States and other selected OECD countries in 1997. Generation amounts are in kilograms per capita. (Both based on statistics from OECD Environmental Data 1999.)

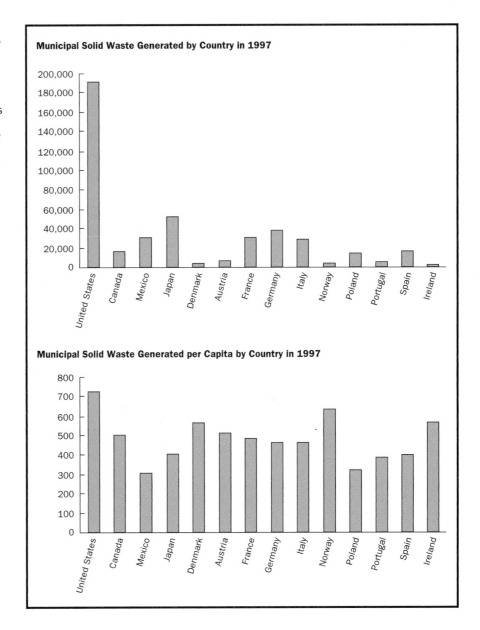

Municipal Solid Waste Generated by Country in 1997

Municipal Solid Waste Generated per Capita by Country in 1997

issues through source reduction, recycling, combustion, and landfill programs. Such community-tailored programs provide possible long-term solutions to decreasing the amount of waste that is produced and ultimately placed in landfills. SEE ALSO COMPOSTING; INCINERATION; LANDFILL; PLASTIC; RECYCLING; REUSE; WASTE; WASTE REDUCTION.

Bibliography

Christiansen, Kim Michael. (1999). *Waste Annual Topic Update: 1998*. Copenhagen: European Environmental Agency.

O'Leary, Philip R., and Walsh, Patrick H. (1995). *Decision Makers Guide to Solid Waste Management*, Vol. II. Washington, D.C.: U.S. Environmental Protection Agency. EPA530-R95-023.

Organisation for Economic Co-operation and Development. (1999). *OECD Environmental Data: Compendium 1999 Edition*. Organisation for Economic Co-operation and Development.

U.S. Environmental Protection Agency. (1994). *Composting Yard Trimmings and Municipal Solid Waste*. Washington, D.C.: U.S. Environmental Protection Agency. EPA530-R-94-003.

U.S. Environmental Protection Agency. (1990). *Environmental Fact Sheet: The Facts on Degradable Plastics*. Washington, D.C.: U.S. Environmental Protection Agency. EPA530-SW-90-017D.

U.S. Environmental Protection Agency. (1990). *Environmental Fact Sheet: The Facts on Recycling Plastics*. Washington, D.C.: U.S. Environmental Protection Agency. EPA530-SW-90-017E.

U.S. Environmental Protection Agency. (1992). *"Green" Advertising Claims*. Washington, D.C.: U.S. Environmental Protection Agency. EPA530-f-92-024.

U.S. Environmental Protection Agency. (1991). *Markets for Scrap Tires*. Washington, D.C.: U.S. Environmental Protection Agency. EPA530-SW-90-074A.

U.S. Environmental Protection Agency. (1997). *Measuring Recycling: A Guide for State and Local Governments*. Washington, D.C.: U.S. Environmental Protection Agency. EPA530-R-97-011.

U.S. Environmental Protection Agency. (2001). *Municipal Solid Waste in the United States: 1999 Facts and Figures*. Washington, D.C.: U.S. Environmental Protection Agency. EPA530-R-01-014. Also available from http://www.epa.gov/epaoswer/non-hw/muncpl/mswfinal.pdf

U.S. Environmental Protection Agency. (1999). *National Source Reduction Characterization Report for Municipal Solid Waste in the United States*. Washington, D.C.: U.S. Environmental Protection Agency. EPA530-R-99-034. Also available from http://www.epa.gov/epaoswer/non-hw/reduce/r99034.pdf.

U.S. Environmental Protection Agency. (1998). *Puzzled About Recycling's Value? Look Beyond the Bin*. Washington, D.C.: U.S. Environmental Protection Agency. EPA530-K-98-008. Also available from http://www.epa.gov/epaoswer/non-hw/recycle/benefits.pdf.

U.S. Environmental Protection Agency. (1998). *RCRA Orientation Manual: 1998 Edition*. Washington, D.C.: U.S. Environmental Protection Agency. EPA530-R-98-004.

U.S. Environmental Protection Agency. (1999). *Recycling Works! State and Local Solutions to Solid Waste Management Problems*. Washington, D.C.: U.S. Environmental Protection Agency. EPA530-K-99-003. Also available from http://www.epa.gov/epaoswer/non-hw/recycle/recycle.pdf.

U.S. Environmental Protection Agency. (1989). *The Solid Waste Dilemma: An Agenda for Action*. Washington, D.C.: U.S. Environmental Protection Agency. EPA530-SW-89-019.

Office of Solid Waste/U.S. Environmental Protection Agency

Space Pollution

In the most general sense, the term space pollution includes both the natural micrometeoroid and man-made orbital debris components of the space environment; however, as "pollution" is generally considered to indicate a despoiling of the natural environment, space pollution here refers to only man-made orbital debris. Orbital debris poses a threat to both manned and unmanned spacecraft as well as the earth's inhabitants.

Environmental and Health Impacts

The effects of debris on other spacecraft range from surface abrasion due to repeated small-particle impact to a catastrophic fragmentation due to a collision with a large object. The relative velocities of orbital objects (10 kilometers per second [km/s] on average, but ranging from meters per second up to 15.5 km/s) allow even very small objects—such as a paint flake—to damage spacecraft components and surfaces. For example, a 3-millimeter (mm) aluminum particle traveling at 10 km/s is equivalent in energy to a bowling ball traveling at 60 miles per hour (or 27 m/s). In this case, all the energy

FRAGMENTATION DEBRIS

	Payloads	Rocket Bodies	Operational Debris	Breakup Debris	Anomalous Debris	Totals
LEO	1,612	758	651	3,232	119	6,372
MEO	126	28	2	0	0	156
GEO	587	116	1	2	0	706
Elliptical	249	515	135	167	0	1,066
Unknown	171	120	185	0	0	476
Totals	2,745	1,537	974	3,401	119	8,776

SOURCE: Anz-Meador, P.D., "History of On-Orbit Satellite Fragmentations", 12th ed., NASA Johnson Space Center Report JSC-29517, 31 July 2001.

would be distributed in an area of the same size as the particle, causing cratering or penetration, depending on the thickness and material properties of the surface being impacted. There has been one accidental collision between cataloged objects to date, but surfaces returned from space and examined in the laboratory confirm a regular bombardment by small particles. Space Shuttle vehicle components, including windows, are regularly replaced due to such damage acquired while in orbit. Debris also poses a hazard to the surface of the Earth. High-melting-point materials such as titanium, steel, ceramics, or large or densely constructed objects can survive atmospheric reentry to strike the earth's surface. Although there have been no recorded fatalities or severe injuries due to debris, reentering objects are regularly observed and occasionally found.

Debris is typically divided into three size ranges, based on the damage it may cause: less than 1 centimeter (cm), 1 to 10 cm, and larger than 10 cm. Objects less than 1 cm may be shielded against, but they still have the potential to damage most satellites. Debris in the 1 to 10 cm range is not shielded against, cannot easily be observed, and could destroy a satellite. Finally, collisions with objects larger than 10 cm can break up a satellite. Of these size ranges, only objects 10 cm and larger are regularly tracked and cataloged by surveillance networks in the United States and the former Soviet Union. The other populations are estimated statistically through the analysis of returned surfaces (sizes less than 1 mm) or special measurement campaigns with sensitive radars (sizes larger than 3 mm). Estimates for the populations are approximately 30 million debris between 1 mm and 1 cm, over 100,000 debris between 1 and 10 cm, and 8,800 objects larger than 10 cm.

The number, nature, and location of objects greater than 10 cm in size are provided in the fragmentation debris table and in the image of space debris around Earth. Low Earth orbit (LEO) is defined as orbital altitudes below 2,000 km above the earth's surface and is the subject of the image of space debris around Earth. Middle Earth orbit (MEO) is the province of the Global Positioning System (GPS) and Russian navigation satellite systems and is located at approximately 20,000-km altitude, whereas the geosynchronous Earth orbit (GEO) "belt" is inhabited primarily by communications and Earth—observation payloads around 35,800 km. The majority of objects in these orbital regions are in circular or near-circular orbits about the earth. In contrast, the elliptical orbit category includes rocket bodies left in their transfer (payload delivery) orbits to MEO and GEO as well as scientific, communications, and Earth-observation payloads. Of all objects listed in the

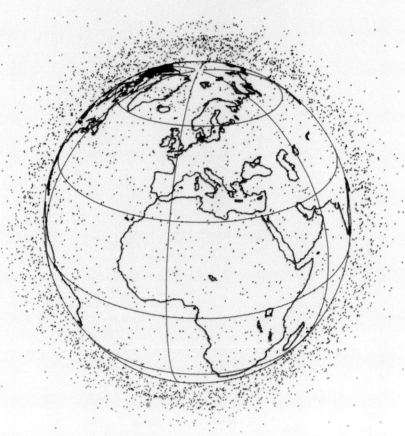

MAN-MADE ORBITAL DEBRIS
IN
LOW EARTH ORBIT

Observer is located 4 Earth radii above Earth's Surface

Objects from NORAD Catalog for May 30, 1987

Only objects larger than 10 cm diameter are shown

fragmentation debris table, the vast majority are "debris"—only about 5 percent of objects in orbit represent operational payloads or spacecraft. Also, of the approximately 28,000 objects that have been tracked, beginning with the launch of *Sputnik 1* in October 1957, those not accounted for in the fragmentation debris table have either reentered the earth's atmosphere or have escaped the earth's influence (to land on Mars, for example). The distribution of debris smaller than 10 cm is predicated on the orbits of the parent objects and is assumed to be very similar to the distributions presented in the image of space debris around Earth.

Remediation Strategies

Remediation takes two courses: protection and mitigation. Protection seeks to shield spacecraft and utilize intelligent design practices to minimize the effects

A NASA map showing man-made orbital debris in low Earth orbit. (©NASA/Roger Ressmeyer/Corbis. Reproduced by permission.)

221

of debris impact. Mitigation attempts to prevent debris from being created. Active mitigation techniques include collision avoidance between tracked and maneuverable objects and the intentional reentry of objects over the oceans. Passive techniques include venting residual fuels or pressurized vessels aboard rockets and spacecraft, retaining operational debris, and placing spacecraft into disposal orbits at the end of a mission. Space salvage or retrieval, while an option, is currently too expensive to employ on a regular basis.

The United States and international space agencies recognize the threat of debris and are cooperating to limit its environmental and health hazards. The Interagency Space Debris Coordination Committee (IADC), sponsored originally by the National Aeronautics and Space Administration (NASA), has grown to include all major space-faring nations. The IADC charter includes the coordination and dissemination of remediation research, and strategies based on research results are being adopted by the worldwide space community.

Remediation strategies have resulted in a decline in the rate of debris growth in the 1990s although the overall population continues to grow. Continued work is necessary, however, to reduce the orbital debris hazard for future generations and continue the safe, economical utilization of space.

Bibliography

Committee on Space Debris, Aeronautics and Space Engineering Board, Commission on Engineering and Technical Systems, National Research Council. (1995). "Orbital Debris: A Technical Assessment." Washington, D.C.: National Academy Press. Also available from http://pompeii.nap.edu/books/0309051258/html/index.html

Johnson, Nicholas L. (1998). "Monitoring and Controlling Debris in Space." *Scientific American* 279(2):62–67.

Internet Resources

Interagency Space Debris Coordination Committee (IADC) Web site. Available from http://www.iadc-online.org.

Phillip Anz-Meador

Sprawl

A term used in debates about urban growth, *sprawl* does not have a precise, academic definition. As a noun, it most often refers to spread-out development that requires people to use a car for every activity, because it strictly separates housing, shopping, schools, offices, and other land uses from each other. The commercial sprawl landscape features wide roads flanked by parking lots that surround mostly single-story buildings; there are usually many cars but few pedestrians. As a verb, sprawl most often refers to metropolitan areas that are consuming land at a faster rate than the population is growing. Sprawl is said to be worst in cities that are spreading out even though their population is stagnant or declining. Some people criticize sprawling growth because it creates traffic congestion, air quality, water pollution, and the revitalization of older neighborhoods harder to address. SEE ALSO SMART GROWTH.

Internet Resource

Smart Growth Online. "The Cost of Sprawl: How Much Does It Cost to Drive to Work?" Available from http://www.smartgrowth.org/news/article.asp?art=3071.

David Goldberg

A sprawling neighborhood in Corona, California. (AP/Wide World Photos. Reproduced by permission.)

Strong, Maurice

CANADIAN ENVIRONMENTAL ADVOCATE; FIRST EXECUTIVE DIRECTOR OF THE UNITED NATIONS ENVIRONMENT PROGRAMME (1929–)

No single international civil servant has contributed more to global attention to environmental problems, including those relating to air and water pollution, than has Maurice F. Strong. Born in Manitoba, Canada, by the age of twenty-two Strong had acquired a small fortune from the Alberta oil boom. His lifelong ambition, however, was public service. After serving as director-general of the Canadian External Aid Office (later the Canadian International

Maurice Strong. (©Robert Patrick/Corbis. Reproduced by permission.)

Development Agency), he agreed to head the United Nations Conference on the Human Environment (1972). As conference secretary-general and United Nations (UN) undersecretary-general for environmental affairs, he began his lifelong quest to focus the world's attention on a future environmental catastrophe. He sought to reconcile environmental concerns with development needs and the need of the present generation for sustainable growth with the necessity of leaving a clean environment for future generations. Knowing that the key to successful conferences was their follow up, he agreed to serve as the first executive director of the United Nations Environment Programme (UNEP) (1973 to 1975). In 1992, the Canadian government nominated him as secretary-general of the UN Conference on Environment and Development (UNCED), commonly known as the Earth Summit, held in Rio de Janeiro, Brazil. In both global environmental conferences, he encouraged participation by nongovernmental organizations.

In his capacity as undersecretary-general and senior advisor to the UN secretary-general, Strong is assisting UN Secretary-General Kofi Annan in reforming the United Nations (something he has long advocated) and serving as a member of the Commission on Global Governance (1992 to 1996). In his 2000 best-selling autobiography, *Where on Earth Are We Going?*, he vowed to continue his lifelong quest. SEE ALSO ACTIVISM.

Bibliography

Strong, Maurice (2000). *Where on Earth Are We Going?* Toronto: Vintage.

Michael G. Schechter

Sulfur Dioxide

anthropogenic human-made; related to or produced by the influence of humans on nature

Sulfur dioxide (SO_2) is an air pollutant known primarily for its role in acid rain. SO_2 is emitted naturally from volcanoes. **Anthropogenic** emissions arise largely from the production of electricity, particularly coal-fired power plants (65%). The sulfur in the coal reacts with oxygen during combustion, converting it to SO_2. Scrubbers, using a slurry of limestone and water, are used to extract the SO_2 before it exits the stack.

Once in the atmosphere, SO_2 is converted to other compounds such as sulfuric acid (H_2SO_4), the primary contributor to acid rain. SO_2 also reacts to form sulfate aerosols. These tiny airborne particles are the major cause of haze in U.S. national parks.

Both SO_2 gas and sulfate aerosols cause breathing problems, particularly for people with existing respiratory illnesses such as asthma. For health reasons, to reduce acid rain, and to improve visibility, SO_2 emissions are regulated by a market-based allowance trading system established by the U.S. Environmental Protection Agency (EPA). SEE ALSO ACID RAIN; COAL; ELECTRIC POWER; SCRUBBERS.

Bibliography

Turco, Richard P. (1997). *Earth under Siege: From Air Pollution to Global Change.* New York: Oxford University Press.

Internet Resource

Agency for Toxic Substances and Disease Registry (ATSDR). "ToxFAQ for Sulfur Dioxide." Available from http://www.atsdr.cdc.gov/tfacts116.html.

U.S. Environmental Protection Agency Web site. Available from http://epa.gov.

Marin Sands Robinson

Superfund

Superfund is a term used for the monies available to the U.S. Environmental Protection Agency (EPA) to clean up abandoned or inactive hazardous waste sites. Such sites may involve soil and/or groundwater contamination, and are often contaminated with **heavy metals,** such as arsenic, cadmium, chromium, lead, mercury, and zinc; pesticides, including aldrin, dieldrin, chlordane, and **DDT**; and chlorinated solvents such as carbon tetrachloride, methylene chloride, and tetra and trichloroethylene. Polychlorinated biphenyls (**PCBs**), cyanide, benzene, toluene, vinyl chloride, and radionuclides, including strontium, plutonium, and uranium are also found at hazardous waste sites. The $1.8 billion Superfund was established in 1980 by federal legislation under the Comprehensive Environmental Response, Compensation, and Liability Act (CERCLA). It was created with taxes imposed by the federal government on major oil and chemical companies. At that time, common belief was that sufficient funds and technology existed to clean up all abandoned hazardous waste sites by 1985.

Historical Perspective

By 1985, although work had started at many sites, only approximately six sites had been completely remediated, and it soon became clear that revisions to legislation were needed to streamline cleanup efforts and additional taxes for Superfund were required to provide funding. In 1986 Superfund was replenished under the Superfund Amendments and Reauthorization Act (SARA). As a result of SARA, Superfund totaled $8.5 billion.

Under CERCLA and SARA, the EPA is given the authority and resources to clean up hazardous waste sites. EPA's priority is to identify responsible parties—those companies that have caused contamination—and require them to clean up, at their own expense, any corresponding hazardous waste sites. EPA thus reserves the use of Superfund monies for sites in which responsible parties are not identified or have claimed bankruptcy. As of 1999, responsible parties have contributed over $16 billion toward the cleanup of hazardous waste sites.

The EPA follows a detailed procedure to evaluate hazardous waste sites and ranks them according to the severity of risk to human health and the environment. The national priorities list (NPL) includes those sites that are deemed eligible for cleanup by Superfund. In 1987 it listed 1,187 sites and nearly 30,000 sites remained to be assessed. As of March 2002, 1,223 sites remained on the NPL and were eligible for cleanup under Superfund. In addition, 810 sites had achieved "construction completed" status which means that all the measures to clean up the sites, as outlined in the EPA Record of Decisions, have been taken.

Site Cleanup Remedies

Technologies employed to clean up sites include procedures that have been used for decades in treating water and air pollution; also, novel techniques

heavy metals metallic elements with high atomic weights (e.g. mercury, chromium, cadmium, arsenic, and lead); can damage living things at low concentrations and tend to accumulate in the food chain

DDT the first chlorinated hydrocarbon insecticide (chemical name: Dichloro-Diphenyl-Trichloroethane); it has a half-life of 15 years and can collect in fatty tissues of certain animals; for virtually all but emergency uses, DDT was banned in the U.S. in 1972

PCBs polychlorinated biphenyls; two-ringed compounds of hydrogen, carbon, and chlorine

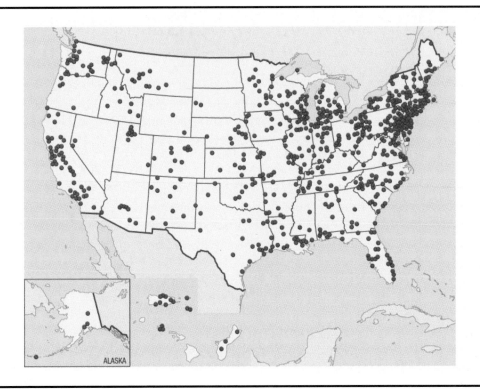

Map illustrating Superfund sites in the United States, illustration. (Gale.)

volatile any substance that evaporates readily

have been developed to clean up specific contaminants in groundwater and soil. Environmental engineers, geologists, chemists, and biologists consider alternatives to clean up sites depending on what medium is contaminated (e.g., groundwater, surface or subsurface soil, surface water, or air), and the nature of the contaminants. Community involvement is also sought as part of the decision process.

Contaminants that are biodegradable may be completely converted to environmentally acceptable products. An example of this would be using microorganisms to biodegrade gasoline components in water or soil to carbon dioxide and water. Alternatively, depending on cost and time constraints, other technologies are employed that transfer the contamination from one medium to another. Air stripping and soil vapor extraction are examples of such technologies. Air stripping involves spraying contaminated water into the top of a vertical tower while air is pumped from the bottom to the top of the tower. Chemicals that are **volatile** will be transferred in the tower from the water to the air. In soil vapor extraction, perforated pipes are drilled into contaminated subsurface soil and a vacuum is applied to encourage volatile chemicals to transfer from the soil to the air. Contaminants transferred to the air by these processes, such as benzene, toluene, and trichloroethylene are sometimes captured with activated carbon or destroyed by a combustion process, such as incineration. Air stripping was employed to clean groundwater contaminated with volatile organic chemicals, including trichloroethylene, benzene, toluene, and xylenes, at the General Mills/Henkel Superfund site, a former technical center and research laboratory in Minneapolis, Minnesota. The contaminants in the ground water have stabilized since the pump and treat system began in the early 1990s, with cleaned water being discharged to the Minneapolis storm sewer system.

Since the inception of SARA, the EPA has expressed a preference for cleanup remedies that destroy contamination rather than transfer it. Contaminants may be destroyed by microorganisms that biodegrade chemicals or by incineration processes that transform the chemical with extreme heat. One billion pounds of contaminated soil were incinerated at the Sikes Disposal Pits near Crosby, Texas, where hazardous waste from petrochemical companies had been dumped in unlined pits during the 1960s. The incineration was completed in 1994 and the site is now planted with local grasses. The excavation of contaminated soil for hauling to a landfill is an example of the removal and transfer of contamination to another area. The concern with "removal technologies" is that the contamination may create a future hazard to human health or the environment. For this reason, the EPA has come to discourage the use of removal technologies.

Pros, Cons, and Other Countries

Superfund's proponents argue that the EPA must have the authority and resources to clean up hazardous waste sites. Otherwise, reluctant responsible parties will have no incentive to bear the burden of cleanup. In such cases, the protection of public health and remediation of damages to the environment would be left for taxpayers to finance. Those against Superfund reauthorization claim that many industries are responsibly handling the matter of hazardous waste sites and have invested sizable resources to clean up such locations. Furthermore, these industries have a vested interest in achieving a cost-effective cleanup in a timely manner.

Many developed countries have implemented hazardous waste remediation programs. Some countries pay for site cleanup from general government revenues (taxes, etc.), whereas others rely on special taxes on industry (similar to Superfund). SEE ALSO ABATEMENT; BIOREMEDIATION; BROWNFIELD; CLEANUP; COMPREHENSIVE ENVIRONMENTAL RESPONSE, COMPENSATION, AND LIABILITY ACT (CERCLA); HAZARDOUS WASTE; RADIOACTIVE WASTE.

Bibliography

LaGrega, Michael D.; Buckingham, Phillip L.; and Evans, Jeffrey C. (2001). *Hazardous Waste Management*, 2nd edition. New York: McGraw-Hill.

Watts, Richard J. (1997). *Hazardous Wastes: Sources, Pathways, Receptors*. New York: John Wiley & Sons.

Internet Resources

Federal Remediation Technologies Roundtable. "Remediation Technology." Available from http://www.frtr.gov.

"Libby, Montana, Groundwater Contamination." Available from http://www.epa.gov/superfund.

Thomas D. DiStefano

SUPERFUND SITE IN LIBBY, MONTANA

In Libby, Montana, the remediation of soil and groundwater contaminated with pentachlorophenol (PCP) and polycyclic aromatic hydrocarbons (PAHs) has been under way since 1985. PCP and PAHs are chemicals used to preserve wood products such as telephone poles and railroad ties. The responsible party, Champion International Corporation, caused soil and groundwater contamination at its lumber and plywood mill in Libby. The EPA determined that wastewater and sludge from the wood-treating process were the sources of contamination. To address the issue of contamination, drinking water from a public water supply was provided to residents of the Libby area, and the use of private wells prohibited. Contaminated soil and groundwater are undergoing cleanup using bioremediation, a technology that employs microorganisms to transform hazardous chemicals into environmentally acceptable products.

Sustainable Development

The term *sustainable development* gained international recognition after the World Commission on Environment and Development (the Brundtland Commission) released its report *Our Common Future* in 1983. In this report, sustainable development was defined as "development that meets the needs of the present without compromising the ability of future generations to meet their own needs." The International Union for Conservation of Nature and Natural Resources had introduced the term earlier in its 1980 publication *World Conservation Strategy*, stating, "Development and conservation operate in the same global context, and the underlying problems that must be overcome if either is to be successful are identical." It thus recommended a strategy entitled, "Towards Sustainable Development."

Development refers to any systematic progress toward some improved or advanced condition. In the international development field, in which the term sustainable development is most often encountered, development refers to the establishment of the physical and social conditions that make economic progress possible. In the past this has at times involved the transformation of forests, wetlands, soil, and other resources in ways that ultimately undermined the capacity of the natural environment to produce conditions able to sustain future advances in the quality of people's lives. The concept of sustainable development thus suggests an alternative strategy in which economic progress and environmental protection go hand in hand.

The negative environmental impacts of some forms of economic development had been recognized long before the term sustainable development was popularized in the 1980s. The earliest settled communities subjected the harvesting of important food and raw materials to rules, customs, and eventually formal laws and regulations designed to protect renewable resources for the future. In his book *Man and Nature* published in 1867, George Perkins Marsh drew attention to the environmental changes he had witnessed in both the United States and the Mediterranean region. His alarm was echoed by early American conservationists Gifford Pinchot and John Muir at the beginning of the twentieth century and again by Rachel Carson in her 1962 book *Silent Spring*. Then in 1972 an environmentally aware group of industrialists known as the Club of Rome issued a report, *The Limits to Growth*, that warned of inadequate natural resource supplies and disruption to global ecosystems if population and economic growth were to continue on their current path. In 1971 the International Institute for Environment and Development (IIED) was established in Britain with a mandate to seek ways to achieve economic progress without destroying the environmental resource base.

In June 1992 the United Nations Conference on Environment and Development (UNCED) further refined the term by developing an agenda for nations to follow that would move the world toward sustainable development. Agenda 21, as it was called, was a three-hundred-page plan for achieving sustainable development in the twenty-first century. To assist in follow up and monitor the progress of Agenda 21, and to report on the implementation of related agreements, the United Nations created the Commission on Sustainable Development (CSD), to report to the UN Economic and Social Council (ECOSO).

Although the concept of sustainable development has received considerable attention in international diplomatic and policy circles, it does have its critics. Many claim sustainable development is an oxymoron. They argue that nothing, least of all economic development, is sustainable forever. For them, the concept of sustainable development is wishful thinking that distracts nations from the necessary transformations of the global economy. Others claim that a determined focus on sustainability is likely to lead to economic stagnation and continued underdevelopment.

The proponents of sustainable development believe that the current mode of economic development is fundamentally destructive and must be radically reformed, and although nothing is absolutely sustainable, the effort to hold development activities accountable for the environmental conditions they produce makes both long-term economic and ethical sense. They argue that this approach, when combined with efforts to reduce population growth rates, reduce consumption among the richest nations of the world, promote the substitution of renewable for nonrenewable natural resources, reduce waste from manufacturing processes, and improve efficiency in the use of materials, is the only approach that offers a positive future outlook for the welfare of the global community.

In the decade since Agenda 21 was accepted as a strategy for sustainable development, progress has been made. International agreements have been promulgated that will have a positive effect on sustainable development. These include, among others, the efforts of the United Nations in formulating a framework convention on climate change, a convention on biological diversity and a global compact that combines concerns for human rights, labor, and the environment. In addition, standards for business activity that consider environmental consequences have been agreed to by the International Organization for Standardization (ISO 14000), and the international business community has created the World Business Council for Sustainable Development. The World Bank has applied the concept of sustainable development with its reformed lending practices requiring recipients to demonstrate sound environmental criteria. Cities around the world are adopting sustainable criteria for land-use planning and zoning, and individuals are making personal consumption choices with sustainable development in mind. Though the problems of a sustainable future are far from solved, there is much about which to be optimistic. SEE ALSO EARTH SUMMIT.

Bibliography

Carson, Rachel. (1963). *Silent Spring.* Boston: Houghton Mifflin.

International Union for the Conservation of Nature and Natural Resources. (1980). *World Conservation Strategy.*

Marsh, George Perkins. (1867). *Man and Nature.* New York: Scribners.

Meadows, Donella H.; Meadows, Dennis L.; Randers, Jorgen; and Behrens, William W. III. (1972). *The Limits to Growth.* New York: Universe Books.

Pinchot, Gifford. (1947). *Breaking New Ground.* New York: Hartcourt, Brace, and Co.

World Commission on Environment and Development. (1987). *Our Common Future.* New York: Oxford University Press.

Internet Resource

U.S. Department of Energy Center for Excellence for Sustainable Development. Available from http://www.sustainable.doe.gov.

Jack Manno and Ross Whaley

Swallow, Ellen

Ellen Swallow Richards (1842–1911) was the first female chemist in the United States and the mother of the science of ecology. As she walked to the Massachusetts Institute of Technology (MIT) each day, this sanitary chemist noticed horse wagons carrying uncovered food over Boston's dirty, unpaved

Ellen Swallow. (Courtesy of the MIT Museum. Reproduced by permission.)

streets, which were often flooded with pools of stagnant waste from the open sewers. She saw filth, disease, suffering, and poverty that took the lives of half the children living in these conditions.

She determined that chemistry should be used to provide a meaningful service to society by improving people's health and environment. Her pioneering work on the effects of industrial pollution and sewage on human health led to the world's first sanitary engineering program and water-purity testing formulas, which are so precise that they are still being used. Her work in food additives led to the creation of the first pure food laws in the United States.

In an 1892 speech to other scientists, she first introduced the word and concept *ecology*, referring to the relationships of organisms to their environments, whether natural, domestic or industrialized and human created. Her book, *Euthenics: The Science of Controllable Environment: A Plea for Better Living Conditions as a First Step toward Higher Human Efficiency*, which was published in 1910, introduced ecology to the public. Swallow's dedication brought environmental concerns about clean air, water, sanitation, and pure food to the societal, home, and individual levels and called for the integration of all the sciences to solve environmental and health problems.

Swallow's difficult but successful entry to higher learning also paved the way for other women. In 1871 she overcame substantial obstacles to become the first woman to attend MIT, where she was denied a doctorate in chemistry because MIT did not allow women to be awarded doctorate degrees. She then became the first female member of the faculty at MIT, resulting in, for example, both women and men having access to the school's new chemistry lab in 1878. SEE ALSO INDUSTRY; ENVIRONMENTAL MOVEMENT; SETTLEMENT HOUSE MOVEMENT.

Bibliography

Internet Resources

Chemical Heritage Foundation. Available from http://www.chemheritage.org/EducationalServices/chemach/hnec/esr.html.

MIT Institute Archives and Special Collections. "Ellen Swallow Richards." Available from http://libraries.mit.edu/archives/exhibits/esr.

Susan L. Senecah

Systems Science

Most traditional science works within a very restricted disciplinary domain requiring a careful and often technically rigorous and demanding approach that includes, at least in theory, the use of the Baconian scientific method of test and control in a restricted laboratory environment. This is how most science operates, and it is often a very successful approach. However, such an approach is very difficult to apply to many real problems, including those in the complex natural or seminatural world outside the laboratory where many interacting variables can render laboratory results rather meaningless. For example, cleaning up sewage in a treatment plant can increase air pollution both directly and through the energy required.

One antidote to this problem is systems science, which seeks to find and use general principals, concepts, and equations that are applicable across, and can integrate, many disciplines. Numerous thinkers throughout history have used some kind of systems approach (e.g., Isaac Newton realized that billiard balls and planets both followed the same laws of motion). General systems theory was formalized and popularized by Ludwig von Bertalanffy in a book by that name; he founded the Society for General Systems and advanced its studies. This society is still active, continuing to attract ecologists, physicians, psychologists, engineers, mathematicians, economists, and others who seek new ideas in other disciplines.

There are, generally speaking, two approaches to systems science. The first undertakes analysis of the properties of systems as a whole. For example, one might ask what is the photosynthesis of an entire ecosystem or indeed of the globe as a whole. The most comprehensive, and some might say most controversial, application of this approach is the Gaia hypothesis of deep ecologists James Lovelock and Lynn Margulis. This hypothesis postulates that the earth as a living system itself regulates the chemical and other characteristics of the atmosphere (and other entities) in order to maintain optimal conditions for life. In other words, life maintains its own environment. This concept, or one somewhat like it, has been called *self-design* by Howard Odum and others, and Odum applies the view especially to ecosystems.

The second general approach is a "systems" analysis of how parts of a system interact and generate the behavior of some entire entity. Such an approach, originally used to link radar, artillery, and aircraft during the Battle of Britain in World War II, has been especially well developed in the engineering sciences. For example, computers are used routinely in designing automobiles to model how springs and shock absorbers interact with wheels and terrain so that automobiles with smoother rides can be designed. Here and elsewhere in a systems approach, the feedback of one motion or operation on the subsequent behavior of the system is of paramount importance. Many systems investigators try to capture the essence of the behavior and other attributes of their ecosystem of interest through the construction of mathematical and/or computer simulation models.

Examples of how systems science has contributed to science include the use of techniques originally designed for measuring photosynthesis and respiration in aquatic ecosystems to understand the metabolism of the Northern Hemisphere. It has also included the application of fisheries analysis techniques to assess the success of drilling for oil. Oil return per unit effort spent in acquiring it, like fishing for fish, decreases with increasing effort.

A systems approach can be applied in many ways, including modeling the fate and transport of pollutants dumped into a river or groundwater. The classic example is the Streeter Phelps model, developed in the 1930s, that predicts the oxygen level in a river as a function of sewage load, dispersion, microbial activity and interactions with the atmosphere. A general systems approach has been most thoroughly developed for the environmental sciences by Odum in *General and Systems Ecology*. Other specific examples include its use in combined hydrological, biological, and economic models to determine the cheapest way to clean up the Delaware estuary; combined atmospheric and pollutant generation models to predict, for example, acid rain deposition downwind; and models to generate groundwater pollution and its impact. More recently, some efforts to integrate economics into traditional systems analyses of natural systems have evolved. An extensive systems approach has been used, for example, to examine the economy of Costa Rica, not just with the conventional tools of economics but also through a biophysical approach originally developed for natural ecosystems (Hall, 2000). In fact, some kind of systems approach is almost a necessity in any sophisticated environmental impact statement. SEE ALSO ENVIRONMENTAL IMPACT STATEMENT; GIS (GEOGRAPHIC INFORMATION SYSTEM); GLOBAL WARMING; GROUNDWATER; RISK.

Bibliography

Hall, C.A.S., and C.J. Cleveland. (1981). "Petroleum Drilling and Production in the United States: Yield per Effort and Net Energy Analysis." *Science* 211:576–579.

Hall, C.A.S., ed. (2000). *Quantifying Sustainable Development: The Future of Tropical Economies.* San Diego: Academic Press.

Odum, Howard T. (1994). *Ecological and General Systems.* Niwot: University of Colorado Press.

von Bertalanffy, Ludwig. (1968). *General Systems Theory.* New York: George Brazillier.

Internet Resource

Principia Cybernetica Project (PCP) Web site. Available from http://pespmc1.vub.ac.be/DEFAULT.

Charles Hall

TCE (Trichloroethylene)　*See Dry Cleaning*

Technology, Pollution Prevention

A pollution prevention (P2) technology is one that creates less pollution in its life cycle than the one it replaces. P2 can be achieved in many ways, from better housekeeping and maintenance to redesign of products and processes. The range of P2 technologies is therefore very broad. It includes relatively cleaner technologies, technologies that help other technologies to be cleaner, and certain mass-market technologies. All of them reduce environmental impacts compared to their alternatives. It is important to understand that P2 technology does not include pollution-control or -treatment technologies that do not make the technology producing the pollution any cleaner itself. They just manage the resulting waste.

Relatively Cleaner Technologies

Technology is always advancing and improving. Many new technologies are naturally more energy efficient and less polluting than the ones they replace. Sometimes, this is because they were designed with environmental improvement in mind. Usually, however, it is simply the result of using newer and better materials and components. Therefore, pollution-preventing technologies can be found in every area of a product's life cycle.

Life cycle analysis (LCA) is needed to determine if a particular technology really pollutes less than its alternatives. LCA is the examination of the environmental impacts of a product, from its origins as raw material through processing and production to use and final disposal. This can be a complex process. For example, fluorescent light bulbs may seem to be less polluting than incandescent light bulbs because they use much less energy. However, they actually use polluting chemicals such as mercury that are not found in incandescent light bulbs. So they use less energy, but more toxic chemicals. The choice of indicators for P2 performance and LCA, such as toxicity or energy efficiency, is important for evaluation.

Facilitative Technologies

Some technologies are important for helping other technologies reduce pollution. For example, process controls such as meters and sensors can make many production processes more efficient and less polluting by providing improved control, which reduces waste and defects. Centrifuges can reduce the amount of solids in wastewaters, thereby reducing water pollution. Catalytic converters on engine exhaust systems can reduce air pollution. There are many such examples of technologies that help other technologies be cleaner. This is important in situations where there is a large investment in an existing technology already installed that cannot be easily or economically replaced with new and cleaner technology.

Technologies Designed to Prevent Pollution

Some technologies are designed specifically for protecting the environment while also improving business performance. For example, recycling technologies can help recover valuable materials from wastes, cutting manufacturing costs, while also preventing pollution. Examples include gene-engineered plants that do not need protection using chemical insecticides and fuel cells for generating electricity. However, it is surprisingly challenging to identify such technologies. Most technologies that stop pollution were usually created to simply reduce costs and save on materials. Technologies designed to prevent pollution usually rely on cost efficiency, rather than pollution prevention, as their main selling point.

One important and fundamental exception is P2 in chemical design. Thousands of chemicals are used in industry, commerce, and daily life. Many of them have environmental impacts, from mild to serious. By developing alternative chemicals with better environmental performance, significant reductions in pollution can be obtained throughout product life cycles. A common application of green chemistry is in the design of environmentally benign solvents. Traditional solvents such as acetone, xylene, and methylene chloride are being replaced by new chemicals designed specifically to be less hazardous or less polluting.

Mass-Market P2 Technologies

Mass-market P2 technologies are those that can be used in many different industries or even in consumer households. These technologies create new markets because their production creates jobs and spin-offs, and they generate ready demand from producers who want to reduce input costs. Each has the following criteria:

1. The technology is widely applicable across a variety of industry types and sizes.

2. The technology does not require very large capital expenditures.

3. The technology's usefulness has been proven through years of implementation experience.

4. The technology has demonstrated free-market feasibility, that is, a positive payback in the productivity of materials, not including reductions in disposal costs.

EU COMMISSION ON THE ENVIRONMENT

Each citizen of the European Union produces an average of 3.5 tonnes (3.85 U.S. tons) of total waste annually. In view of this, on May 27, 2003, the European Union Commission announced a formal communication or policy statement aimed at reducing waste generation and the use of natural resources, and developing a coherent policy on recycling. Current recycling regulations in the EU are inconsistent. For instance, cardboard and paper packaging are recycled but office paper and newsprint are not. Recycling also often costs more than landfilling or incineration. More industry involvement, tradable environmental permits, national landfill bans and taxes, pay-as-you-throw schemes, and producer responsibility initiatives are among the communication's proposals.

5. The technology can be supported in the field by local technicians with basic competence.

6. Parts for repair are locally available at reasonable cost.

Example mass-market technologies for P2 include household water-conservation fixtures, variable-speed motors, programmable heating and air conditioning controls, citrus-based solvent cleaners, plastic films for reducing heat transmission through windows, and many others.

International P2 Technologies

The major differences in P2 technologies among countries lie in the age of the technology and the level of process control. In less developed countries, much of the technology is old and would be considered out of date and uncompetitive in developed countries. Consequently, it usually produces much more pollution per unit of output. Less developed countries also tend to use fewer process controls and instrumentation. Much of the operation is controlled by hand or based on experience, rather than real-time data. Human error thus potentially creates more waste and pollution in such situations. But there are no hard and fast rules for differences in P2 technologies between countries. In Thailand, for example, there has been significant investment in new factories in the electronics and auto parts industries. These plants use the latest technology and management practices and are much less polluting than older plants in the same industries operating nearby. SEE ALSO CATALYTIC CONVERTER; ENERGY EFFICIENCY; GREEN CHEMISTRY; LIFE CYCLE ANALYSIS.

Bibliography

European Environment Agency. (1997). *Comparing Environmental Impact Data on Cleaner Technologies*, Copenhagen: European Environment Agency.

U.S. Environmental Protection Agency. (2001). *Cleaner Technologies Substitutes Assessment*, Washington: U.S. Environmental Protection Agency.

Burt Hamner

Terrorism

Terrorism, as defined by the Federal Bureau of Investigation (FBI), is "the unlawful use of force or violence against persons or property to intimidate or coerce a government, the civilian population, or any segment thereof, in furtherance of political or social objectives."

The destruction inherent in any act of mass terrorism inevitably causes secondary environmental pollution effects, many of them serious. Acts of terrorism can also be directed against the environment itself, or specific natural resources such as freshwater, oil, or agricultural products.

Terrorist Attack on the World Trade Center

The secondary environmental effects of terrorism can often be as significant as its primary effects. The attack on the World Trade Center (WTC) in New York City on September 11, 2001, had negative health consequences beyond the staggering loss of life. The collapse of the structures and subsequent fires spewed an enormous cloud of dust and toxins into the air over the city.

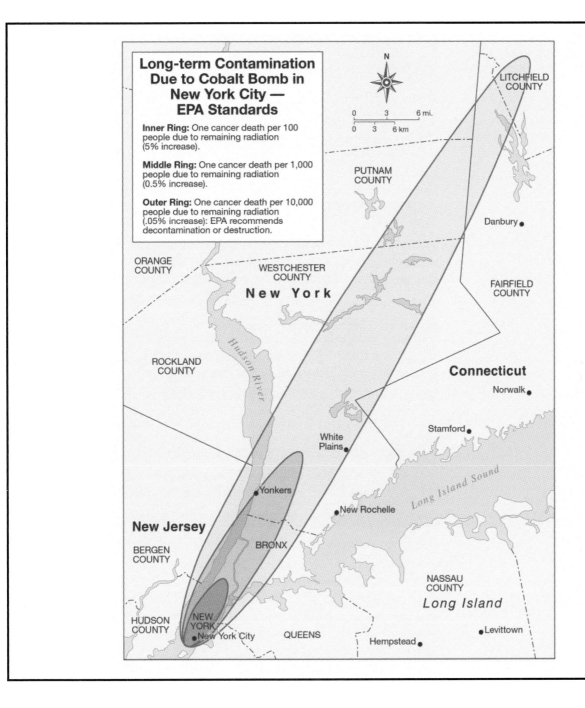

Long-term Contamination Due to Cobalt Bomb in New York City — EPA Standards

Inner Ring: One cancer death per 100 people due to remaining radiation (5% increase).

Middle Ring: One cancer death per 1,000 people due to remaining radiation (0.5% increase).

Outer Ring: One cancer death per 10,000 people due to remaining radiation (.05% increase): EPA recommends decontamination or destruction.

Pulverized concrete, building materials, heavy metals, and human remains were inhaled by residents and rescue workers in lower Manhattan until a heavy rain three days later washed away most of the dust.

The immediate environmental fallout from the WTC collapse contained asbestos and fibrous glass from the building structure; mercury, dioxins, furans, and other cancer-causing toxins from the burning of fluorescent light bulbs and computer screens; heavy metals such as cadmium and lead and volatile organic compounds like benzene. Federal, state, and local agencies went right to work monitoring air quality and cleaning up dust and debris from the WTC collapse, but these actions themselves have serious environmental consequences. One in four cleanup workers at Ground Zero report-

Map showing long-term contamination due to detonation of cobalt bomb in New York City, illustration. (Gale.)

South tower of the World Trade Center collapsing as black smoke billows from the burning north tower following the terrorist attack on September 11, 2001, by members of al-Qaeda. (AP/Wide World Photos. Reproduced by permission.)

edly suffer from asthma and respiratory illness brought about by dust inhaled at the site. Some airborne pollutants and dust were resuspended as a result of ongoing cleanup efforts.

The secondary pollution concerns include possible contamination of waterways around lower Manhattan as well as the challenge of where to dispose of the catastrophe's 1.2 million tons of waste. Fresh Kills landfill on Staten Island has been accepting WTC debris, some containing asbestos and other toxic materials, despite being slated to close December 31, 2001. Since Fresh Kills was not designed to accept hazardous waste, there is concern about whether or not contaminants could leach from the landfill into surrounding groundwater.

COMMON POLLUTANTS FROM SEPTEMBER 11, 2001, ATTACK AND THEIR HEALTH EFFECTS

Contaminant	Health Effects	Source
Asbestos	Carcinogenic. Causes tissue damage in the lungs when inhaled over long periods and can lead to asbestosis, mesothelioma, and lung cancer.	Used as an insulator and fire retardant, applied to steel beams.
Benzene	Flammable and carcinogenic. Short-term effects include dizziness, headaches, and tremors. Long-term exposure can lead to leukemia.	Combustion of plastics.
Biohazards	Exposure to blood and body parts can transmit infectious diseases such as hepatitis and AIDS.	Human remains.
Chromium	Carcinogenic when inhaled at high concentrations; can cause skin ulcers.	Video and computer monitors.
Copper	Can cause dizziness, headaches, vomiting, liver and kidney damage.	Electrical wiring and cables.
Diesel fumes	Asthma trigger. Can aggravate symptoms in asthmatics.	Truck traffic and heavy machinery.
Dioxins	Chloracne is a short-term effect of exposure. Strong evidence for carcinogenic, teratogenic, reproductive, and immuno-suppressive effects.	Combustion of polyvinyl chloride found in electrical cables and other insulating materials.
Freon	Damages the ozone layer. When burned, can produce phosgene, a potent cause of severe and life-threatening pulmonary edema.	Refrigeration and air-conditioning equipment.
Lead	Neurotoxin. Damages the central nervous system, especially in children. Can also cause kidney damage and reproductive damage in adults.	Video and computer monitors, rustproofing paint used on steel beams.
Mercury	Neurotoxin. Damages the peripheral nervous system, especially in children.	Thermometers and other precision instruments.
Particulate matter	Asthma trigger. Can also aggravate cardio-vascular disease.	Pulverized concrete and other materials, smoke, dust and soot.
Polychlorinated biphenyls	Carcinogen. May also cause reproductive and developmental abnormalities.	Electrical equipment.
Sulfur dioxide	Pulmonary toxicant. Can cause severe airway obstruction when inhaled at high concentrations.	Combustion.

SOURCE: Adapted from *Environmental Health Perspectives,* Vol. 109, No. 11, November 2001.

With the passage of time, and through the cleansing effect of rainfall and the specialized cleanup efforts of the U.S. Environmental Protection Agency (EPA), air quality in lower Manhattan has now returned roughly to pre-9/11 levels. However, despite reassurances from the EPA and the Occupational Safety and Health Administration (OSHA), residents of lower Manhattan worry about the long-term health effects of dust and particulates deposited on rooftops and windowsills, and in the ventilation systems of nearby buildings. Only now are the long-term effects of exposure to Ground Zero being studied.

Renewed Efforts to Protect Environmental Infrastructure

After the attacks of September 11, 2001, federal and state authorities began to wonder what else might offer a tempting target for terror attacks. New

York City and other large cities immediately took steps to protect their water systems by guarding the infrastructure and testing the water for known contaminants. In 2002 President George W. Bush's administration passed the Public Health and Bioterrorism Preparedness and Response Act, which required, among other actions, that all water utilities across the country conduct security assessments to gauge possible vulnerability and take steps to protect their water.

Bioterrorism

The environment can also be a conduit for terrorism. Biological elements such as disease-causing bacteria and viruses can become potent weapons when taken out of their natural environment. Shortly after the attack on the WTC, several pieces of mail in and around Florida, Washington, D.C., and New York City tested positive for the biocontaminant anthrax. Anthrax is a bacterium that, in its most potent inhaled form, has a fatality rate of over 90 percent. Over ten thousand people may have been exposed, and five people died of **inhalational** anthrax before the contaminated mail was quarantined. The FBI and the Postal Service have offered a $2.5 million reward for information leading to those responsible, and medical researchers have been working on a cure. Authorities have not yet determined if the anthrax-contaminated mail is connected to al-Qaeda and the events of September 11, but the combined effect of these two attacks occurring in close proximity served to heighten the perception that America is under siege.

inhalation drawing into the lungs by breathing

Nuclear Terrorism

Biocontamination is not the only threat to safety in the United States. One of the most frightening terror scenarios that government officials must consider is the possibility of a nuclear device, or "dirty bomb," being detonated in a U.S. city. Quite separate from the direct human health consequences, the environmental effects of even a low-yield (five kiloton) nuclear weapon are severe: The shock wave will disperse radioactive fallout over a wide area, poisoning wildlife and groundwater. The heat (thermal radiation) will destroy plants and trees. And although the global nuclear winter theory (cooling of the earth's surface due to airborne fallout, thus blocking sunlight) has largely been discredited, this phenomenon can have devastating effects on local agriculture and ecosystems.

Relationship between Resource Competition and Terrorism

The United States is often a target of **asymmetrical warfare**, such as terrorism, because of its military superiority and worldwide economic interests. Many scholars studying peace have reasoned that, in order to defeat terrorism, we must remedy the conditions that give rise to it.

asymmetrical warfare conflict between two forces of greatly different sizes; e.g., terrorists versus superpower

One of the most pressing American national security interests is ensuring continued global access to natural commodities such as oil, minerals, and timber. However, the United States already consumes approximately 30 percent of all raw materials consumed by humans in a given year and is perceived as a nation that seeks more than its fair share of the world's resources. One concern is that as the world population grows and resources are stretched to cover its needs, supplies will fall and prices will rise, making necessary commodities

ENVIRONMENTAL CONSEQUENCES OF THE GULF WAR

Although there is some debate among scholars about the difference between war and terrorism, the retreating Iraqi army committed two particularly wanton acts of environmental destruction during the 1991 Persian Gulf War. First, they released six million barrels of oil from the Kuwaiti Sea Island offshore loading terminal, and scuttled five fully loaded oil tankers at the Mina Ahmadi terminal. They also set fire to 732 oil wells across Kuwait. These burned for months before they were extinguished. The combined oil pollution output from these acts totaled 1.5 billion barrels, or 6,000 times the amount spilled from the *Exxon Valdez*.

The environmental effects of these acts were clearly immense. A plume of soot and oil droplets spread over 1.3 million square miles, contaminating the air with pollutants such as nitrous oxides, sulfur dioxide, polycyclic aromatic hydrocarbons, and vast amounts of CO_2. The oil lakes on land have contaminated the fragile desert ecosystem, virtually guaranteeing that it will not regenerate for decades. The oil in the Gulf itself destroyed mangrove thickets, fish, shrimp, marine mammals, and sea birds. Ten years after the war, this region is still environmentally degraded.

available only to wealthy countries or the upper class within a country. This means that the rich would get richer and the poor poorer, and such inequity of supply and distribution might give rise to unilateral actions on the part of those who feel they are on the losing end of this globalization gap.

To reduce this potential for conflict, developed societies are being encouraged to recognize that global resource consumption and international security are connected, and that obtaining resources cooperatively rather than competitively will enhance long-term security. International agencies can help to ensure the equitable distribution of critical resources both between and within countries. In addition, nations can contribute their relative expertise to finding new sources of natural resources, to developing substitutes for commodities such as oil and natural gas, and to enhancing conservation and efficiency technology to certify that existing resources are used to their maximum benefit. If poorer citizens can be assured they have access to the resources needed to live, they are less likely to adopt combative ideologies that lead to terrorism. SEE ALSO ECOTERRORISM; WAR.

Bibliography

Gugliotta, Guy, and Matsumoto, Gary. (2002). "FBI's Theory on Anthrax Is Doubted." *Washington Post*, October 28, 2002, A1.

Hawley, T.M. (1992). *Against the Fires of Hell: The Environmental Disaster of the Gulf War*. New York: Harcourt Brace Jovanovich.

Klare, Michael T. (2001). *Resource Wars: The New Landscape of Global Conflict*. New York: Henry Holt.

Makhijani, Arjun; Hu, Howard; and Yih, Katherine, eds. (2000). *Nuclear Wastelands: A Global Guide to Nuclear Weapons Production and Its Health and Environmental Effects*. Special Commission of International Physicians for the Prevention of Nuclear War and the Institute for Energy and Environmental Research. Cambridge, MA: MIT Press.

Nordgren, Megan D.; Goldstein, Eric A.; and Izeman, Mark A. (2002). *The Environmental Impacts of the World Trade Center Attacks: A Preliminary Assessment*. New York: Natural Resources Defense Council.

Internet Resources

Federation of American Scientists. (2002). "Special Weapons Primer: Biological Warfare Agents." Available from http://www.fas.org/nuke.

New York State Department of Environmental Protection. (2001). "Statement on Water Supply Security." Available from http://www.nyc.gov/html.

U.S. Environmental Protection Agency Web site. Further information on the environmental and human health effects of 9/11 available from http://www.epa.gov/wtc.

Elizabeth L. Chalecki

Thermal Pollution

The broadest definition of thermal pollution is the degradation of water quality by any process that changes **ambient** water temperature. Thermal pollution is usually associated with increases of water temperatures in a stream, lake, or ocean due to the discharge of heated water from industrial processes, such as the generation of electricity. Increases in ambient water temperature also occur in streams where shading vegetation along the banks is removed or where sediments have made the water more **turbid**. Both of these effects allow more energy from the sun to be absorbed by the water and thereby increase its temperature. There are also situations in which the effects of colder-than-normal water temperatures may be observed. For example, the discharge of cold bottom water from deep-water reservoirs behind large dams has changed the downstream biological communities in systems such as the Colorado River.

Sources

The production of energy from a fuel source can be direct, such as the burning of wood in a fireplace to create heat, or by the conversion of heat energy into mechanical energy by the use of a heat engine. Examples of heat engines include steam engines, **turbines**, and internal combustion engines. Heat engines work on the principal of heating and pressuring a fluid, the performance of mechanical work, and the rejection of unused or waste heat to a **sink**. Heat engines can only convert 30 to 40 percent of the available input energy in the fuel source into mechanical energy, and the highest efficiencies are obtained when the input temperature is as high as possible and the sink temperature is as low as possible. Water is a very efficient and economical sink for heat engines and it is commonly used in electrical generating stations.

The waste heat from electrical generating stations is transferred to cooling water obtained from local water bodies such as a river, lake, or ocean. Large amounts of water are used to keep the sink temperature as low as possible to maintain a high thermal efficiency. The San Onofre Nuclear Generating Station between Los Angeles and San Diego, California, for example, has two main reactors that have a total operating capacity of 2,200 **megawatts** (MW). These reactors circulate a total of 2,400 million gallons per day (MGD) of ocean water at a flow rate of 830,000 gallons per minute for each unit. The cooling water enters the station from two intake structures located 3,000 feet offshore in water 32 feet deep. The water is heated to approximately 19°F above ambient as it flows through the **condensers** and is discharged back into the ocean through a series of **diffuser**-type discharges that have a series of sixty-three exit pipes spread over a distance of 2,450 feet. The discharge water is rapidly mixed with ambient seawater by the diffusers and the average rise in temperature after mixing is less than 2°F.

ambient surrounding or unconfined; air: usually but not always referring to outdoor air

turbid containing suspended particles

turbine machine that uses a moving fluid (liquid or gas) to gas to turn a rotor, creating mechanical energy

sink hole or depression where a compound or material collects; thermodynamics: part of a system used to collect or remove heat

megawatt one million watts

condenser apparatus used to condense vapors

diffuser something that spreads out or dissipates another substance over a wide area

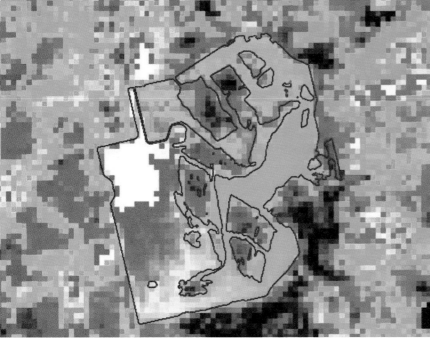

These ASTER false-color images were acquired over Joliet 29, a coal-burning power plant in Illinois. Joliet 29 can be seen in the VNIR image (top) as the bright blue-white pixels just above the large cooling pond. Like many power plants, Joliet 29 uses a cooling pond to discharge heated effluent water. In the bottom image a single ASTER Thermal Infrared band was color-coded to represent heat emitted from the surface. The progression from warmest to coolest is shown with the following colors: white, red, orange, yellow, green, blue, and black. (Image courtesy NASA/GSFC/MITI/ERSDAC/JAROS, and U.S./Japan Aster Science Team. Reproduced by permission.)

Environmental Effects

The primary effects of thermal pollution are direct **thermal shock**, changes in dissolved oxygen, and the redistribution of organisms in the local community. Because water can absorb thermal energy with only small changes in temperature, most aquatic organisms have developed enzyme systems that operate in only narrow ranges of temperature. These **stenothermic** organisms can be killed by sudden temperature changes that are beyond the tolerance limits of their metabolic systems. The cooling water discharges of power plants are designed to minimize heat effects on local fish communities. However, periodic heat treatments used to keep the cooling system clear of

thermal shock rapid temperature change beyond an organism's ability to adapt

stenothermic living or growing within a narrow temperature range

fouling organisms that clog the intake pipes can cause fish mortality. A heat treatment reverses the flow and increases the temperature of the discharge to kill the mussels and other fouling organisms in the intake pipes. Southern California Edison had developed a "fish-chase" procedure in which the water temperature of the heat treatment is increased gradually, instead of rapidly, to drive fish away from the intake pipes before the temperature reaches lethal levels. The fish chase procedure has significantly reduced fish kills related to heat treatments.

Small chronic changes in temperature can also adversely affect the reproductive systems of these organisms and also make them more susceptible to disease. Cold water contains more oxygen than hot water so increases in temperature also decrease the oxygen-carrying capacity of water. In addition, raising the water temperature increases the decomposition rate of organic matter in water, which also depletes dissolved oxygen. These decreases in the oxygen content of the water occur at the same time that the metabolic rates of the aquatic organisms, which are dependent on a sufficient oxygen supply, are rising because of the increasing temperature.

The composition and diversity of communities in the vicinity of cooling water discharges from power plants can be adversely affected by the direct mortality of organisms or movement of organisms away from unfavorable temperature or oxygen environments. A nuclear power-generating station on Nanwan Bay in Taiwan caused bleaching of corals in the vicinity of the discharge channel when the plant first began operation in 1988. Studies of the coral *Acropora grandis* in 1988 showed that the coral was bleached within two days of exposure to temperatures of 91.4°F. In 1990 samples of coral taken from the same area did not start bleaching until six days after exposure to the same temperature. It appears that the **thermotolerance** of these corals was enhanced by the production of heat-shock **proteins** that help to protect many organisms from potentially damaging changes in temperature. The populations of some species can also be enhanced by the presence of cooling water discharges. The only large population of sea turtles in California, for example, is found in the southern portion of San Diego Bay near the discharge of an electrical generating station.

Abatement

The dilution of cooling water discharges can be effectively accomplished by various types of diffuser systems in large bodies of water such as lakes or the ocean. The only thermal effects seen at the San Onofre nuclear generating station are the direct mortality of **planktonic** organisms during the twenty-five-minute transit through the cooling water system. The effectiveness of the dilution systems can be monitored by **thermal infrared imaging** using either satellite or airborne imaging systems. The use of cooling towers has been effective for generating stations located on smaller rivers and streams that do not have the capacity to absorb the waste heat from the cooling water **effluent**. The cooling towers operate by means of a recirculating **cascade** of water inside a tower, with a large column of upwardly rising air that carries the heat to the atmosphere through **evaporative** cooling. Cooling towers have been used extensively at nuclear generating stations in both the United States and France. The disadvantages of cooling towers are the potential for local changes in meteorological conditions due to large amounts of warm air

thermotolerance ability to withstand temperature change

protein complex nitrogenous organic compound of high molecular weight made of amino acids; essential for growth and repair of animal tissue; many, but not all, proteins are enzymes

planktonic that portion of the plankton community comprised of tiny plants; e.g. algae, diatoms

thermal infrared imaging photographs in which contrast depends on differences in temperature

effluent discharge, typically wastewater—treated or untreated—that flows out of a treatment plant, sewer, or industrial outfall; generally refers to wastes discharged into surface waters

cascade waterfall; a system that serves to increase the surface area of the water to speed cooling

evaporative relating to transition from liquid to gas

entering the atmosphere and the visual impact of the large towers. SEE ALSO ELECTRIC POWER; ENERGY; FISH KILLS; VISUAL POLLUTION; WATER POLLUTION

Bibliography

Brown, Richard D.; Ouellette, Robert P.; and Chermisinoff, Paul N. (1983). *Pollution Control at Electric Power Stations: Comparisons for U.S. and Europe.* Boston: Butterworth-Heinemann.

Henry, J. Glenn, and Heinke, Gary W. (1996). *Environmental Science and Engineering.* Upper Saddle River, NJ: Prentice-Hall.

Hinrichs, Roger A., and Kleinbach, Merlin. (2001). *Energy: Its Use and the Environment,* 3rd edition. Monterey, CA: Brooks/Cole Publishing Company.

Langford, Terry E. (1990). *Ecological Effects of Thermal Discharges.* New York: Elsevier Applied Science.

Larminie, James, and Dicks, Andrew. (2000). *Fuel Cell Systems Explained.* New York: John Wiley & Sons.

Liu, Paul Ih-fei. (1997). *Introduction to Energy and the Environment.* New York: John Wiley & Sons.

Ristinen, Robert A., and Kraushaar, Jack J. (1998). *Energy and the Environment.* New York: John Wiley & Sons.

Slovic, Paul. (2000). *The Perception of Risk.* London: Earthscan Publications Ltd.

Other Resources

California Energy Commission. "Energy-Related Environmental Research." Available from http://www.energy.ca.gov/pier/energy/energy_aquatic.html.

Larry Deysher

Thermal pollution from power plants in Florida turned out to be a lifesaver for the state's threatened manatee population. The ecology changed when irrigation wells and diversion channels that support Florida's agricultural development severely impacted the natural springs that moderate river-water temperatures. Manatees cannot survive in cold water and naturalists feared that irregular cold snaps would put the sea mammals at risk. Manatees, however. discovered the power-plant discharge zones and today, naturalists take advantage of cold weather to tally manatee population as the herds gather at local power plants.

Times Beach, Missouri

According to former mayor Marilyn Leistner, the 2,000 residents of Times Beach, Missouri, a community located along the Meramec River, endured a lasting toxic waste episode throughout the Christmas holiday season of 1982. In 1974 the U.S. Centers for Disease Control (CDC) identified dioxin-contaminated waste oil as the cause of death for an unspecified number of dogs and songbirds in Times Beach. In the early 1970s, many municipalities, including Times Beach, commissioned the use of waste oil to control dust on unpaved roads.

On December 3, 1982, in response to local complaints spurred by the CDC's earlier findings, the U.S. Environmental Protection Agency (EPA) conducted soil tests in Times Beach. Floodwaters from the Meramec forced the evacuation of the entire community the very next day. On December 23 the CDC received the EPA's Times Beach soil test results. Because dioxin levels in the soil significantly exceeded public health standards, officials recommended that Times Beach residents not return home.

The Times Beach episode exemplifies how agencies can use their financial and legal resources to address environmental risks to public health. On February 23, 1983, for instance, former EPA Director Anne Burford announced a $25 million plan to buy out the homes and businesses of Times Beach through the Superfund program. Later, a presidential commission on this environmental pollution episode fined Syntex Agribusiness $200 million for their culpability. Syntex Agribusiness produced the dioxin as a waste product in manufacturing pesticides. Russell Bliss, a commercial waste hauler, transported the dioxin from Syntex Agribusiness, mixed the chemical with waste oil, and then, for a fee, sprayed the oil on Times Beach roads. Between 1996 and 1997 Missouri officials, using an environmentally controversial incineration technique, restored the former Times Beach site and turned what was Times Beach into a state park on and dedicated to U.S. Route 66. SEE ALSO DIOXIN; SUPERFUND; U.S. ENVIRONMENTAL PROTECTION AGENCY.

Bibliography

Humphrey, Craig R.; Lewis, Tammy L.; and Buttel, Frederick H. (2002). *Environment, Energy, and Society: A New Synthesis.* Belmont, CA: Wadsworth.

Internet Resources

Leistner, Marilyn. (1985). "The Times Beach Story." In *Proceedings of the 3rd Annual Hazardous Materials Management Conference*, Philadelphia, PA, June 1985. Available at www.greens.org/s-r/078/07-09.html.

U.S. Environmental Protection Agency Web site. "History." Available from http://www.epa.gov/history.

Craig R. Humphrey

Tobacco Smoke

Tobacco smoke has long been recognized as a major cause of mortality and morbidity, responsible for an estimated 434,000 deaths per year in the United States. It is also a source of indoor air pollution due to the release of harmful chemicals, particles, and carcinogens. Exposure to tobacco smoke affects everybody. Children are more vulnerable than any other age group because they are still growing and developing.

Chemical Composition and Health Effects

Tobacco smoke from cigarettes, cigars, and pipes is composed of more than 4,000 different chemicals including carbon monoxide and formaldehyde. More than forty of these compounds are known to cause cancer in humans or animals, and many of them are strong irritants.

The U.S. Environmental Protection Agency (EPA) has concluded that exposure to tobacco smoke in the United States poses a serious and significant public health threat. New long-term studies estimate that about half of all regular cigarette smokers die of smoking-related diseases. However, controversy still surrounds the exact extent of such health effects.

Attempts have been made to study the effect of tobacco smoke on individuals exposed to other toxic chemicals. The risk of developing lung cancer among asbestos workers grows when they smoke an increasing number of cigarettes per day and their cumulative asbestos exposure increases. Cigarette-smoking asbestos workers tend to develop both restrictive lung disease (decreased lung capacities) and chronic obstructive lung disease, as compared to nonsmoking asbestos workers who have a tendency to develop only restrictive lung disease.

In recent years, there has been great concern that nonsmokers may also be at risk for some of the above health effects as a result of their exposure to the tobacco smoke (known as secondhand smoke) that occurs in various environments occupied by smokers.

The tobacco industry has denied the claim of such health hazards and has legally challenged the EPA over its secondhand smoke findings. In addition, some researchers argue that a number of the studies involve flawed data or the selective interpretation of findings. Many of these critics contend that the health risks involved with secondhand smoke are not as extensive as reported.

Regulations on Smoke-free Environment

As of December 31, 1999, smoke-free indoor air laws of one type or another had been enacted in forty-five states and the District of Columbia. Smoking in private work sites is limited in twenty states and the District of Columbia. Forty-one states and the District of Columbia have laws restricting smoking in state government work sites. Thirty-one states have enacted laws that regulate smoking in restaurants, and out of these, only Utah and Vermont completely prohibit smoking in restaurants.

Most European countries have regulations that either ban or restrict smoking to designated areas in public places such as government/private work sites, health care facilities, and educational facilities. Japan and Singapore also have enacted laws that restrict smoking to designated areas, whereas other Asian countries such as India have no regulations in place. South Africa introduced a ban on tobacco smoking in public places, including the workplace, in 1999. SEE ALSO ASBESTOS; ASTHMA; CANCER; HEALTH, HUMAN; INDOOR AIR POLLUTION.

Bibliography

American Conference of Governmental Industrial Hygienists. (1998). *Industrial Ventilation: A Manual of Recommended Practice*, 23rd edition. Cincinnati, OH: Author.

Wadden, Richard A., and Scheff, Peter A. (1983). *Indoor Air Pollution: Characterization, Prediction and Control*. New York: Wiley.

Internet Resources

American Lung Association. "Trends in Tobacco Use." Available from http://www.lungusa.org/data.

National Cancer Institute. "Health Effects Associated with Tobacco Smoke." Available from http://cis.nci.nih.gov/fact.

National Tobacco Information Online System. "Laws and Regulations." Available from http://apps.nccd.cdc.gov/nations.

Ashok Kumar and Sunil Ojha

Todd, John

INNOVATIVE ECOLOGICAL DESIGNER
(1939–)

John Todd is an internationally recognized biologist and pioneer in ecological design. He has been a practical activist in the ecology movement since 1969 when he cofounded the New Alchemy Institute in order to explore science and engineering based on ecological principles. Todd developed earth-based technologies to grow food, generate fuel, transform waste, and purify water.

Todd is best known for his wastewater treatment systems in which floating structures support plants whose roots grow in the wastewater, becoming home to a variety of introduced creatures, including bacteria, fungi, snails,

John Todd. (Courtesy of Ocean Arks International. Reproduced by permission.)

insects, and fish. Underwater baffles direct water through the plant roots, and bubbled air increases oxygen and microbial activity. As the bacteria and other organisms feed off the waste and purify the water, they reproduce to form an efficient ecological living machine. Because they utilize natural processes, these systems require little energy to operate, minimize the use of chemicals, produce only small amounts of sludge, and cost less to install than traditional wastewater treatment plants. More than one hundred such systems are currently operating worldwide.

Todd is president of Ocean Arks International, a nonprofit organization that is dedicated to ecological research, education, and development. A professor at the University of Vermont, he has authored over two hundred articles on biology and earth stewardship. He is involved in developing a zero-emissions community food center in Burlington, Vermont, where wastes from food production are recycled as resources. Spent grain, for example, can be used to grow mushrooms. He has received a number of awards for his innovative wastewater treatment system, including the U.S. Environmental Protection Agency's Environmental Achievement Award in 1996. SEE ALSO WASTEWATER TREATMENT.

Bibliography

Todd, Nancy Jack, and Todd, John. (1994). *From Eco-Cities to Living Machines: Principles of Ecological Design.* Berkeley, CA: North Atlantic Books.

Internet Resource

Ocean Arks Web site. Available from http://www.oceanarks.org.

Patricia Hemminger

Toxic Release Inventory

In 1986 the U.S. Congress passed a federal law called the Emergency Planning and Community Right-to-Know Act (EPCRA), which gives the public the right to know about industrial toxic chemicals that are released into the environment. At present this law, which is also known as Title III of the Superfund Amendment and Reauthorization Act, requires businesses in certain industries that manufacture, process, or otherwise use any chemical from a list of 651 designated chemicals or chemical groups in amounts greater than a certain threshold to report annually to the U.S. Environmental Protection Agency (EPA) on their releases of these chemicals. The EPA maintains this information in a database called the Toxics Release Inventory (TRI), which is available to the public over the Internet.

TRI Reporting Requirements

A plant, factory, or other facility must report chemical releases if it has ten or more full-time employees and manufactures, processes, or imports any of the listed chemicals in amounts greater than 25,000 pounds per year—or 10,000 pounds per year if any of the listed chemicals are otherwise used but not incorporated into a final product. The TRI classifies the chemicals according to their chemical and physical characteristics and contains information on release location. The TRI reports amounts that are released each year to the air, water, and land, as well as information on chemicals sent to waste-management facilities. Air emissions are separated into passive emissions

TOP TEN TRI CHEMICAL ON-SITE AND OFF-SITE REPORTED RELEASES (IN POUNDS) FOR FACILITIES IN ALL INDUSTRIES, UNITED STATES, 2000

Chemical	Air	Water	Underground Injection	Land	Total On- and Off-Site
Copper compounds	1,656,106	426,419	1,737,251	1,346,061,845	1,367,338,006
Zinc compounds	7,513,386	1,276,151	22,580,44	828,086,567	1,037,602,367
Hydrochloric acid	645,632,582	96,763	54,125	15,549	647,112,538
Manganese compounds	2,214,810	5,696,403	10,829,146	4,048,797,705	479,942,409
Arsenic compounds	240,956	166,482	1,809,735	469,413,711	476,640,941
Lead compounds	1,225,794	80,510	8,512,731	328,875,879	357,844,917
Nitrate compounds	336,731	232,960,319	57,203,694	13,041,063	317,119,741
Barium compounds	2,850,794	1,749,324	2,099,443	243,702,122	299,780,394
Methanol	183,176,226	3,753,931	18,353,232	1,828,212	208,566,348
Ammonia	139,047,851	7,560,654	27,335,270	5,772,773	184,124,675

SOURCE: U.S. Environmental Protection Agency

TRI TOTAL RELEASES BY INDUSTRY, 1998–2000. (DOES NOT INCLUDE PBT CHEMICALS.)

Industry	Total On- and Off-Site Releases, 2000, in Pounds	Change 1998–2000, in Pounds and Percentage	
Manufacturing Industries	2,267,118,555	–154,218,664;	–6.4
Metal Mining	3,310,956,485	–252,183,558;	–7.1
Coal Mining	15,327,860	19,334,956;	14.4
Electric Utilities	1,120,615,348	–9,834,598;	–0.9
Chemical Wholesale Distributors	1,611,790	91,350;	6.0
Petroleum Terminals/Bulk Storage	3,725,152	–786,620;	–17.4
Hazardous Waste/Solvent Recovery	7,001,138,027	–409,262,569;	–5.5

SOURCE: U.S. Environmental Protection Agency

from storage or production and "stack" or point emissions. Releases to water include the name of the receiving water body.

Businesses required to report to TRI have expanded from the original manufacturing facilities and now include manufacturing, metal mining, coal mining, electric utilities that combust coal and/or oil, chemical wholesale distributors, petroleum terminals, bulk-storage facilities, hazardous-waste treatment and disposal facilities, solvent-recovery services, and federal facilities.

PBT Emissions

Persistent bioaccumulative toxic (PBT) chemicals are a class of compounds that persist and bioaccumulate in the environment. They have the potential to result in greater exposure to humans and the environment over a longer period of time, making even smaller quantities of these chemicals of concern. In 2000 the TRI was expanded to include new PBT chemicals, and the reporting threshold was lowered for both the newly added chemicals and certain PBT chemicals already on the TRI list. The reporting criteria for most PBT chemicals was lowered to a threshold of one hundred pounds if manufactured, used, or processed. A threshold of ten pounds was established for another subset of PBT chemicals that are highly persistent and highly bioaccumulative, including mercury compounds, pesticides such as chlordane,

COMPANIES WITH LARGEST EMISSIONS; TOTAL SURFACE WATER DISCHARGES

Companies with Largest Discharges	Surface Water Discharges, in Pounds, 2000	Principal Chemical Releases
AK Steel Corp., PA	28,048,653	Nitrate Compounds
BASF Corp., TX	21,515,040	Nitrate Compounds
AK Steel, IN	12,211,850	Nitrate Compounds
Vicksburg Chemical Co., MS	7,966,805	Nitrate Compounds
IBP Inc., NE	6,700,250	Nitrate Compounds
Smithfield Packing Co., NC	5,129,795	Nitrate Compounds

SOURCE: U.S. Environmental Protection Agency

COMPANIES WITH LARGEST EMISSIONS; TOTAL AIR EMISSIONS

Companies with Largest Emissions	Total Air Emissions, in Pounds, 2000	Principal Chemical Releases
Magnesium Corp. of America, UT	43,932,001	Chlorine, Hydrochloric Acid
CP&L Roxboro Steam Electric Plant, NC	19,247,325	Hydrochloric Acid
Reliant Energies Inc., Keystone Power Plant, PA	18,460,972	Hydrochloric Acid, Sulfuric Acid
Bowen Steam Electric Plant, GA	17,807,778	Sulfuric Acid, Hydrogen Fluoride
Lenzing Fibers Corp., TN	17,345,982	Carbon Disulfide
Gulf Power Co. Crist Plant, FL	16,621,882	Hydrochloric Acid, Sulfuric Acid

SOURCE: U.S. Environmental Protection Agency

COMPANIES WITH LARGEST EMISSIONS; TOTAL LAND RELEASES

Companies with Largest Releases	Total Releases to Land, in Pounds, 2000	Principal Chemical Releases
Kennecott Utah Copper Mine, UT	813,758,255	Copper, Zinc, Antimony
Red Dog OPS Mine Facility, AK	445,322,528	Cadmium, Lead
Barrick Goldstrike Mines Inc., NV	346,539,178	Arsenic, Manganese, Zinc
Newmont Mining Corp., Twin Creeks Mine, NV	219,922,901	Arsenic, Antimony
ASARCO Inc. Ray Complex Mine, AZ	155,098,189	Copper
Newmont Mining Corp., Carlin, NV	154,157,564	Arsenic, Zinc, Antimony

SOURCE: U.S. Environmental Protection Agency

heptachlor, methoxychlor, and toxaphene, polychlorinated biphenyls (PCBs), and polycyclic aromatic compounds (PACs). Since dioxins are highly persistent but are produced in extremely small amounts, the threshold for dioxin and dioxin-like compounds was set at 0.1 grams, with the provision that reporting include dioxin and dioxin-like compounds that are present as contaminants in a chemical or that are created during the manufacture of another chemical.

Reporting Trends

From 1998 to 2000, total TRI releases by all industries fell by 409.3 million pounds, or more than 5 percent. The largest decrease from 1999 to 2000 occurred in the metal mining industry. SEE ALSO COMPREHENSIVE

ENVIRONMENTAL RESPONSE, COMPENSATION, AND LIABILITY ACT (CERCLA); HAZARDOUS WASTE; RESOURCE CONSERVATION AND RECOVERY ACT.

Internet Resource

U.S. Environmental Protection Agency. "Toxics Release Inventory Program." Available from http://www.epa.gov/tri.

Joan Rothlein

Toxic Substances Control Act (TSCA)

The Toxic Substances Control Act (TSCA), enacted by Congress in 1976, gives the U.S. Environmental Protection Agency (EPA) the responsibility for checking the relative safety of all chemical substances not already covered under other federal laws. The EPA can control or ban a chemical if it poses an unreasonable risk to human or environmental health. Manufacturers must give the EPA information about new chemicals before they are commercially produced or marketed. The EPA then reviews the information and can order further testing to determine, for instance, whether the substance is persistent, carcinogenic, or otherwise acutely toxic. The acute toxicity or short term poisoning effects of chemicals can be evaluated by the LD50 test that determines the lethal dose required to kill fifty percent of test animals, usually rats or mice. Microbial biotechnology products for use in industry have been subject to EPA review under TSCA since 1997. Over 70,000 chemicals were in use in the United States in 2002 according to the TSCA Chemical Substances Inventory.

Pesticides, and substances used in cosmetics, food, and drugs are regulated under other federal laws, but many chemicals, including polychlorinated biphenyls (PCBs), were not subject to review or regulation until TSCA was passed. Studies showing PCBs to be dangerous to human health were an impetus for TSCA. In 1977 the EPA outlawed PCBs and subsequently regulated their disposal with strict safety requirements.

Amendments to TSCA in 1986, 1988, and 1992 were aimed at reducing the health threats from asbestos, radon, and lead exposure. The amendments required the EPA to test schools and federal buildings for radon contamination and establish state programs for monitoring and reducing lead exposure levels. The Asbestos Hazard Emergency Response Amendment (AHERA) imposed stricter standards on the reduction of asbestos contamination in schools.

Any person or company not complying with TSCA can be fined or jailed. Many landlords have been fined and required to remove lead-based paint as a result of TSCA's enforcement. In 2002 two landlords were also sentenced to prison terms for noncompliance.

In Europe regulations for assessing the safety of new chemical substances were established in 1981 and for all existing chemicals in 1993. In 2001 the European Commission proposed a new policy called "Strategy for a Future Chemicals Policy" aimed at determining the environmental risk posed by thousands of chemicals that came on the European market before 1981. SEE ALSO Asbestos; Lead; Radon.

Internet Resources

EPA's New Chemical Program Web site. Available from http://www.epa.gov/opptintr.

European Commission Environment Web site. Available from http://www.europa.eu.int/comm.

Patricia Hemminger

Toxicology

Toxicology is the science of poisons, which are sometimes referred to as toxins or toxicants. The former term applies to all natural poisons produced by organisms, such as the botulinum toxin produced by the bacteria *Clostridium botulinum*. The latter more generic term includes both natural and anthropogenic (human-made) toxicants like dichlorodiphenyl trichloroethane (DDT), which is perhaps the most commonly recognized toxicant.

Even though the botulinum toxin is extremely toxic to humans, and DDT is relatively toxic to insects, it is important to recognize that virtually any element or compound will become toxic at some concentration. For example, iron, which is an essential component of **hemoglobin**, can cause vomiting, liver damage, and even death if it is **ingested** in excess. This concept of toxicity was recognized five centuries ago by the Swiss alchemist and physician Paracelsus (1493–1541), who stated that, "The right dose differentiates a poison from a remedy." How much of the toxicant an organism receives depends on both the exposure and dose. Exposure is a measure of the amount of a toxicant that comes into contact with the organism through air, water, soil, and/or food. Dose is a measure of the amount of toxicant that comes into contact with the target organ or tissue, within the organism, where it exerts a toxic effect. The dose is largely determined by how effectively the toxicant is absorbed, distributed, **metabolized**, and eliminated by the body.

As a consequence, basic toxicological studies include measurements of the effects of increasing doses of a toxicant on an organism or some component of that organism (e.g., tissue, cell, subcellular structure, or compound). The measurements are commonly plotted as dose–response curves. A dose–response curve typically ranges from relatively low concentrations that do not elicit a toxic effect to higher concentrations that are increasingly toxic.

One of the great challenges to the science of toxicology is the prediction and discovery of chronic, sublethal responses. For example, in the 1920s, excessive exposure of workers to tetraethyl lead (the lead in leaded gasoline) in several United States gasoline production facilities caused approximately fifteen deaths, and over three hundred cases of psychosis. Despite this discovery of the apparent hazard of lead in gasoline, and the concerns of many at the time, rigorous scientific studies were required to demonstrate the subtle, sublethal dangers of chronic lead exposure, including adverse neurological effects in children, which eventually led to the ban of lead additives in gasoline in the United States.

Characterizing Toxicity

One measure of response is acute toxicity, which is the amount of a toxicant that will cause an adverse effect within a relatively short period of time (e.g., from instantaneous to within a few days). Another measure of response is chronic toxicity, which is the long-term response to a toxicant. Although the same types of dose–response curves are used to measure the chronic toxicity

hemoglobin oxygen-carrying protein complex in red blood cells

ingest take in through the mouth

metabolize chemically transform within an organism

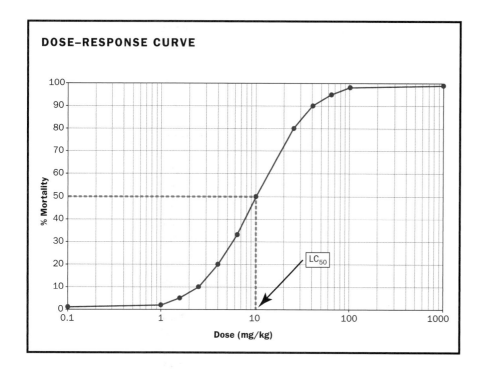

DOSE–RESPONSE CURVE

of toxicants, those measurements are more difficult to quantify because the responses are often less absolute and more complex. For example, chronic benzene toxicity causes lung cancer, but it may be years before that benzene-induced cancer appears, and many other factors may retard the development of that cancer (**antagonistic** effect), contribute to its development (**synergistic** effect), or independently cause lung cancer (e.g., smoking cigarettes).

Forms of toxicity can also be characterized by the type of adverse response they create. Carcinogens cause cancer, either by the initiation or promotion of an uncontrolled growth of cells. Mutagens cause mutations by altering the DNA sequences of chromosomes. Teratogens cause mutations in the DNA structure of developing fetuses that can result in developmental abnormalities. The latter form of toxicity includes the infamous teratogen thalidomide, which was prescribed as a **sedative** for pregnant women before it was found to cause severe birth defects in their children.

Differences in Sensitivities

Resolving the adverse effects of a toxicant are further complicated by the variations in those effects in different species. Some species are more sensitive to certain toxicants than others, and the effects of toxicants on different tissues often vary between species. Because such variations occur between humans and rodents, in spite of the similarity (95%) in their DNA, extrapolations of laboratory studies on the effects of toxicants on rats and mice to human health must always take this into account. Moreover, the toxic effects of a pollutant on the gall bladder of humans cannot be determined in studies involving rats because rats do not have gall bladders.

There are also relatively large differences in the sensitivities and effects of toxicants between individuals of the same species. Fetuses, **neonates**, and infants are more sensitive to the **neurotoxic** effects of lead than older individuals, because lead interferes with the development of the central nervous

antagonistic working against

synergistic combination of effects greater than the sum of the parts

sedative substance that reduces consciousness or anxiety

neonate newborn

neurotoxic harmful to nerve cells

system, which is formed during the first few years of life. Finally, healthy individuals are generally less sensitive to pollutants than individuals with weakened immune systems who are less capable of responding to additional threats to their health.

Genetics also plays a major role in the sensitivities of individuals. Although some differences have been observed in humans, the most commonly recognized genetic differences in toxic responses have been observed in other species. These include the acquired genetic resistance of some mosquitoes to DDT and some bacteria to antibiotics. However, the development of molecular techniques to genotype humans has now made it possible to identify individual sensitivities to different toxicants.

Risk Assessment

Another important aspect of toxicology is risk assessment, which is a characterization of the potential adverse effects resulting from exposure to a toxicant. Risk is the probability of an adverse outcome. The basic steps involved in risk assessment are the identification of the magnitude of the hazard, which is the potential for harm of a toxicant, and the resultant characterization of risk, which is the probability of realizing that harm. The results of risk assessments are routinely used by regulators to establish acceptable concentrations of toxicants in the environment.

Environmental Toxicology

Environmental toxicology is a relatively recent field that examines the occurrence of, exposure to, and form of toxicants in the environment, and the comparative effects of these toxicants on different organisms. DDT, for example, is a pesticide that has been used to control mosquitoes responsible for spreading malaria. Although this pesticide is effective in combating the spread of malaria, DDT and its chemical products have also been found to affect reproduction in birds by causing egg shell thinning, and in other organisms (e.g., alligators) by altering their estrogen balance. Consequently, studies of toxicology now extend well beyond dose–response assays of toxicants on specific target organisms to analyses of their impact on entire ecosystems.

In addition to anthropogenic toxicants like pesticides, environmental toxicologists also study naturally occurring toxicants, such as metals and metalloids. Selenium, for example, is a naturally occurring element that is essential at low concentrations in the diet of many animals. Excessive intake of selenium, however, can be toxic to organisms. In the 1980s scientists working at Kesterson Slough in the San Joaquin Valley, California, observed a large number of deformed and dying waterfowl. The slough was part of a water project designed to receive and evaporate excess irrigation water and remove pesticides from the highly productive agriculture regions in the San Joaquin Valley. The observed effects on the waterfowl were eventually linked to an excess of selenium in the water. The selenium accumulated in the slough because the soils and runoff from the valley were naturally rich in selenium, and because evaporation in the slough further increased its concentration in the water. In this example, it was discovered that a rare, but naturally occurring and essential element was unwittingly concentrated to toxic levels in the environment by human activity. SEE ALSO CANCER; DDT (DICHLORODIPHENYL TRICHLOROETHANE); HAZARDOUS WASTE; HEALTH, HUMAN; LEAD; RISK.

Bibliography

Crosby, Donald G. (1998). *Environmental Toxicology and Chemistry.* New York: Oxford University Press.

Needleman, H.L. (1998). "Clair Patterson and Robert Kehoe: Two Views of Lead Toxicity." *Environmental Research* 78(2):79–85.

Ohlendorf, H.M.; Hoffman, D.J.; Daiki, M.K.; and Aldrich, T.W. (1986). "Embryonic Mortality and Abnormalities of Aquatic Birds—Apprent Impacts of Selenium from Irrigation Drainwater." *Science of the Total Environment* 52(44).

Williams, P.L.; James, R.C.; and Roberts, S.M., eds. (2000). *Principles of Toxicology: Environmental and Industrial Applications,* 2nd edition. New York: John Wiley & Sons.

Internet Resource

Society of Toxicology. Available from http://www.toxicology.org.

A. Russell Flegal and Christopher H. Conaway

Tragedy of the Commons

The term *tragedy of the commons* was coined by Garrett Hardin who hypothesized in 1968 that, as the size of the human population increased, there would be mounting pressures on resources at the local and global levels, leading to overexploitation and ruin. Partly the tragedy would occur because some "commoners" (or users of common resources) would reap the full benefit of a particular course of action while incurring only a small cost, while others would have to share the cost but receive none of the benefits. The classic examples of such overexploitation are grazing, fishing, and logging, where grasslands, fish stocks, and trees have declined from overuse. Hardin suggested that governmental intervention and laws could become the major method of solving such overexploitation. More recently, the concept of the commons has been expanded to include air, water, the Internet, and medical care.

Much controversy has developed over whether commoners are caught in an inevitable cycle of overexploitation and destruction of resources, or whether the wise use and management of natural resources are possible. Although many examples of overexploitation exist, particularly in fisheries, Elinor Ostrom, Bonnie McCay, Joanna Burger, and others have argued that there are also examples of local groups effectively managing commonly held resources, and that such local control requires accepted rules, with appropriate sanctions and some governmental control to prevent exploitation by outside interests. That is, a fishing cooperative can succeed only if outside fishermen agree to adhere to existing rules or laws. In an age with increasing populations, understanding how different societies and groups have managed a common pool of resources allows us to apply successful methods in managing these resources. SEE ALSO EHRLICH, PAUL; LIMITS TO GROWTH, THE; MALTHUS, THOMAS ROBERT.

Bibliography

Burger, Joanna, and Gochfeld, Michael. (1998). "The Tragedy of the Commons—30 Years Later." *Environment* 41:4–13, 26–28.

Ostrom, Elinor; Burger, Joanna; Field, Christopher B.; Norgaard, Richard B.; and Policansky, David. (1999). "Revisiting the Commons: Local Lessons, Global Challenges." *Science* 284:278–282.

Internet Resource

Hardin, Garret. (1968). "The Tragedy of the Commons." *Science* 162:12–13. Also available from http://dieoff.org/page95.htm.

Joanna Burger

Treaties and Conferences

Treaties, conventions, protocols, and conferences are tools for creating and shaping international law, and for establishing sanctions in the event of noncompliance.

A treaty is a compact, or contract, made between or among sovereign nations, involving matters of each country's public interest. It has the force of law within each signing nation. Treaties are the formal conclusion of the negotiating process rather than an intermediate step. Ideally, they include both the formal commitment of nations and mechanisms for enforcement, although many international environmental treaties fall short on the adequacy of enforcement mechanisms.

A convention is also an international agreement, although it often has a narrower scope and is less politically motivated than a treaty. In addition, a convention may consist of agreed-upon arrangements that precede a formal treaty or that serve as the basis for an anticipated treaty.

A protocol is an agreed-upon document or instrument that provides the template for subsequent diplomatic transactions, serving, in a manner of speaking, as a first draft that is subject to further refinement.

Conferences are diplomatic meetings conducted in order to agree upon policy statements in lieu of formal, and more time-consuming, international negotiations. In addition to such bilateral or even multilateral agreements between nations, international organizations may create mechanisms for examining and resolving international disputes and other issues. Most notably, the United Nations, through its Environmental Programme, and joined by the World Meteorological Organization, was instrumental in establishing the Intergovernmental Panel on Climate Change (IPCC) in 1988. The IPCC created working groups and special committees that assessed the scientific information related to various components of climate change, including, specifically, data regarding the emissions of major greenhouse gases, analyzed that information in environmental and socioeconomic contexts, and then formulated realistic response strategies for the management of climate change. The IPCC's analysis and recommendations thus became the template for subsequent attempts to draft international agreements. The Climate Change Convention, discussed below, was one such result.

These various tools and arrangements are unique to international law, which imposes constraints not typically present in national, or "domestic," law. International law has traditionally differed from the domestic law of nation-states in that it is fundamentally voluntary, notwithstanding the fact that political or military pressures may have prompted the parties to negotiate or enter into any compacts in the first place. Hence, ultimate enforcement, short of political or military responses, can be problematic.

The contrast with a nation's own regulatory law is instructive. When a country enacts and enforces laws that have an effect within its boundaries or

with respect to its citizens or other residents, it does so by virtue of its internationally recognized sovereign power to coercively regulate within its own boundaries. However, with the possible exception of an evolving European Union, no supranational authority exists to regulate conduct among countries. Thus, treaties and other international agreements historically evolved as a means of controlling the behavior of nations and, by less direct means, their businesses and citizens. These compacts then operate between nations at a level that traditionally was often beyond the realm of domestic regulation. As such, they have traditionally been more akin to contracts, into which countries voluntarily enter, rather than manifest regulatory authority. However, in recent decades, particularly on matters where international environmental law and trade law intersect, there has been an interesting convergence. The distinctions between these contractual remedies that characterize treaties and the enforcement remedies that characterize regulations in which a tribunal issues supposedly binding decisions on disputes, have been merging in interesting ways. For instance, the World Trade Organization (WTO), of which the U.S. is a member, and whose rule the U.S. has agreed to submit, ruled against efforts by the United States to protect dolphins and sea turtles. The WTO decision had included trade sanctions, and found that such unilateral American efforts violated the WTO's free trade rules. Hence, the U.S. was faced with a choice between submitting to this suprasanctional "regulation," supposedly binding, or pursuing conservation-oriented treaty-making that was inconsistent with some of the tenets of free trade.

Treaty making extends to all aspects of international dealings. It is most often associated with a declaration of war or the end of an armed conflict, or the allocation of resources between or among countries, with those two broad areas historically related. Interestingly, international environmental treaties and law have usually addressed resource conservation rather than resource

Hiroshi Oki, Japanese Environment Minister, addressing the final plenary session of the United Nations Global Warming Conference, Kyoto, Japan, December 1–10, 1997. (Photograph by Aizawa Toshiyuki. Hulton/Archive. Reproduced by permission.)

exploitation, although, naturally, various parties may seek to maximize their use of resources within the scope of international environmental treaties. In terms of transnational pollution, international sensitivities have evolved that recognize the drastic regional and potentially global impact of uncontrolled or poorly regulated disposal practices. Thus, it is also interesting that although treaties traditionally acknowledged the political and jurisdictional sovereignty of nation-states, present international environmental treaty making effectively underscores the geographic and climate-related commonality of nation-states. Therefore, in an era when globalism is much discussed as an emerging economic paradigm, environmental factors highlight the global consequences of many environmental events and policies formerly viewed through only local or domestic legal prisms. Perhaps as a consequence, treaties and conventions are assuming significantly wider reach, and are entered into, or participated in, by many more numerous parties than was historically typical in international law.

Despite concerns about the ultimate effectiveness of international environmental law developed by means of treaties and other agreements, recent history provides the basis for some optimism. As of 2002 approximately 140 multinational agreements on numerous international environmental issues (or including environmental provisions) have been reached. Preliminary studies indicate that notwithstanding weak or even nonexistent enforcement mechanisms, the general trend has been one of compliance. However, it has also been pointed out that nations often enter such agreements when it appears the price of their cooperation will be low, with the result that many such agreements are successfully negotiated by only minimally addressing the particular environmental problems at hand.

Modern treaty making has also demonstrated a greater tendency to accommodate equitable goals, although the ability to effectively resolve disputes on the basis of common principles has not been an easy matter. The best treaties, as with contracts, are pragmatic. And, of course, all is not equitable. The usual methods used to influence other parties still prevail.

The influence of the United States is a case in point on how national attributes may affect the outcome of international environmental agreements. It is hard to generalize about the direction of U.S. policy. The economic clout of the United States, especially the manner in which American business interests mesh with other national economies, and the sheer dimension of U.S. consumption of the world's resources, continue to influence the outcome of international environmental negotiations. Recent official U.S. policy has demonstrated a reluctance to enter into agreements that would require a significant reduction in resource or energy use by or within the United States. On the other hand, U.S. environmental activists and national policy have often taken the lead in highlighting significant international environmental issues. Nongovernmental organizations, such as Environmental Defense, the Natural Resources Defense Council (NRDC), and the Sierra Club, all headquartered in the United States, not only assert a global presence but often shape both international and U.S. domestic policies. The manner in which the United States formulates its own environmental policies has especially complicated its international role on environmental matters. With power divided between a powerful presidency and a strong but often fragmented legislative branch, with various leaders elected in different regions from different political parties at different times, a balance on domestic issues may be reached, but often at the expense of a coherent and consistent foreign policy. This dynamic, resulting in a diffuse and often contradictory U.S. voice on international environmental issues, still requires some development in the relatively young field of international environmental law.

International agreements in recent years have paralleled the recognition of international environmental problems, even if they have not always effectively mitigated those issues. Early examples include the Convention on International Trade in Endangered Species of Wild Fauna and Flora (CITES), which enumerates specific compliance mechanisms and has generally been successful in banning the international trade of endangered species or products derived from such species; the Intergovernmental Conference on the Dumping of Wastes at Sea, also known as the London Dumping Convention, which banned the maritime dumping of radioactive wastes and spurred enactment of the U.S. Ocean Dumping Ban Act; the Convention on Wetlands of International Importance Especially as Waterfowl Habitat, also known as the Ramsar Convention, which addressed the loss of migratory waterfowl habitat.

More recent agreements include the Montréal Protocol on Substances That Deplete the Ozone Layer, known generally as the Montréal Protocol, which entails formal compliance mechanisms and has enjoyed significant success in reducing the use of CFCs (chloroflurocarbons) despite the absence of actual sanctions; the 1994 protocol to the 1979 Convention on Long-Range Transboundary Air Pollution on Further Reduction of Sulphur Emissions, known generally as the Sulfur Protocol, largely directed against acid rain, for which compliance provisions have been negotiated; and the previously mentioned Intergovernmental Panel on Climate Change, which matured into the United Nations Framework Convention on Climate Change, effective in 1994, now known as the Climate Change Convention.

Other regional and global environmental crises have been addressed under the auspices of the United Nations Commission on Environment and Development, including the Convention on Biological Diversity; the Convention to Combat Desertification in Those Countries Experiencing Serious Drought and/or Desertification; and the Convention on the Prior Informed Consent Procedure for Certain Hazardous Chemicals and Pesticides in International Trade, also known as the Rotterdam Convention. This latter convention sought to transfer the responsibility for regulating hazardous materials trade to exporting nations rather than leaving regulatory control only to importing nations that, for various reasons, might be ineffective regulators.

Most recently, the Kyoto Protocol has been the subject of much negotiation and perhaps even greater controversy. An outgrowth of the Climate Change Convention, it seeks the global reduction of greenhouse gases below 1990 levels by targeted dates and provides for an international emissions trading program. The actual reductions proposed are not evenly distributed, and equitable factors have been cited as the justification for the reductions required of, and exemptions afforded to, particular nations. Although the need for an effective mechanism to reduce hydrocarbon emissions as a means of addressing incipient global warming has been widely acknowledged, the best mechanisms for achieving this goal have been sharply debated. The Kyoto Protocol is not yet in force, and the United States is one of several large industrialized nations that have not signed it. SEE ALSO AGENDA 21; EARTH SUMMIT;

257

LAWS AND REGULATIONS, INTERNATIONAL; MONTRÉAL PROTOCOL; NAFTA (NORTH AMERICAN FREE TRADE AGREEMENT); PRECAUTIONARY PRINCIPLE.

Kevin Anthony Reilly

Trichloroethylene   *See Dry Cleaning*

TSCA   *See Toxic Substances Control Act*

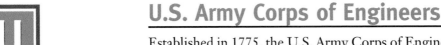

U.S. Army Corps of Engineers

Established in 1775, the U.S. Army Corps of Engineers (otherwise known as the corps) is the world's largest public, engineering, design, and construction management agency. The corps obtains its authority from the secretary of the army and is a division serving the chief of engineers within the Department of the Army. Funded by Congress, the corps' primary responsibilities include the management and execution of civil works programs in or adjacent to the nation's waterways (e.g., rivers, harbors, and wetlands), administration of environmental laws to protect and preserve these waterways, and the review of applications and issuance of permits for proposed projects affecting such bodies of water. As part of its responsibility, the corps assesses the consequences of proposed activities on water bodies, balancing environmental and developmental need and concerns. This often brings environmental and business groups into conflict such as in the case of dredging. Environmental groups oppose dredging due to its adverse effects on aquatic species whereas industry asserts that such dredging reduces the costs of river transportation by allowing larger ships to pass through waterways with fuller cargo loads. The corps reviews and issues permits under the Clean Water Act or Rivers and Harbor Act, ensuring that proposed activities do not adversely affect or impede U.S. waterways. Under the Clean Water Act, the corps primarily issues permits for the discharge of excavated material or fill, whereas under the Rivers and Harbor Act, the agency issues permits for the construction of structures such as bridges, dams, dikes, or causeways. With respect to both laws, the corps considers reasonable and alternative locations and methods for a proposed project, potential effects on private and public uses, and the need for a specified project. During the past several years, however, senators have introduced legislation such as the Corps of Engineers Modernization and Improvement Act of 2002, in an effort to reform the corps' project review and authorization procedures. These procedures have been criticized for allowing a number of projects to go forward that have had few economic benefits and high environmental costs. Agencies similar in purpose to the corps exist in countries such as Australia, Britain, and Canada, but they function on a much smaller scale in comparison.

Bibliography

National Research Council, Committee to Assess the U.S. Army Corps of Water Resources Planning Procedures. (1999). *New Directions in Water Resources Planning for the U.S. Army Corps of Engineers.* Washington, D.C.: National Academy Press.

Internet Resource

Services for the Public. Available at http://www.usace.army.mil/public.html #environmental.

Robert F. Gruenig

U.S. Coast Guard

Established in 1790 as the Revenue Marine Service but named as such after combination with the U.S. Lifesaving Service, the U.S. Coast Guard provides support for the protection and preservation of the United States' marine and natural resources. Although a branch of the armed forces, the agency operates under the jurisdiction of the Department of Transportation during times of peace. The agency is responsible for managing the nation's seas and coastal waters, with environmental issues primarily handled by two offices: Marine Safety, Security, and Environmental Protection, and Law Enforcement and Defense Operations. Two primary agency functions are the enforcement of environmental laws (e.g., Clean Water Act, Marine Protection, Research, and Sanctuaries Act, and Oil Pollution Act) and the provision of an emergency response system to mitigate the release of pollution (e.g., garbage discharges, hazardous substance releases, and oil spills) into seas and coastal waters. With respect to enforcement, the agency enforces U.S. environmental laws along with all treaties and international agreements that allow the Coast Guard to assess penalties for violations under the law. With respect to its emergency response system, the agency is proactive by serving as a lead agency under the National Oil and Hazardous Pollution Plan by coordinating federal, state, local, and responsible party resources in conducting spill response efforts in the containment, removal and disposal of oil, and hazardous substance discharges in the country's coastal zone areas. The agency also assesses movements of potential pollutants (e.g., discharges and spills), accounts for wind and ocean currents, and evaluates potential chemistry changes due to those caused by evaporation, mixing, and sunlight. SEE ALSO CLEAN WATER ACT; MARINE PROTECTION, RESEARCH, AND SANCTUARIES ACT; PETROLEUM.

Bibliography

Goldsteen, Joel B. (1999). *The ABCs of Environmental Regulation.* Rockville, MD: Government Institutes.

Internet Resource

Marine Safety, Security and Environmental Protection. Available from http://www.uscg.mil/hq/g-m/gmhome.htm.

Robert F. Gruenig

U.S. Department of Agriculture

Established in 1862, the U.S. Department of Agriculture (DOA) works with landowners to maintain the productive capacity of their land while helping them to protect soil, water, forests and other natural resources. The department conducts a large part of this work through two of its agencies: the Forest Service and Natural Resources Conservation Service (NRCS). The Forest Service is charged with the oversight of 191 million acres of federal land. In advancing its pollution-control efforts, the Forest Service relies on a number of practices to inhibit air, land, and water pollution, including erosion and flood control, timber-harvesting methods to protect water bodies, and the minimization of pollution created by natural resource extraction. It also invokes a number of laws (the Clean Air Act, Clean Water Act, and National Forest Management Act) to penalize individuals or industries operating contrary to its efforts. The NRCS oversees pollution management of

U.S. agricultural and range lands. Such management is conducted cooperatively with farmers, ranchers, and landowners who utilize technical assistance provided by the NRCS to address such things as the environmental effects of pesticides on agricultural and ranch lands. Among the programs that the NRCS has jurisdiction over are the Natural Resources Inventory, Rural Abandoned Mines, and Wetlands Reserve Program. A number of countries, including Australia, Britain, Canada, France, Germany, and Spain, have taken the DOA's lead in conducting similar pollution control activities. SEE ALSO AGRICULTURE; PESTICIDES.

Internet Resources

U.S. Department of Agriculture Web site. Available at www.usda.gov/energyandenvironment/faq.html.

U.S. Forest Service Web site. Available at www.fs.fed.us.

Robert F. Gruenig

U.S. Department of the Interior

Established in 1849, the U.S. Department of the Interior has primary management and conservation responsibility for all federal lands and minerals, national parks, water resources, and wildlife refuges. Its secretary reports directly to the president, and the department's responsibilities are divided among a number of agencies, including the Bureau of Land Management, Bureau of Mines, Bureau of Reclamation, Fish and Wildlife Service, Geological Survey, National Park Service, and Office of Surface Mining Reclamation and Enforcement. Among its primary objectives are the wise use of land and natural resources, the protection of animal and plant species, the promotion of environmental values among U.S. citizens, and environmental protection balanced with mineral resource needs. Its responsibilities include the coordination of its agencies' activities, data collection and analysis concerning natural resources, and minimization and mitigation of mining and other human activities adversely affecting public lands. Serving in a complementary role to the department's management responsibilities, the U.S. Environmental Protection Agency enforces a number of environmental laws (e.g., the Clean Air Act, Clean Water Act, Endangered Species Act, National Environmental Policy Act, Surface Mining Control and Reclamation Act, Wild and Scenic Rivers Act, and Wilderness Act) which help to protect the resources under the department's jurisdiction. SEE ALSO MINING; NATIONAL PARK SERVICE.

Bibliography

Goldsteen, Joel B. (1999). *The ABCs of Environmental Regulation.* Rockville, MD: Government Institutes.

Internet Resource

"Orientation to the U.S. Department of the Interior." Available at http://www.doiu.nbc.gov/orientation.

Robert F. Gruenig

U.S. Environmental Protection Agency

The U.S. Environmental Protection Agency (EPA) is the primary regulatory agency of the federal government responsible for pollution control. EPA's

A worker is undergoing a decontamination process. (U.S. EPA. Reproduced by permission.)

stated mission is to protect human health and to safeguard the natural environment—air, water, and land—on which life depends. The EPA was created in 1970 as an outgrowth of the burgeoning environmental movement in the United States during the 1960s. President Richard M. Nixon signed the Reorganization Plan No. 3 of 1970, the legal document that established the EPA. Although at that time a number of federal environmental programs already existed, they were scattered throughout several different federal agencies. For example, the Federal Water Quality Administration of the Department of the Interior was responsible for certain water pollution programs, the Department of Agriculture was responsible for the regulation of pesticides, and the Department of Health, Education and Welfare was responsible for air pollution and solid waste management. The creation of the EPA was an attempt to consolidate these environmental programs in a coordinated way under the control of one agency with clear-cut responsibility for environmental protection. The EPA opened its doors for business on December 2, 1970, less than eight months after the first Earth Day celebration.

Organization and Administration

The EPA is one of many independent agencies of the executive branch of the U.S. government. It derives its authority to carry out pollution-control

Seal of the U.S. Environmental Protection Agency (U.S. EPA. Reproduced by permission.)

programs through statutes passed by Congress. Although there have been several unsuccessful efforts over the years, especially during the late 1980s and early 1990s, to make the EPA a cabinet-level department, it remains an independent agency. EPA's administrator is appointed by the president, but must be confirmed by the Senate. Although not a member of the cabinet, the administrator is directly responsible to the president. The EPA has a number of assistant administrators who oversee offices with responsibility for EPA's primary programs, including air and radiation; enforcement and compliance assurance; international affairs; prevention, pesticides, and toxic substances; research and development; solid waste and emergency response; and water.

In addition, the EPA has ten regional offices throughout the United States. Each of these is responsible for working with the states in its region to implement and enforce EPA's regulations. Within these various offices and regional centers, the EPA carries out wide-ranging duties related to environmental protection, including:

- Researching the causes and effects of specific environmental problems

- Monitoring environmental conditions

- Determining how to best regulate activities causing environmental harm

- Setting specific standards for particular pollutants of concern

- Administering environmental permitting programs

- Providing financial and technical assistance to states

- Coordinating and supporting research activities of states and other private and public organizations

- Providing oversight of states that have assumed responsibility for federal environmental program

- Enforcing environmental laws

The EPA receives its funding through congressional appropriation. In 1970 EPA's annual budget was slightly over $1 billion. In 2002 its annual budget was in excess of $7.3 billion. EPA's workforce has grown from approximately 4,000 employees in 1970 to more than 17,000 employees in 2002.

Activities and Accomplishments

The EPA is responsible for implementing and enforcing more than twenty-four major environmental statutes. Some of the most significant environmental statutes include the Clean Air Act; the Clean Water Act; the Comprehensive Environmental Response, Compensation, and Liability Act (Superfund); the Toxic Substances Control Act; the Federal Insecticide, Fungicide and Rodenticide Act; and the Safe Drinking Water Act.

The EPA has achieved many significant successes in implementing these programs. One of the agency's earliest accomplishments was banning the pesticide DDT in 1972 after it was found to accumulate in the food chain, where it threatened wildlife populations. This ban, enacted fewer than two years after the formation of the EPA, had particular significance because the environmental risks associated with DDT, about which Rachel Carson warned the public about in her 1962 book *Silent Spring*, were in large part publicized and disseminated by the environmental movement. Some of EPA's other early accomplishments include the 1973 ban on lead in gasoline; funding to build an advanced network of sewage-treatment facilities to prevent raw sewage from flowing into the nation's waters; establishing discharge limitations for industrial water pollution under the Clean Water Act; the establishment of health-based standards to protect the public water supply under the Safe Drinking Water Act; the 1978 ban on the use of chlorofluorocarbons (CFCs) as propellants in most aerosol cans to protect the ozone layer; and the 1980 establishment of the Superfund program for hazardous waste cleanup.

In the 1990s some of EPA's major accomplishments included the annual release of information on the location and nature of toxic chemical releases in communities throughout the country through the Toxics Release Inventory (TRI); the 1990 establishment of the first public–private partnership to reduce industrial emissions under the Pollution Prevention Act; obtaining the largest environmental criminal damage settlement in history in 1991 (totaling over $1 billion) for the 1989 *Exxon Valdez* oil spill; and establishing pollution-control standards under the Clean Air Act to reduce toxic air pollutants by 90 percent. During this same decade, in response to the rapid development of biotechnology products, the EPA established new regulatory programs to address the risks from the release of genetically modified organisms into the environment. In addition, one of EPA's most significant roles remains that of enforcer of the nation's environmental laws. In 1997 alone, the EPA levied nearly $170 million in administrative penalties, and referred 278 criminal cases to the Department of Justice (DOJ) for prosecution. In that same year the EPA referred 426 civil cases to the DOJ and assessed $95 million in civil penalties.

Relationship to Other Environmental Agencies

Although the EPA is the primary federal agency responsible for environmental protection in the United States, there are several other federal agencies that bear some responsibility for environmental protection in specified areas. The EPA is primarily concerned with regulatory programs, such as pollution-control programs, designed to protect human health and the environment. Other federal agencies are responsible for other types of programs, such as the management of public lands and natural resources and the protection of threatened and endangered species.

In addition to other federal agencies with environmental responsibility, virtually every one of the fifty states has an agency responsible for pollution control. The type and extent of state regulation vary widely. The EPA has delegated the majority of federal environmental laws it administers to state environmental agencies. However, when the EPA delegates a program to a state, it retains oversight authority over that program.

Outside the United States, many other developing countries, particularly those in the West, have agencies responsible for environmental protection that are very similar in scope and structure to the EPA. For example, Germany, France, and Great Britain all have national environmental agencies with primary responsibility for the regulation of air and water pollution and public and hazardous waste disposal. Other countries have taken somewhat different approaches. For example, in Japan, although a national Environmental

> "It seemed to me important to demonstrate to the public that the government was capable of being responsive to their expressed concerns; namely, that we would do something about the environment. Therefore, it was important for us to *advocate* strong environmental compliance, back it up, and *do* it; to actually show we were willing to take on the large institutions in the society which hadn't been paying much attention to the environment."
>
> —William D. Ruckelshaus, on his expectations when named the first EPA Administrator in December 1970.

Agency was established in 1971, the national government initially did not play an active role in environmental regulation. Instead, many of Japan's large cities developed their own environmental protection programs. Rather than rely on formal laws and regulations, these cities sought to achieve environmental protection through agreements between the local governments and industry there. The pollution-control agreements resulting in environmental protection within Japan stand in contrast to the regulatory systems of many Western countries.

Conclusion

During its existence, despite its numerous successes and accomplishments, EPA often has been the target of criticism, both by industries asserting that EPA's regulations are too stringent and are imposing too great an economic cost, and by environmentalists who claim that EPA is not doing enough to protect public health and the environment. Despite these debates, public opinion polls consistently show a strong support for environmental protection programs. Nevertheless, controversy continues over the appropriate direction and scope of EPA's specific regulatory programs. SEE ALSO AGENCIES, REGULATORY; U.S. ARMY CORPS OF ENGINEERS; ENVIRONMENT CANADA; ENVIRONMENT MEXICO; U.S. COAST GUARD.

Bibliography

Antista, James V.; Boardman, Dorothy Lowe; Cloud, Thomas A.; et al. (2001). "Federal, State, and Local Environmental Control Agencies." In *Treatise on Florida Environmental and Land Use Law*, Vol. 1. Tallahassee, FL: The Florida Bar.

Carson, Rachel. (1962). *Silent Spring*. New York: Houghton Mifflin.

Ferrey, Steven. (2001). *Environmental Law: Examples and Explanations*, 2nd edition. New York: Aspen Publishers.

Lovei, Magda, and Weiss, Charles, Jr. (1998). *Management and Institutions in OECD Countries: Lessons from Experience*. Washington, D.C.: World Bank.

Moya, Olga L., and Fono, Andrew L. (2001). *Federal Environmental Law: The User's Guide*, 2nd edition. St. Paul, MN: West Publishing Company.

Rodgers, William H., Jr. (1994). *Environmental Law*, 2nd edition. St. Paul, MN: West Publishing Company.

U.S. Environmental Protection Agency. (1995/1996). *Information Resource Management*. Access EPA 220-B-95-004. Washington, D.C.: U.S. Government Printing Office.

Internet Resource

Government Institutes. (1994). "How EPA Works: A Guide to EPA Organization and Functions." Rockville, MD. Available from http://www.epa.gov/html.

Mary Jane Angelo

U.S. Food and Drug Administration (FDA)

Established in 1927, the U.S. Food and Drug Administration (FDA) protects public health by guarding against impure and unsafe foods, drugs, cosmetics, and other potential hazards. The FDA carries out this role through regulation, testing, studies, and consumer advisories. In addition, the FDA actively enforces a number of laws, including the Food Quality Protection Act and Lead-Based Paint Poisoning Prevention Act, to protect the public against unsafe foods and other products. Foods can be adversely affected by dioxins, mercury, and lead that are ingested or absorbed by, or adhere to animals and plants intended for human consumption. The FDA takes the necessary measures to ensure that these substances do not make the food supply unsafe. It monitors dairy and seafood products for dioxin residues created by fuel burning and material incineration. With respect to mercury, the FDA tests for its bioaccumulation in fish because fetuses and infants are especially sensitive to and can be adversely affected by its presence. Lead, existing in food cans (often imported from foreign countries), plumbing, solder, and brass faucets, has led to the FDA's establishment of a contaminants branch in the office of plant and dairy foods and beverages and the creation of a test kit to screen for the presence of lead. SEE ALSO DIOXIN; LEAD; MERCURY.

Bibliography

Parisian, Suzanne. (2001). *FDA Inside and Out.* Front Royal, VA: Fast Horse Press.

Internet Resource

"Dioxins: FDA Strategy for Monitoring, Method Development, and Reducing Human Exposure." Available at http://www.cfsan.fda.gov/~lrd/dioxstra.html.

Robert F. Gruenig

U.S. Geological Survey

Established as part of the Department of the Interior in 1879 and funded by Congress, the U.S. Geological Survey (USGS) provides support to federal agencies (e.g., the Environmental Protection Agency or EPA, the National Oceanographic and Atmospheric Administration or NOAA, and the U.S. Coast Guard) in the form of useful information for decision-making purposes concerning the management of U.S. environmental and natural resources. As part of this support, the USGS examines the relationship between humans and the environment by conducting data collection, long-term research assessments, and ecosystem analyses, and providing forecast changes and their implications. One example of this support is the provision of information about earthquake and seismic activities that is used to assess the potential impact of such activities on water quality. In addition to its federal agency support, the USGS also manages some of the following programs that address the problems of environmental pollution: (1) coastal and marine geology program; (2) contaminants program; (3) energy program; (4) fisheries and aquatic resources; and (5) global change/wetland ecology program. These external support activities and internal programs have been similarly adopted by countries such as Australia, Britain, Finland, and Japan, although not to the same degree as provided by the USGS. SEE ALSO ENVIRONMENTAL PROTECTION AGENCY; INTERIOR DEPARTMENT, UNITED STATES; NATIONAL OCEANOGRAPHIC AND ATMOSPHERIC ADMINISTRATION (NOAA); U.S. COAST GUARD.

Bibliography

Natural Research Council, Committee on Geosciences, Environment and Resources. (2001). *Future Roles and Opportunities for the U.S. Geological Survey.* Washington, D.C.: National Academy Press.

Internet Resource

Coastal and Marine Geology Program Site. Available from http://marine.usgs.gov.

Robert F. Gruenig

Ultraviolet Radiation

Ultraviolet (UV) radiation is a form of electromagnetic radiation that lies between visible light and x rays in its energy and wavelength. It is a component of the radiation that reaches the Earth from the sun. The broad UV band, having wavelengths between 190 nanometers (nm) and 400 nm, is conventionally divided into three parts: UV-A or near-UV (315 to 400 nm), UV-B or mid-UV (280 to 315 nm), and UV-C or far-UV (190 to 280 nm). Much of the **incident solar** UV radiation is absorbed by gases in the earth's atmosphere and never reaches the earth's surface. This is fortunate, because UV radiation can chemically alter important biological molecules, including proteins and deoxyribonucleic acid (DNA), and thereby cause damage to living systems. The most familiar effect on humans is sunburn, which is the manifestation of UV's damage to outer skin cells. Long-term effects of excessive UV exposure include skin cancer, eye damage (cataracts), and **suppression** of the immune system.

Among the atmospheric gases that are the major absorbers of UV radiation is ozone (O_3), which lies predominantly in the upper atmospheric region known as the stratosphere. Stratospheric ozone is particularly important in absorbing UV-B radiation. A current environmental issue concerns the depletion of stratospheric ozone (the ozone layer) by human-made chemicals such as chlorofluorocarbons (CFCs) and halons. With even small percentages of ozone depletion, more UV-B radiation reaches the surface of the earth and the harmful effects of UV increase. SEE ALSO CFCs (CHLOROFLUOROCARBONS); HALON; OZONE.

Bibliography

World Meteorological Organization. (2003). *Scientific Assessment of Ozone Depletion: 2002.* Global Ozone Research and Monitoring Project, Report No. 47. Geneva: Author.

Internet Resource

NASA Advanced Supercomputing Division Web site. "Ultraviolet Radiation." Available from http://www.nas.nasa.gov/About/Education/Ozone/radiation.html.

United Nations Environment Programme. (1998). "Environmental Effects of Ozone Depletion 1998 Assessment." In the Global Change Research Information Office Web site. Available from http://www.gcrio.org/ozone/toc.html.

World Meteorological Organization. "UV Radiation Page." Available from http://www.srrb.noaa.gov/UV.

Christine A. Ennis

Underground Storage Tank

Leaking underground storage tanks (LUSTs) containing hazardous liquids, primarily petroleum products such as gasoline, diesel, kerosene, or oil have contaminated the **groundwater** and **drinking water** of thousands of communities across the United States.

Following the boom in automobile sales after World War II, gasoline stations mushroomed across the county to meet the demand for personal mobility. At these new stations, gasoline was stored underground in tanks made of bare steel, which were not protected from corrosion—the oxidation, or rusting, of other metals as well as iron metal in steel that can cause metals to

incident solar sun energy that hits a particular spot

suppression reduction in or prevention of an effect

groundwater the supply of freshwater found beneath the Earth's surface includes; aquifers, which supply wells and springs

drinking water water used or with the potential to be used for human consumption

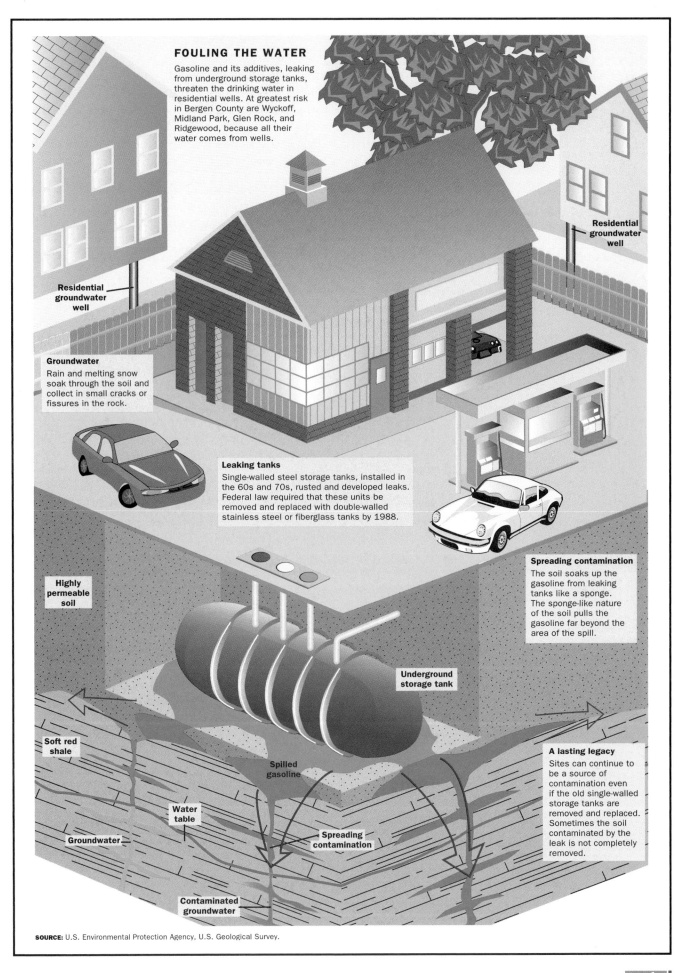

FOULING THE WATER

Gasoline and its additives, leaking from underground storage tanks, threaten the drinking water in residential wells. At greatest risk in Bergen County are Wyckoff, Midland Park, Glen Rock, and Ridgewood, because all their water comes from wells.

Residential groundwater well

Residential groundwater well

Groundwater
Rain and melting snow soak through the soil and collect in small cracks or fissures in the rock.

Leaking tanks
Single-walled steel storage tanks, installed in the 60s and 70s, rusted and developed leaks. Federal law required that these units be removed and replaced with double-walled stainless steel or fiberglass tanks by 1988.

Spreading contamination
The soil soaks up the gasoline from leaking tanks like a sponge. The sponge-like nature of the soil pulls the gasoline far beyond the area of the spill.

Highly permeable soil

Underground storage tank

Soft red shale

Spilled gasoline

A lasting legacy
Sites can continue to be a source of contamination even if the old single-walled storage tanks are removed and replaced. Sometimes the soil contaminated by the leak is not completely removed.

Water table

Groundwater

Spreading contamination

Contaminated groundwater

SOURCE: U.S. Environmental Protection Agency, U.S. Geological Survey.

crack or disintegrate and leak. The average life expectancy of steel tanks installed in the 1950s and 1960s was thirty to fifty years. This statistic suggests that petroleum products have been leaking from these tanks, spread throughout the country, since the early 1980s. By 2001, the U.S. Environmental Protection Agency (EPA) was dealing with cleanups at 379,243 LUST sites in the United States.

Corrosion usually causes tanks to leak slowly. Leaks from older tanks are often difficult to detect because inventory control is imprecise. Once released from a tank, gasoline sinks through **unsaturated** soil and, because gasoline is less dense than water, floats on the surface of the **water table**. Because most components of gasoline are fairly **volatile**—they readily become a vapor at a relatively low temperature—leaks often go undetected until the vapors are present at the ground's surface. In addition to the risk to water supplies, leaking gasoline also presents risk of fire and explosion when vapors from leaking tanks can travel through sewer lines and soils into buildings.

unsaturated capable of dissolving more solute, i.e., water

water table the level of water in the soil

volatile any substance that evaporates readily

Because nearly half of all Americans depend on groundwater for their drinking water, leaking underground storage tanks represent a significant public health threat. The most hazardous components of petroleum products are the BTEX compounds—benzene, toluene, ethylbenzene, and xylenes. Benzene is the most hazardous of these compounds due to the risk of cancer from drinking and bathing in water containing benzene. The **maximum contaminant level** set by the EPA is 5 parts per billion (ppb).

maximum contaminant level in water: the maximum permissible level of a contaminant in water delivered to any user of a public system; MCLs are enforceable standards

Another potentially hazardous compound in gasoline is methyl tertiary butyl ether (MTBE). Ironically, MTBE is added to gasoline to combat air pollution by making the fuel burn cleaner. At concentrations as low as 20 ppb, MTBE makes drinking water unfit for human consumption. (This assessment is based on standards that correlate unfitness with the taste and odor left in the water by MTBE.) MTBE is currently classified as a potential human **carcinogen**, but there is no maximum contaminant level for MTBE in drinking water. As many as 9,000 community water wells in thirty-one states may be affected by MTBE contamination, and many states are phasing out its use in gasoline.

carcinogen any substance that can cause or aggravate cancer

Once LUSTs are identified and the extent of soil and groundwater contamination is determined, remediation can include removal of the leaking tanks, the contaminated soil, and the contaminants from the groundwater. Tank and contaminated soil removal is accomplished by excavation. Removal of groundwater contaminants is accomplished by the suction pumping of gasoline floating on the water table by air stripping, a process in which air is pumped through the water to cause the volatile compounds to evaporate, and by natural attenuation, the **biodegradation** of contaminants by microorganisms. Remediating contaminated groundwater can take decades, and some waters will never be made safe enough to drink. Dozens of communities have had to find alternative sources of drinking water because of gasoline contamination.

biodegradation decomposition due to the action of bacteria and other organisms

The primary responsibility for the licensing, operation, and regulation of underground storage tanks (USTs) and the cleanup of LUSTs falls to the state. Most states fund **remediation** of LUST sites through licensing fees and surcharges on most petroleum products. The EPA oversees the state programs and augments their remediation efforts through grants to support

remediation cleanup or other methods used to remove or contain a toxic spill or hazardous materials from a Superfund site or for the Asbestos Hazard Emergency Response program

LUST program staffing, and through direct assistance with emergency responses and cleanup.

To prevent future problems, the EPA established UST standards in 1988 and gave tank owners ten years to upgrade or replace old tanks. New tanks must have corrosion protection and improved leak-detection systems. Nearly 1.5 million USTs and LUSTs have been closed. SEE ALSO PETROLEUM; SUPERFUND; WATER POLLUTION.

Bibliography

American Petroleum Institute. (1989). "Recommended Practices on Underground Petroleum Storage Tank Management." RP 1650. Washington, D.C.: Author.

American Society for Testing and Materials. (1998). "Standard Guide for Performing Evaluations of Underground Storage Tank Systems for Operational Conformance with 40 CFR, Part 280 Regulations." Report ASTM E 1990-98. West Conshohocken, PA: ASTM.

Internet Resources

Office of Solid Waste and Emergency Response. (1998). "Technical Standards and Corrective Action Requirements for Owners and Operators of Underground Storage Tanks (Section 610 Review)." 63 FR 22709. Washington, D.C.: U.S. Environmental Protection Agency. Also available from http://www.epa.gov/swerust1.

Office of Solid Waste and Emergency Response. (2000). Catalog of EPA Materials on Underground Storage Tanks. EPA Report 510-B-00-001. Washington, D.C.: U.S. Environmental Protection Agency. Also available from http://www.epa.gov/swerust1.

Joseph N. Ryan

Unintended Consequences

Solutions to environmental problems occasionally create unintended consequences, that is, solving one problem creates another. Scientists and engineers must carefully evaluate potential negative results before implementing new remediation programs. For example, burying wastes in landfills may cause groundwater contamination, incinerating wastes reduces waste volumes but can cause air pollution, and excavating abandoned waste sites as part of a remediation effort may expose workers to contamination. Recycling can have a net negative environmental impact if air pollution associated with transportation outweighs environmental benefits. Stimulating the **biodegradation** of trichloroethylene (TCE) in contaminated groundwater can lead to the formation of vinyl chloride, a more hazardous chemical. Two examples are described here: MTBE and disinfection by-products.

biodegradation decomposition due to the action of bacteria and other organisms

MTBE

Methyl tertiary-butyl ether (MTBE) is a fuel additive that has improved air, but degraded groundwater. Its primary use in the United States began in the 1990s as a fuel **oxygenate** added to gasoline to help meet the requirements of the Clean Air Act. By providing a source of oxygen during gasoline combustion, MTBE reduces carbon monoxide levels. It has been used in a number of localities to help combat significant air pollution problems, and studies have identified important air quality and public health benefits from its use.

oxygenate increase the concentration of oxygen within an area

Unfortunately, the addition of MTBE to fuels resulted in unintended consequences. MTBE is highly soluble in water and relatively nonbiodegradable. It has been detected in groundwater across the United States, primarily from fuel leaks and spills. For example, the U.S. Geological Survey (USGS) analyzed drinking water information from over one thousand community water systems (CWS) in the Northeast and Mid-Atlantic regions of the United States for the period from 1993 to 1998. MTBE was found in drinking water from 8.9 percent of the CWSs. Levels over 20 µg/l were determined in 1 percent of those same cases. Once introduced to groundwater, MTBE's high solubility makes it very mobile. The U.S. Environmental Protection Agency (EPA) did issue a drinking water advisory for MTBE in 1997. Although there are no data on the effects of drinking MTBE-contaminated water on humans, cancer and other deleterious effects occur in animals at high exposure levels. Furthermore, MTBE has an unpleasant taste and odor.

Disinfection By-Products

Disinfection, one of the primary tools of water treatment, is the removal and inactivation of pathogenic microbes, that is, small organisms such as viruses, bacteria, and protozoa, that can cause disease. Disinfection has historically been accomplished using chlorination, the destruction of microbes by hypochlorous acid and the hypochlorous ion, formed by the reaction of chlorine gas and water or added directly as hypochlorite salts. Large improvements in public health occur when pathogen-free waters are available for human consumption, and significant portion of the life span increase achieved in the modern era is the result of safe drinking water.

However, there have been unintended consequences of disinfection by chlorination. If organic compounds are present in the water, halogenated disinfection by-products (DBPs) may be formed. Two halogenated DBPs regulated by U.S. drinking water standards are trihalomethanes (THM) and haloacetic acids. Both can increase the risk of cancer. THMs can also cause liver, kidney, and central nervous system problems. A USGS study found THMs in the drinking water of 45 percent of some 2,000 CWSs randomly selected in the Northeast and Mid-Atlantic regions of the United States. Fortunately, there are a number of ways CWSs can limit the generation of halogenated DBPs, including using water sources with low organic content, removing organic compounds before chlorination, and using disinfectants that produce fewer or no halogenated DBPs, such as ozone or chloramines. SEE ALSO ABATEMENT; DISINFECTION; VEHICULAR POLLUTION.

Bibliography

Grady, S., and Casey, G. (2001). "Occurrence and Distribution of Methyl *tert*-Butyl Ether and Other Volatile Organic Compounds in Drinking Water in the Northeast and Mid-Atlantic Regions of the United States, 1993–98." Washington, D.C.: U.S. Geological Survey.

U.S. Environmental Protection Agency. (1997). "Drinking Water Advisory: Consumer Acceptability Advice and Health Effects Analysis on Methyl Tertiary-Butyl Ether (MtBE)." EPA-822-F-97-009. Washington, D.C.: U.S. Environmental Protection Agency.

U.S. Geological Survey, Water Resources Investigations Report 00-4228.

Internet Resources

Davis, J. Michael. "How to Avert the Problems of MTBE." Available from http://www.epa.gov/ord.

Reshkin, K. "EPA Student Center." Available from http://www.epa.gov/students.

Jess Everett

Union of Concerned Scientists

The Union of Concerned Scientists (UCS) is a nonprofit alliance of some fifty thousand scientists and citizens across the United States. The group's stated goal is to combine rigorous scientific analysis with committed citizen advocacy in order to build a cleaner environment and a safer world. The group focuses on issues such as global warming and the environmental impact of vehicles and various energy sources.

The UCS was formed in 1969 at the Massachusetts Institute of Technology, where a number of faculty members and students banded together to protest what they saw as the abuse of science and technology for military purposes. The new group called for greater emphasis on the application of scientific research to solve social and environmental problems. In its early years, the organization issued statements urging an end to the nuclear arms race and a ban on space weapons research. In recent years, the group has focused more on environmental issues.

In 1992, seventeen hundred of the world's leading scientists, including many Nobel prize winners, issued an emotional appeal through the UCS. Their statement, titled "World Scientists' Warning to Humanity," noted that "human activities inflict harsh and often irreversible damage on the environment and on critical resources." It urged the world community to take action by moving away from fossil fuels and giving high priority to more efficient use of natural resources such as water.

In 1997, the UCS issued another statement at the Kyoto Climate Summit in Japan. This statement, which addressed the threat of global warming, was signed by more than fifteen hundred scientists from sixty-three countries, including sixty U.S. National Medal of Science winners. UCS efforts helped set the stage for the adoption of an international treaty on climate change. Such joint appeals are influential, because they show world leaders that there is growing agreement among scientists on key issues.

In the United States, the UCS has been a force for social change as well. For example, in California, the UCS and other environmental and public health groups helped convince the state to begin requiring sport utility vehicles, light trucks, and diesel cars to meet the same tailpipe emissions standards as gasoline-powered cars. In Connecticut, the UCS and its allies helped persuade the legislature to pass a law that included strong support for clean, renewable energy sources. In short, the UCS continues to be a powerful voice for concerned scientists and citizens. SEE ALSO ENVIRONMENTAL MOVEMENT; GLOBAL WARMING; TREATIES AND CONFERENCES.

Bibliography

Brown, Michael, and Leon, Warren. (1999). *The Consumer's Guide to Effective Environmental Choices: Practical Advice from the Union of Concerned Scientists*. New York: Three Rivers Press.

Internet Resources

Union of Concerned Scientists. "World Scientists' Call for Action" and "World Scientists' Warning to Humanity." Available from http://www.ucsusa.org.

Linda Wasmer Andrews

Urban Sprawl *See Sprawl*

Vehicular Pollution

The large majority of today's cars and trucks travel by using internal combustion engines that burn gasoline or other fossil fuels. The process of burning gasoline to power cars and trucks contributes to air pollution by releasing a variety of emissions into the atmosphere. Emissions that are released directly into the atmosphere from the tailpipes of cars and trucks are the primary source of vehicular pollution. But motor vehicles also pollute the air during the processes of manufacturing, refueling, and from the emissions associated with oil refining and distribution of the fuel they burn.

Primary pollution from motor vehicles is pollution that is emitted directly into the atmosphere, whereas secondary pollution results from chemical reactions between pollutants after they have been released into the air.

Despite decades of efforts to control air pollution, at least 92 million Americans still live in areas with chronic smog problems. The U.S. Environmental Protection Agency (EPA) predicts that by 2010, even with the benefit of current and anticipated pollution control programs, more than 93 million people will live in areas that violate health standards for ozone (urban smog), and more than 55 million Americans will suffer from unhealthy levels of fine-particle pollution, which is especially harmful to children and senior citizens.

While new cars and light trucks emit about 90 percent fewer pollutants than they did three decades ago, total annual vehicle-miles driven have increased by more than 140 percent since 1970 and are expected to increase another 25 percent by 2010. The emission reductions from individual vehicles have not adequately kept pace with the increase in miles driven and the market trend toward more-polluting light trucks, a category that includes sports utility vehicles (SUVs). As a result, cars and light trucks are still the largest single source of air pollution in most urban areas, accounting for one-quarter of emissions of smog-forming pollutants nationwide.

Ingredients of Vehicular Pollution

The following are the major pollutants associated with motor vehicles:

- Ozone (O_3). The primary ingredient in urban smog, ozone is created when hydrocarbons and nitrogen oxides (NO_x)—both of which are chemicals released by automobile fuel combustion—react with sunlight. Though beneficial in the upper atmosphere, at the ground level ozone can irritate the respiratory system, causing coughing, choking, and reduced lung capacity.

- Particulate matter (PM). These particles of soot, metals, and pollen give smog its murky color. Among vehicular pollution, fine particles (those less than one-tenth the diameter of a human hair) pose the most serious threat to human health by penetrating deep into lungs. In addition to direct emissions of fine particles, automobiles release nitrogen oxides, hydrocarbons, and sulfur dioxide, which generate additional fine particles as secondary pollution.

- Nitrogen oxides (NO_x). These vehicular pollutants can cause lung irritation and weaken the body's defenses against respiratory infections such as pneumonia and influenza. In addition, they assist in the for-

During the morning rush hour, the Miguel Hidalgo area of Mexico City is clogged with traffic and smog. (©Stephanie Maze/Corbis. Reproduced by permission.)

mation of ozone and particulate matter. In many cities, NO_x pollution accounts for one-third of the fine particulate pollution in the air.

- Carbon monoxide (CO). This odorless, colorless gas is formed by the combustion of fossil fuels such as gasoline. Cars and trucks are the source of nearly two-thirds of this pollutant. When inhaled, CO blocks the transport of oxygen to the brain, heart, and other vital organs in the human body. Newborn children and people with chronic illnesses are especially susceptible to the effects of CO.

- Sulfur dioxide (SO_2). Motor vehicles create this pollutant by burning sulfur-containing fuels, especially diesel. It can react in the atmosphere to form fine particles and can pose a health risk to young children and asthmatics.

- Hazardous air pollutants (toxics). These chemical compounds, which are emitted by cars, trucks, refineries, gas pumps, and related sources, have been linked to birth defects, cancer, and other serious illnesses. The EPA estimates that the air toxics emitted from cars and trucks account for half of all cancers caused by air pollution.

Vehicular Emissions That Contribute to Global Warming

Carbon monoxide, ozone, particulate matter, and the other forms of pollution listed above can cause smog and other air quality concerns, but there are vehicular emissions that contribute to a completely different pollution issue: global warming.

Morning rush hour traffic waiting to pay the toll to cross the Oakland Bay Bridge in August 1989. (©James A. Sugar/Corbis. Reproduced by permission.)

The gases that contribute to global warming are related to the chemical composition of the Earth's atmosphere. Some of the gases in the atmosphere function like the panes of a greenhouse. They let some radiation (heat) in from the sun but do not let it all back out, thereby helping to keep the Earth warm. The past century has seen a dramatic increase in the atmospheric concentration of heat-trapping gasses, due to human activity. If this trend continues, scientists project that the earth's average surface temperature will increase between 2.5°F and 10.4°F by the year 2100.

One of these important heat-trapping gasses is carbon dioxide (CO_2). Motor vehicles are responsible for almost one-quarter of annual U.S. emissions of CO_2. The U.S. transportation sector emits more CO_2 than all but three other countries' emissions from all sources combined.

Curbing Vehicular Pollution

Vehicular emissions that contribute to air quality problems, smog, and global warming can be reduced by putting better pollution-control technologies on

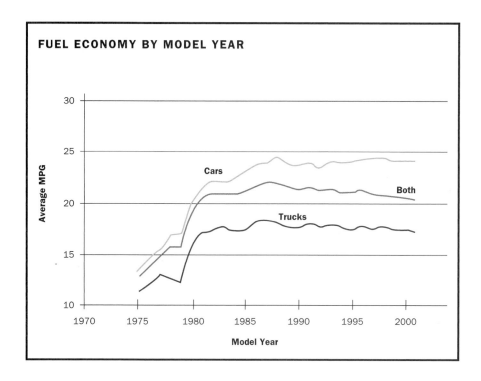

FUEL ECONOMY BY MODEL YEAR

cars and trucks, burning less fuel, switching to cleaner fuels, using technologies that reduce or eliminate emissions, and reducing the number of vehicle-miles traveled.

Pollution Control Technology

Federal and California regulations require the use of technologies that have dramatically reduced the amount of smog-forming pollution and carbon monoxide coming from a vehicle's tailpipe. For gasoline vehicles, "three-way" catalysts, precise engine and fuel controls, and evaporative emission controls have been quite successful. More advanced versions of these technologies are in some cars and can reduce smog-forming emissions from new vehicles by a factor of ten. For diesel vehicles, "two-way" catalysts and engine controls have been able to reduce hydrocarbon and carbon monoxide emissions, but nitrogen oxide and toxic particulate-matter emissions remain very high. More advanced diesel-control technologies are under development, but it is unlikely that they will be able to clean up diesel to the degree already achieved in the cleanest gasoline vehicles.

Added concerns surround the difference between new vehicle emissions and the emissions of a car or truck over a lifetime of actual use. Vehicles with good emission-control technology that is not properly maintained can become "gross polluters" that are responsible for a significant amount of existing air-quality problems. New technologies have also been developed to identify emission-equipment control failures, and can be used to help reduce the "gross polluter" problem.

Burning Less Fuel

The key to burning less fuel is making cars and trucks more efficient and putting that efficiency to work in improving fuel economy. The U.S. federal government sets a fuel-economy standard for all passenger vehicles. However, these standards have remained mostly constant for the past decade. In addition, sales of lower-fuel-economy light trucks, such as SUVs, pickups, and minivans, have increased dramatically. As a result, on average, the U.S. passenger-vehicle fleet actually travels less distance on a gallon of gas than it did twenty years ago. This has led to an increase in heat-trapping gas emissions from cars and trucks and to an increase in smog-forming and toxic emissions resulting from the production and transportation of gasoline to the fuel pump.

This trend can be reversed through the use of existing technologies that help cars and trucks go farther on a gallon of gasoline. These include more efficient engines and transmissions, improved aerodynamics, better tires, and high strength steel and aluminum. More advanced technologies, such as hybrid-electric vehicles that use a gasoline engine and an electric motor plus a battery, can cut fuel use even further. These technologies carry with them additional costs, but pay for themselves through savings at the gasoline pump.

Zero-Emission Vehicles

As more cars and trucks are sold and total annual mileage increases, improving pollution-control technology and burning less fuel continues to be vital, especially in rapidly growing urban areas. However, eliminating emissions from the tailpipe goes even further to cut down on harmful air pollutants.

Hydrogen fuel-cell and electric vehicles move away from burning fuel and use electrochemical processes instead to produce the needed energy to drive a car down the road. Fuel-cell vehicles run on electricity that is produced directly from the reaction of hydrogen and oxygen. The only by-product is water—which is why fuel-cell cars and trucks are called zero-emission vehicles. Electric vehicles store energy in an onboard battery, emitting nothing from the tailpipe.

The hydrogen for the fuel cell and the electricity for the battery must still be produced somewhere, so there will still be upstream emissions associated with these vehicles. These stationary sources, however, are easier to control and can ultimately be converted to use wind, solar, and other renewable energy sources to come as close as possible to true zero-emission vehicles.

Cleaner Fuels

The gasoline and diesel fuel in use today contains significant amounts of sulfur and other compounds that make it harder for existing control technology to keep vehicles clean. Removing the sulfur from the fuel and cutting down on the amount of light hydrocarbons helps pollution-control technology to work better and cuts down on evaporative and refueling emissions.

Further large-scale reductions of other tailpipe pollution and CO_2 can be accomplished with a shift away from conventional fuels. Alternative fuels such as natural gas, methanol, ethanol, and hydrogen can deliver benefits to the environment while helping to move the United States away from its dependence on oil. All of these fuels inherently burn cleaner than diesel and gasoline, and they have a lower carbon content—resulting in less CO_2. Most of these fuels are also more easily made from renewable resources, and fuels such as natural gas and methanol help provide a bridge to producing hydrogen for fuel-cell vehicles.

Reducing Driving

Because we are still dependent on fossil fuels and the number of cars on the road is expected to double, a significant reduction in vehicular pollution requires more than gains in fuel efficiency. Measures that encourage us to drive less can help curb vehicular pollution and protect natural resources and public health.

Alternatives that can reduce the number of vehicle-miles traveled include

- providing transportation alternatives to cars, including mass transit, bicycle, and pedestrian routes;

- promoting transit-oriented, compact developments in and around cities and towns; and adopting policies to improve existing roads and infrastructure.

Personal Contributions

Individuals can also make a difference in the effort to reduce pollution from cars and trucks. How we drive and how we take care of our vehicles affects fuel economy and pollution emissions. The following are several ways people can reduce the harmful environmental impact of cars.

- Driving as little as possible is the best way to reduce the harmful environmental impact of transportation needs. Carpooling, mass transit, biking, and walking are ways to limit the number of miles we drive. Choosing a place to live that reduces the need to drive is another way.

- Driving moderately and avoiding high-speed driving and frequent stopping and starting can reduce both fuel use and pollutant emissions.

- Simple vehicle maintenance—such as regular oil changes, air-filter changes, and spark plug replacements—can lengthen the life of your car as well as improve fuel economy and minimize emissions.

- Keeping tires properly inflated saves fuel by reducing the amount of drag a car's engine must overcome.

- During start-up, a car's engine burns extra gasoline. However, letting an engine idle for more than a minute burns more fuel than turning off the engine and restarting it.

- During warm periods with strong sunlight, parking in the shade keeps a car cooler and can minimize the evaporation of fuel.

SEE ALSO AIR POLLUTION; CARBON DIOXIDE; CARBON MONOXIDE; LEAD; NO_x; OZONE; SMOG; VOCs (VOLATILE ORGANIC COMPOUNDS).

Internet Resources

American Automobile Association. "Daily Fuel Gauge Report." Available from http://198.6.95.31/index.asp.

American Council for an Energy Efficient Economy. "GreenerCars.com." Available from http://www.greenercars.com/indexplus.html.

How Stuff Works. "How Ozone Pollution Works." Available from http://science.howstuffworks.com/ozone-pollution.htm.

U.S. Department of Energy. "Fuel Economy." Available from http://www.fueleconomy.gov.

U.S. Environmental Protection Agency. "Green Vehicle Guide." Available from http://www.epa.gov/greenvehicles.

U.S. Environmental Protection Agency. "Light-Duty Automotive Technology and Fuel Economy Trends: 1975 through 2003." Available from http://www.epa.gov/otaq/fetrends.htm.

U.S. Environmental Protection Agency. "Motor Vehicle Emissions." Available from http://www.epa.gov/otaq/ld-hwy.htm.

Weather Channel. "Health Forecast Search." Available from http://www.weather.com/activities/health/search.html?from=tabset.

David Friedman

The top of a ninety-five-foot-tall wireless phone antenna made to look like a cypress tree, blending with the other cypress trees in a Metairie, Louisiana, neighborhood. (AP/Wide World Photos. Reproduced by permission.)

deuterium a hydrogen atom with an extra neutron, making it unstable and radioactive

Visual Pollution

Visual pollution is an aesthetic issue, referring to the impacts of pollution that impair one's ability to enjoy a vista or view. The term is used broadly to cover visibility, limits on the ability to view distant objects, as well as the more subjective issue of visual clutter, structures that intrude upon otherwise "pretty" scenes, as well as graffiti and other visual defacement.

Visibility is a measure of how far and how well people can see into the distance. Haze obscures visibility. It is caused when light is absorbed or scattered by pollution particles such as sulfates, nitrates, organic carbon compounds, soot, and soil dust. Nitrogen dioxide and other pollution gases also contribute to haze. Haze increases with summer humidity because sulfate and other particles absorb moisture and increase in size. The larger the particles, the more light they scatter.

Haze is most dramatically seen as a brownish-grey cloud hovering over cities, but it also obscures many beautiful vistas in U.S. national parks. At Acadia National Park in Maine, visual range on a clear day can be 199 miles. On a hazy day, that can be reduced to 30 miles. At its worst, haze at Grand Canyon National Park was so severe that people could not see across the 10-mile wide canyon. An enormous coal-fired electric plant, the Navajo Power Generating Station, about 80 miles north of the Grand Canyon, was thought to be the source of the pollution causing canyon haze. In 1985 researchers at Colorado State University injected methane-containing **deuterium** into the power plant's smoke emissions. Deuterium is not normally present in the air. When monitors determined the presence of deuterium in canyon air, researchers were able to demonstrate that the plant was responsible for much of the canyon haze. The result was a landmark settlement in which Navajo's owners agreed to a 90-percent cutback in sulfur dioxide emissions by 1999.

Utility boilers and vehicular emissions are both major sources of haze-causing pollution. The haze problem is greatest on the east coast of the United States because of the higher levels of pollution and humidity in that region. The pollution that causes haze can travel thousands of miles, and improving regional visibility requires interstate cooperation. Wood smoke is a contributor in the west, and forest fire smoke and windblown dust are natural sources of haze.

The pollutants that cause haze are also a health concern because they often result in respiratory problems among humans and other species. Controls designed to reduce the pollution from vehicular and smokestack emissions will also reduce visual pollution. In addition, the U.S. Environmental Protection Agency (EPA) has issued regional haze regulations that call on

The Los Angeles skyline with mountain peaks visible in the background. (© Mark L. Stephenson/Corbis. Reproduced by permission.)

states to establish goals and strategies and to work together in regional groups to improve visibility in 156 national parks and wilderness areas.

In Southeast Asia, haze caused by massive forest fires cost billions of dollars in health care and lost tourist revenue in the last decade. Fires in Sumatra and Borneo affected not only Indonesia, but also Malaysia, Singapore, and Thailand. Most fires were set deliberately, and often illegally, to clear land for planting and development and to cover up illegal logging. Some of the fires spread to peat deposits beneath the forest, and these may continue to burn for years.

Visual blight—billboards, power lines, cell towers, even ugly buildings—is literally in the eye of the beholder. It is subjective. To the businessman, a well-placed billboard may be a thing of beauty. But to the traveler whose view of the rolling hills or the rustic village is obstructed, it is visual pollution.

Billboards proliferated in the 1940s and 1950s, spurred by the growth of automobile traffic and construction of interstate highway system, but in 1965 Lady Bird Johnson, wife of President Lyndon Johnson, attacked their growing presence on our nation's roadways. "Ugliness is so grim," the first lady proclaimed, and she fought for and won passage of the Highway Beautification Act of 1965. This groundbreaking law prompted a number of states, including Alaska, Hawaii, Maine and Vermont, to ban billboards totally; there were loopholes, however.

Sensitivity to visual pollution has led utility companies to bury power and telephone lines in some communities. The latest fight against visual pollution centers on cell towers, needed to provide cellular telephone service. One solution has been to disguise cell towers as trees or cacti. Graffiti, spray-painted

A similar perspective of the Los Angeles skyline, but with much of the scenery obscured by smog. (© Robert Landau/Corbis. Reproduced by permission.)

names and messages, are a form of urban visual blight. Attempts to curb graffiti by banning the sale of spray paint to minors have had little effect.

Bibliography

Gudis, Catherine (2003). *Buyways: Automobility, Billboards and the American Cultural Landscape.* New York: Routledge.

National Research Council Board on Environmental Studies and Toxicology. (1991). *Haze in the Grand Canyon: An Evaluation of the Winter Haze Intensive Tracer Experiment.* Washington, D.C.: National Academy Press.

National Research Council Environment and Resources Commission on Geosciences. (1993). *Protecting Visibility in National Parks and Wilderness Areas.* Washington, D.C.: National Academy Press.

Internet Resources

Malm, William (National Park Service and Colorado State Institute for Research on the Atmosphere). "Introduction to Visibility." Available from http://www.epa.gov/oar/visibility.

Scenic America Web site. Available from http://www.scenic.org/billboards.htm.

Richard M. Stapleton

VOCs (Volatile Organic Compounds)

volatile of any substance that evaporates readily

photochemical light-induced chemical effects

Volatile organic compounds (VOCs) are small organic molecules that take part in **photochemical** reactions in the atmosphere, resulting in smog. They have low boiling points and vaporize easily. When present in the atmosphere,

VOCs, such as benzene and ethylbenzene, are not removed by passing the air through a filter. The atmosphere also contains nonvolatile organic compounds and semivolatile species such as anthracene and nicotine. The latter separate partly on a filter and partly in the gas phase, depending on temperature. VOCs (isoprene and pinene) are emitted by living trees and decomposition of vegetation. The process of refining crude oil to various fuels and the use, spillage, and incomplete combustion of those fuels in vehicles is another major source of VOCs. When mixed with nitric oxide emissions, mainly from combustion sources, and allowed to stagnate in intense sunlight, this mix forms ozone (a colorless gas) and oxidizes much of the VOCs to involatile particulate matter that scatters and absorbs light. This combination is termed *photochemical smog*. SEE ALSO AIR POLLUTION; HEALTH, HUMAN; RISK; SMOG.

Internet Resource

U.S. Environmental Protection Agency. "Organic Gases (Volatile Organic Compounds—VOCs)." Available from http://www.epa.gov/iaq.

Donald Stedman

Volatile Organic Compounds   *See VOCs*

War

War, defined as armed conflict between nations or between opposing factions within a nation, can have grave consequences for the environment, public health, and natural resources. The impact of military tactics and weaponry extends beyond military targets to affect civilian populations and their infrastructure, air and water; armed forces directly target forests, jungles, and other ecosystems in order to deprive enemy troops of cover, shelter, and food; mass refugee movements and other disruptions caused by armed conflict can deplete nearby sources of timber and wildlife; and the general atmosphere of lawlessness that often prevails during or after conflict can make it difficult to prevent illegal logging, mining, and poaching. Even peacetime military activities and preparation for war can be extraordinarily harmful to the environment.

Although wartime environmental damage is as old as war itself, it is modern, industrial warfare that has raised the possibility of destruction on an **ecosystem** or global scale. From the use of poison gases in World War I and atomic bombs in World War II to the use of chemical **defoliants** in Vietnam and land mines in numerous internal conflicts, war now leaves a legacy that extends far beyond the battlefield and long past the duration of the original conflict. This problem has resulted in international treaties that attempt to constrain the adverse impacts of warfare on civilian populations and the environment. It also has ensured that environmental issues are closely monitored during wartime by the international community, in much the same way as humanitarian or refugee issues.

History

Wartime environmental impacts were noted as far back as the ancient world, when the Romans salted the earth around Carthage to keep the Carthaginians from replanting their fields. Medieval sieges took a heavy toll on soldiers and civilians alike. During the U.S. Civil War, General William Tecumseh

Environmental destruction as a tool of war is not new. In 146 B.C.E., at the end of the Third Punic War, Roman soldiers reportedly plowed salt into the fields of Carthage, leaving them infertile and ensuring that the North African city would never again be a challenge to the Roman Empire.

ecosystem the interacting system of a biological community and its non-living environmental surroundings

defoliant an herbicide that removes leaves from trees and growing plants

A U.S. Air Force jet spraying Agent Orange along the Cambodian border during the Vietnam War. Bettmann/Corbis. Reproduced by permission.)

Geneva Conventions humanitarian rules governing treatment of soldiers and civilians during war

Hague Conventions international agreements governing legal disputes between private parties

Sherman's "March to the Sea" laid waste to large areas of the South, including civilian settlements and farms. In World War I, British forces deliberately set Romanian oilfields afire; in World War II, both Germany and the Soviet Union engaged in "scorched earth" tactics; and in the Korean War, the United States intentionally bombed North Korean dams to cause floods.

Such tactics have always been controversial and led to periodic attempts to regulate them. The Old Testament (Deuteronomy 20:19–20) prohibits armies from cutting down fruit-bearing trees, and the Qur'an similarly commands against cutting trees or killing animals unless necessary for food. In 1863 the U.S. Army adopted the Lieber Code, which limited the actions of Union troops and was a precursor of modern military manuals. Since the twentieth century, international armed conflict has been governed by a series of treaties, the **Geneva Conventions** and the **Hague Conventions**, that have progressively restricted military tactics and weaponry, such as banning the targeting of civilian property or the use of poisonous gases. Occasionally, this body of law was directed toward environmental damage. For example, at the Nuremberg Trials, German General Alfred Jödl was found guilty of war crimes for his scorched earth tactics in occupied territory (although another general who used similar tactics, Lothar Rendulic, was found not guilty on the grounds that his actions were dictated by military necessity). However, the primary purpose of the international law of war remained humanitarian, aimed at eliminating inhumane weapons and reducing civilian casualties.

Burning oil wells in Kuwait, which were sabotaged by retreating Iraqi troops at the end of the Persian Gulf War, 1991. (©Peter Turnley/Corbis. Reproduced by permission.)

Vietnam War

The Vietnam War was the first conflict to highlight the devastating effects of modern warfare on entire ecosystems. There, U.S. forces adopted a strategy of defoliating jungle canopy, ultimately spraying "Agent Orange" and other toxic herbicides over 10 percent of South Vietnam. In addition to destroying vegetation, the public health implications of these actions—primarily birth defects, diseases, and premature deaths—have since become apparent, both in the Vietnamese population and U.S. war veterans. In his memoir *My Father, My Son*, Admiral Elmo Zumwalt Sr., the commander of U.S. naval forces in Vietnam, defended his order to defoliate Vietnamese river banks as

This Vietnamese infant was born with deformed arms and legs caused by his parents' exposure to Agent Orange. (©Owen Franken/Corbis. Reproduced by permission.)

tribunal committee or board appointed to hear and settle an issue

necessary to save American sailors from ambush, even though he acknowledged that it ultimately may have caused cancer in his own son, who was serving there at the time. U.S. veterans eventually were compensated for illnesses resulting from their exposure to Agent Orange, but proposals to compensate the Vietnamese victims have remained controversial.

The defoliation campaign and other U.S. tactics in Vietnam led to an international movement for treaties that specifically protect the environment during wartime. This resulted in adoption of the Environmental Modification Convention (1976), which prohibits manipulating the environment as a weapon of war, and of Protocol Additional I to the Geneva Conventions (1977), which includes a prohibition against "widespread, long-term and severe damage to the natural environment." However, many critics have called these treaties vague and impractical, and in fact they have yet to be applied to a specific case of wartime environmental damage. The U.S. government signed both treaties, but has never formally ratified Protocol Additional I.

Persian Gulf War

Wartime environmental damage again came to the fore during the 1990 to 1991 Persian Gulf War, in which Iraq invaded and occupied neighboring Kuwait. Driven from Kuwait by a U.S.–led military alliance, Iraqi troops deliberately ignited hundreds of Kuwaiti oil wells and diverted pipelines directly into the Persian Gulf. The resulting smoke plumes and oil slicks caused enormous harm to the Kuwaiti population and to desert and marine ecosystems and wildlife. Smoke from the oil fires was reported as far away as the Himalayas and was visible from space.

As images of the devastation circulated around the globe, the United Nations Security Council passed Resolution 687, which held Iraq liable for all damage, including environmental damage, resulting from the occupation and liberation of Kuwait. This unprecedented action resulted in the establishment of a special commission, the United Nations Compensation Commission, to verify damage claims and issue awards. Kuwait and other Gulf countries filed more than sixty billion dollars in environmental, natural resource, and public health claims against Iraq, which a decade later were still being resolved. The extraordinary nature of the Security Council's action led to renewed calls for an international treaty or institution to regulate the environmental impacts of armed conflict. Subsequently, prohibitions against environmental damage were included in the charter for the International Criminal Court, a new **tribunal** that will have global jurisdiction over war crimes.

Internal Conflicts

Although the best-known examples of wartime environmental damage occurred during international conflicts, the vast majority of recent conflicts have been civil wars or other internal strife, in places such as Angola, Cambodia, Colombia, Congo, El Salvador, Ethiopia, Liberia, Nicaragua, Rwanda, Sierra Leone, and the former Yugoslavia. These conflicts often take the form of low-level guerrilla warfare that continues for years, with the same territory changing hands several times. In addition to the tragic toll on civilian

HERBICIDE SPRAY MAP

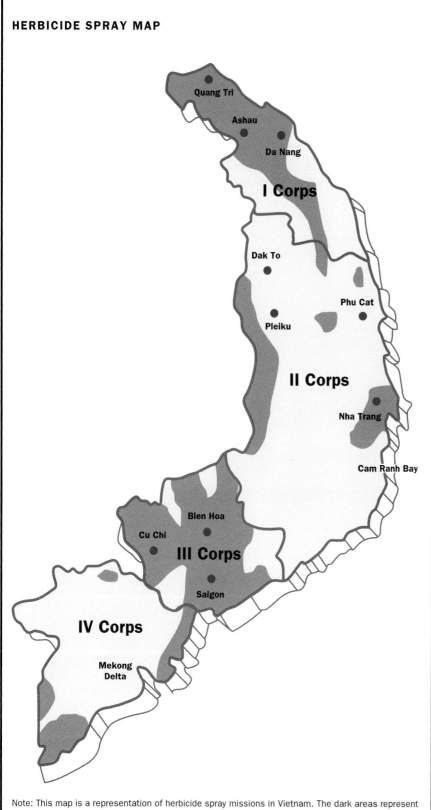

Note: This map is a representation of herbicide spray missions in Vietnam. The dark areas represent concentrated spraying areas. This map only represents fixed-wing aircraft spraying, and does not include helicopter spraying of perimeters, or other spray methods.

The III Corps area received the heaviest concentrations of spraying, followed by I Corps, II Corps, and IV Corps.

The pollution associated with military preparedness is substantial, ranging from the effects of housing, feeding, supplying, and moving large bodies of people, to the impacts of weapons practice and war games. The closure, under protest, of the U.S. Navy's live-fire bombing and artillery ranges on Vieques Island off the coast of Puerto Rico will require the cleanup of nearly sixty years of accumulation of bomb fragments, unexploded ordinance, waste munitions, and landfills. The Navy is conducting an environmental investigation under a **consent order** signed with the U.S. Environmental Protection Agency.

The Comprehensive Environmental Response, Compensation, and Liability Act, or Superfund, requires the military to clean up hazardous waste on its bases. In particular, this is required at bases being closed. The scope and cost of these cleanups are staggering, even for the Department of Defense. A RAND research study of the closure of six California bases recommended setting interim cleanup goals, concluding that "cleanup too long delayed—in the interest of fulfilling a total cleanup program—is cleanup never realized."

consent order a legal agreement requiring specific actions to remedy a violation of law

half-life the time required for a pollutant to lose one-half of its original concentration; for example, the biochemical half-life of DDT in the environment is fifteen years

populations, such conflicts have considerable environmental impacts: Opposing armies engage in deforestation and defoliation, hunt wildlife for food, lay thousands of antipersonnel land mines, and clash over valuable natural resources (such as timber and diamonds) to finance their arms purchases.

Because sovereign nations generally control their own affairs, it has been very difficult for the international community to address internal conflicts and their human and environmental consequences. Most international treaties governing wartime environmental damage do not apply to internal conflict, and even where they do, they are difficult to apply to loosely organized guerilla forces. Armed intervention or peacekeeping missions can solve some humanitarian and environmental problems while creating others. For example, the 1999 NATO bombing of Kosovo ignited a petrochemical plant in the city of Pancevo, exposing thousands of civilians to a cloud of toxic fumes; during the Rwandan civil war, United Nations refugee camps stressed natural resources and wildlife reserves in neighboring Congo. Another attempted solution has been global consumer boycotts of tropical timber, diamonds, and other commodities that originate in war-torn countries and give rise to or finance armed conflict.

The Cold War Legacy

Military activities and preparations for war can have enormous environmental impacts even without a shot being fired. The development of the atomic bomb during the early 1940s, referred to as the Manhattan Project, not only had devastating consequences in Hiroshima and Nagasaki, but also produced a long-lasting legacy of deadly radioactive pollution in the United States. In 1939 Nobel Prize physicist Niels Bohr warned that although it was possible for the United States to build an atom bomb, it could not be done without "turning the country into a gigantic factory." Following the end of the Cold War in 1991, it became apparent to what extent that factory had contaminated such diverse sites as Hanford, Washington; Oak Ridge, Tennessee; and Rocky Flats, Colorado; where the air, groundwater, surface water, soil, vegetation, and wildlife all show signs of radioactivity. The Soviet Union's nuclear program created similar problems, concentrating production in "secret cities" such as Chelyabinsk-7, which many have called the most polluted city on earth. Given the highly toxic nature and extremely long **half-life** of most radioactive waste, cleanup and containment of these sites will pose problems for generations.

The Cold War legacy brings into focus the "necessity" and "proportionality" calculations that underlie most reasoned decisions about environmentally damaging wartime actions: whether there are alternatives to taking a particular action, and whether the military advantage gained from taking such an action outweighs the environmental and other harm that potentially may result. Most scholars would agree that the development of the atomic bomb was justifiable as a means of defeating fascism and winning World War II; they similarly agree that Iraq's actions in retreating from Kuwait were indefensible, even on military grounds. Other cases, such as the United States' defoliation campaign in Vietnam or bombing of civilian infrastructure in Kosovo, are more controversial. In any case, the historical record, the continued development of international treaties and institutions, and the increasing awareness that environmental issues must be considered even during wartime, all should provide a basis for improved military tactics and more environmentally aware decision making in the future. SEE ALSO TERRORISM.

Bibliography

Austin, Jay E., and Bruch, Carl, eds. (2000). *The Environmental Consequences of War: Legal, Economic, and Scientific Perspectives.* Cambridge, UK: Cambridge University Press.

Bloom, Saul; Miller, John M.; Warner, James; and Winkler, Philippa, eds. (1994). *Hidden Casualties: The Environmental, Health and Political Consequences of the Persian Gulf War.* Berkeley, CA: North Atlantic Books.

Browne, Malcolm W. (1991). "War and the Environment." *Audubon* 93:89.

Dycus, Stephen. (1996). *National Defense and the Environment.* Hanover, NH: University Press of New England.

Earle, Sylvia A. (1992). "Persian Gulf Pollution: Assessing the Damage One Year Later." *National Geographic* 181:122.

Feshbach, Murray, and Friendly, Albert. (1992). *Ecocide in the U.S.S.R.: The Looming Disaster in Soviet Health & Environment.* New York: Basic Books.

Hawley, T.M. (1992). *Against the Fires of Hell: The Environmental Disaster of the Gulf War.* New York: Harcourt Brace Jovanovich.

Lanier-Graham, Susan. (1993). *The Ecology of War: Environmental Impacts of Weaponry and Warfare.* New York: Walker & Co.

Levy, Barry S., and Sidel, Victor W., eds. (1997). *War and Public Health.* New York: Oxford University Press.

Rhodes, Richard. (1986). *The Making of the Atomic Bomb.* New York: Simon & Schuster.

Roberts, Guy B. (1991). "Military Victory, Ecological Defeat." In *Worldwatch*, July/Aug. 1991.

Webster, Donovan. (1996). *Aftermath: The Landscape of War.* New York: Pantheon.

Weinberg, William J. (1992). *War on the Land: Ecology and Politics in Central America.* London: Zed Press.

Zumwalt, Elmo Jr.; Zumwalt, Elmo III; and Pekkanen, John. (1986). *My Father, My Son.* New York: Macmillan.

Internet Resources

Environmental Change and Security Project. "Bibliographic Guide to the Literature." Available from http://wwics.si.edu/PROGRAMS.

Environmental Law Institute. (1998). "Addressing Environmental Consequences of War: Background Paper for the First International Conference on Addressing Environmental Consequences of War: Legal, Economic, and Scientific Perspectives." Washington, D.C.: Environmental Law Institute. Available from http://www.eli.org/pdf.

Environmental Law Institute. (1998). "Annotated Bibliography: First International Conference on Addressing Environmental Consequences of War: Legal, Economic, and Scientific Perspectives." Washington, D.C.: Environmental Law Institute. Available from http://www.eli.org/pdf.

Hoffman, Leslie. "Saving the Ghost Ship." *Albuquerque Tribune*, July 31, 1998. Available from http://www.abqtrib.com/arc1.

United Nations Environment Programme. (1999). "The Kosovo Conflict: Consequences for the Environment & Human Settlements." Geneva: United Nations Environment Programme. Available from http://www.grid.unep.ch/btf.

Jay Austin

Ever since the U.S.S. *Arizona* sank in Pearl Harbor on December 7, 1941, a slow trickle of fuel oil has seeped toward the surface, casting a rainbow sheen on the now-still waters. The *Arizona* had 1.5 million gallons of oil in its tanks when it was attacked, and it is unknown how much remains. Although the current 2.5-gallon-per-day leak does not present much of an environmental hazard, the caretakers of what is now the Pearl Harbor National Monument have made plans to minimize impacts if the *Arizona's* hull collapses and releases the remainder into the harbor's fragile marine ecosystem.

Warren County, North Carolina

In 1982 residents of the predominantly African-American Warren County, North Carolina, began to protest the construction of a hazardous waste land-fill near Warrenton in which the state planned to bury 400,000 cubic yards of soil contaminated with polychlorinated biphenyls (PCBs). The contamination occurred when a disposal contractor dripped approximately 12,850 gallons of PCB-tainted fluids along 210 miles of roads in fourteen counties in North Carolina in 1978. Soon after the spill was discovered, the state acquired a 142.3-acre tract of land on which it proposed building a 19.3-acre landfill to bury the wastes. Opponents of the Warren County site filed two lawsuits in 1979 in their attempts to halt plans for the landfill.

At the time, the Warren County site, chosen from ninety sites considered, had a higher percentage of African-American residents of any county in the state. It was 64 percent black and the unincorporated Shocco Township, site of the landfill, was 75 percent black. Warren County ranked ninety-seventh in per capita income out of North Carolina's one hundred counties.

In November 1981 the district courts ruled against landfill opponents. Shortly thereafter protests began; these received widespread national attention. Local police and soldiers from the U.S. Army base at Fort Bragg (which was also contaminated with PCBs) were called in to quell the protests. In total, 523 people were arrested, including local congressman Walter Fauntroy and members of the United Church of Christ Commission for Racial Justice. Fauntroy and other protesters urged the General Accounting Office (USGAO) to examine the relationship between the location of landfills in the Southeast and the demographics of host communities. This led to the publication of the well-known 1983 USGAO study.

Four years later, the United Church of Christ (UCC) Commission for Racial Justice published a national study examining the siting of hazardous facilities and waste sites. Both of these widely cited studies had a significant impact on mobilizing minority communities around environmental issues and the growth of the environmental justice movement. They were among the earliest studies to link race with the increased likelihood of close proximity to hazardous facilities and toxic waste sites. Unlike other studies of the same genre, they were widely circulated among minority activists and in minority communities. SEE ALSO ENVIRONMENTAL JUSTICE.

Bibliography

LaBalme, Jenny. (1988). "Dumping on Warren County." In *Environmental Politics: Lessons from the Grassroots*, edited by Bob Hall. Durham, NC: Institute for Southern Studies, pp. 23–30.

Twitty v. *State of North Carolina*. (1981). 527 F. Supp. 778; 1981 U.S. District, Nov. 25.

UCC. (1987). *Toxic Waste and Race in the United States*. New York: United Church of Christ.

U.S. General Accounting Office (USGAO). (1983). *Siting of Hazardous Waste Landfills and Their Correlation with the Racial and Socio-economic Status of Surrounding Communities*. Washington, D.C.: General Accounting Office.

Warren County v. *State of North Carolina*. (1981). 528 F. Supp. 276; 1981 U.S. District, Nov. 25.

Dorceta E. Taylor

Waste

Waste has been defined as a moveable object with no direct use that is discarded permanently. There are many different kinds of waste, including solid, liquid, gaseous, hazardous, radioactive, and medical. Wastes can also be

defined by generator, for example, municipal, commercial, industrial, or agricultural.

Garbage in bags and containers accumulating at curbside. (U.S. EPA. Reproduced by permission.)

Waste Types

A solid waste does not flow like water or gas. Examples include paper, wood, metals, glass, plastic, and contaminated soil. Solid wastes can be hazardous or nonhazardous. Problems associated with nonhazardous solid waste include aesthetic problems (litter and odors), leachate from the infiltration of water through the waste, and off-gases resulting from biodegradation. Nonhazardous solid wastes are commonly handled by recycling, combustion, landfilling, and composting.

Liquid wastes must be transported in containers or through pipes. Examples include sewage, contaminated groundwater, and industrial liquid discharges. In some cases, direct discharge to the environment may be allowed. However, depending on the waste's characteristics, direct discharge may cause unacceptable environmental harm. For example, large amounts of sewage discharged into a stream can result in fish kills. Liquid wastes containing excreta can contain pathogenic organisms. Other liquid wastes may be toxic. Liquid wastes are often handled at wastewater treatment plants, followed by discharge to the environment.

Sludges contain various ratios of liquid and solid material. They generally result from liquid waste-treatment operations, such as sedimentation tanks. Depending on the percent of solids, sludge may have the characteristics of a liquid or solid. Biological sludge can contain pathogenic organisms. Some sludges contain heavy metals or other toxins. Sludges are commonly handled with treatment, combustion, landfilling, and land application.

Gaseous wastes, of course, consist of gases. They are primarily generated by combustion (e.g., internal combustion engines, incinerators, coal-fired electrical generating plants) and industrial processes. Depending on their characteristics, gaseous wastes can be odiferous or toxic. Some are implicated in global warming, ozone depletion, and smog. Gaseous wastes may be released to the atmosphere or captured/treated with pollution control equipment.

Hazardous wastes pose a substantial present or potential danger to human health or the environment. They can be solid, sludge, liquid, or gas. Hazardous wastes have at least one of the following characteristics: corrosivity, ignitability, reactivity, and toxicity. Hazardous wastes are commonly handled by recycling, combustion, stabilization, chemical-physical-biological treatment, and landfilling.

half-life the time required for a pollutant to lose one-half of its original concentration; for example, the biochemical half-life of DDT in the environment is 15 years

Radioactive wastes emit particles or electromagnetic radiation (e.g., alpha particles, beta particles, gamma rays, and x rays). Radioactive wastes can be high level, transuranic, or low level. High-level radioactive wastes are from spent or reprocessed nuclear reactor fuel. Transuranic wastes are from isotopes above uranium in the periodic table. They are generally low in radioactivity, but have long **half-lives**. Low-level wastes have little radioactivity and can often be handled with little or no shielding. Radiation can damage living cells and cause cancer. Although recycling and incineration may reduce waste amounts, the primary method for handling radioactive wastes is long-term storage.

Medical wastes, that is, wastes generated at medical facilities, can be infectious, toxic, and/or radioactive. Though they may have hazardous characteristics, they are not regulated as hazardous wastes. Some medical wastes are sterilized, disinfected, or incinerated, especially infectious wastes. Recycling and landfilling are also used to dispose of them.

Waste Amounts

The amount of waste generated by a given household is directly related to lifestyle, culture, and economic status. Climate can also increase generation rates (e.g., yard waste). General differences are great enough to produce different country-wide generation rates. The United States has the highest rate, 2.0 kilograms per person per day—probably the result of high economic status, a culture of consumption, and a lifestyle that includes large amounts of disposable items. However, the United States also has a relatively high recycling rate, 27.8 percent in 1999. Some European countries have generation rates varying from 0.9 to 1.7 kilograms per person per day. Developing regions tend to have still lower rates, ranging from 0.3 to 1. SEE ALSO AIR POLLUTION; HAZARDOUS WASTE; LIFESTYLE; MEDICAL WASTE; OZONE; RADIOACTIVE WASTE; SOLID WASTE; WASTE TO ENERGY; WASTE, TRANSPORTATION OF; WASTEWATER TREATMENT.

Bibliography

Davis, M., and Cornwell, D. (1998). *Introduction to Environmental Engineering*, 3rd edition. New York: WCB McGraw-Hill.

Reinhardt, P., and Gordon, J. (1991). *Infectious and Medical Waste Management*. Chelsea, MN: Lewis Publishers.

Siegel, M. (1993). "Garbage and Other Pollution—How Do We Live with All the Trash?" *Information Plus*. Detroit, MI: Gale.

Tchobanoglous, G.; Theisen, H.; and Vigil, S. (1993). *Integrated Solid Waste Management*. New York: McGraw-Hill.

Jess Everett

Waste, Hazardous   *See Hazardous Waste*

Waste, International Trade in

During the past three decades, one of the most persistent international environmental issues has been the toxic waste trade between industrialized countries and less developed nations. From 1968 to 1988 alone, more than 3.6 million tons of toxic waste—solvents, acetone, cobalt, cadmium, chemical and pharmaceutical waste, and perhaps some low-level radioactive waste—were shipped to less developed nations. The saga of the freighter *Khian Sea* is a graphic example of this trade. In 1988 the ship departed from Philadelphia loaded with toxic incinerator ash. Four thousand tons of the waste, which contained dioxin and furans, two of the most toxic chemicals known to humans, were dumped on a beach in Haiti. No effort has ever been made to clean it up. Another ten thousand tons were later dumped illegally at sea.

That same year, another international toxic waste shipment led to a major diplomatic row in New Guinea. The government there jailed a Norwegian consul and fined him $600 after a Norwegian ship transported fifteen thousand tons of incinerator ash from the United States and dumped it in New Guinea.

Ironically, the growing clout of environmentalists in the United States has driven much of this trade. Strict U.S. laws now regulating toxic waste disposal have considerably increased the cost of disposing of toxic waste. In 2001 one U.S. official estimated that it cost from $250 to $300 a ton to dispose of toxic wastes in the United States, whereas some developing countries have accepted the same wastes for as little as $40 per ton.

Officials of some developing nations have called the trade "toxic terrorism" and "garbage imperialism," and others worry that the developing world will change from "the industrialized world's backyard to its outhouse," as an African official said. Many developing countries have had little appreciation of both the short- and long-term health and environmental risks of toxic waste and the dangers it can create.

Those developing countries willing to accept shipments of toxic waste are usually enticed by the prospect of millions of dollars that can be made for their struggling economies. In one deal, for example, the local government of Oro, New Guinea, negotiated a deal with Global Telesis Corporation, a firm from California, to build in that province a $38 million detoxification plant, which would process six million metric tons of toxic waste a month from the West Coast. The deal fell through under pressure by the national government and because of concerns that Global Telesis would not be able to raise the necessary funding.

Since the 1980s, when the issue of international trade in toxic waste first came to widespread attention, there has been a global movement to ban it. In March 1989, 105 countries met under the auspices of United Nations Environment Program (UNEP) in Basel, Switzerland, and passed the Basel Convention on the Control of Transboundary Movement of Hazardous Wastes and Their Disposal. When the convention went into effect in 1992, eighty-eight countries signed it. In 1995 parties to the 1989 Basel Convention agreed to make legally binding a voluntary export ban that was agreed to in 1994. Surprisingly, the United States, one of the original proponents of the Basel Convention, has not thus far ratified the agreement. Supporters of stronger international toxic trade laws insist that, for the Basel Convention to have full force, the United States, as the world's leading producer of toxic waste, must be convinced to sign it.

Bibliography

French, Hilary. (2001). "Can Globalization Support the Export of Hazard." In *USA Today (Magazine)*, May 2001, p. 20ff.

"Ratifying Global Toxic Treaties: The United States must Provide Leadership." (2002) In *SAIS Review*, Winter-Spring, 2002, p. 109ff.

Internet Resource

Lewis, Deana L., and Chepesiuk, Ron. "The International Trade in Toxic Waste; A Select Bibliography." Available from http://egj.lib.uidaho.edu/egj02/lewis01.htm.

Ron Chepesiuk

Waste, Transportation of

The transportation of waste is the movement of waste over a specific area by trains, tankers, trucks, barges, or other vehicles. The types of wastes that may be transported range from municipal garbage to radioactive or hazardous wastes.

Hazardous wastes may be transported to be treated, stored, or disposed of. Facilities that generate hazardous waste are required to prepare a shipping document, or "manifest," to accompany the waste as it is transported from the site of generation. This manifest must accompany the waste until its final destination and is used to track the wastes from cradle-to-grave.

The potential for pollution releases during the transportation of waste varies; the more hazardous the waste and the larger the volume that is transported, the more devastating the environmental/human health impact if an accident occurs. Traffic accidents or train wrecks can result in waste spills and releases of pollutants that may contaminate the air, water, and soil. Wastes may also be released while being loaded or unloaded during transportation.

Approximately four billion tons of regulated hazardous materials are shipped within the United States each year with more 250,000 shipments entering the U.S. transportation system daily. The Emergency Response Notification System (ERNS) database of the Environmental Protection Agency (EPA) shows that from 1988 to 1992 an average of nineteen transportation accidents involving toxic chemicals occurred each day.

DOT Regulations

The U.S. Department of Transportation (DOT) requires that placards identifying the type of hazardous material being transported be placed on the outside of any vehicle transporting hazardous materials or wastes. Placards are used to determine potential hazards in the event of a spill and are placed on all four sides of a vehicle so that **HAZMAT teams**, fire, emergency, medical, and other personnel who respond to accidents may quickly identify the contents

HAZMAT team hazardous materials response group

and associated hazards. Placards are required if one thousand pounds or more of a hazardous material is transported and if any amount of material classified as explosive, poisonous, radioactive, or a flammable solid is transported. The DOT classifies materials based on nine hazard classes represented by symbols. The classes are explosives, gases, **flammable** liquids, flammable solids, oxidizers, poisonous materials, biohazards, radioactive materials, corrosives, or other regulated materials.

The routes that transporters of hazardous waste use must be carefully considered to minimize the risk of an accidental release. If possible, densely populated areas should be avoided. The type of highway or road and the weather conditions along the route must also be considered. Risk analysis may become important in selecting routes for hazardous waste transport in order to minimize adverse impacts to human health in case of an accidental release.

flammable any material that ignites easily and will burn rapidly

Municipal Waste

Due to rapidly decreasing space in urban landfills, officials have been forced to find alternate locations for municipal waste disposal. This has created significant financial incentives for rural communities to accept garbage from urban areas. Depending on the location of these rural facilities, it may be necessary to transport large quantities of wastes by a variety of methods, most often by truck, railway, or barge. Many citizens are concerned about the transportation of the waste through their communities and the risks involved. People are also concerned that the municipal waste from urban areas may be contaminated with toxic chemicals or substances that could contaminate local drinking water supplies.

Disposal of hazardous wastes in the United States can cost up to $2,500 per ton. This has led to the practice of selling waste to developing countries for disposal at a much lower cost. This international waste trade may be illegal in some instances, but the hefty sum paid to those who accept the wastes remains tempting to developing countries. However, the actual composition of the wastes received by developing countries is often misrepresented by those selling the waste. In addition, most developing countries lack the resources and technical expertise to safely manage these hazardous wastes.

Trade in hazardous wastes is a global issue. About ten percent of all hazardous wastes generated around the world cross international boundaries. A large portion goes from industrialized countries to developing countries where disposal costs are lower. Although developing countries may lack the financial and technical capacities to clean up hazardous waste releases in their countries, these countries nevertheless are sites for treatment, recycling, and disposal of wastes from abroad.

The Basel Convention on the Control of the Transboundary Movement of Hazardous Wastes and Their Disposal is the first global environmental treaty to control the international trade of waste. Under the Convention, trade in hazardous wastes cannot take place without the consent of the importing country and cannot occur under conditions that are assessed as not environmentally sound. As of April 2002, 150 countries had ratified the convention. A new protocol adopted by the convention in 2000 provides the first international framework establishing liability for damages that may result from the transportation or disposal of hazardous wastes across foreign

MOBRO BARGE ACCOUNT

Due to overcapacity at the Islip landfills, New York, officials negotiated with Jones County, North Carolina, to accept 3,200 tons of municipal garbage in March 1987. The garbage was transported on the *Mobro* barge. When officials discovered hospital wastes in the garbage, North Carolina refused to accept it for fear that it might contaminate local water supplies. Louisiana, Mexico, Belize, British Honduras, and the Bahamas all refused to accept the contaminated garbage and the *Mobro* returned to New York. The *Mobro* then began a six-thousand-mile, six-month voyage looking for some place to take the garbage. After several court battles, the controversy ended when numerous flatbed trucks were used to transport the garbage to a Brooklyn incinerator where the volume was reduced and the ash was landfilled.

—Goff, Liz. "The Old Disaster: Queens' Garbage Standoff." The *Queens Tribune*. Available from http://queenstribune.com/archives /featurearchive/feature2001/0208 /feature_story.html

—"The Voyage of the Mobro." Available from http://www.grace-space.com/Hamilton/recycle.htm.

borders. SEE ALSO ECONOMICS; HAZARDOUS WASTE; LAWS AND REGULATIONS, UNITED STATES; RADIOACTIVE WASTE; SEWAGE SLUDGE; SOLID WASTE.

Bibliography

La Grega, Michael D.; Buckingham, Philip L.; Evans, Jeffrey C., and Environmental Resources Management. (2001). *Hazardous Waste Management.* Boston: McGraw Hill.

Watts, Richard J. (1998). *Hazardous Wastes: Sources, Pathways, Receptors.* New York: John Wiley & Sons.

Internet Resources

U.S. Department of Transportation. "HAZMAT Safety." Available from http://hazmat.dot.gov.

U.S. Environmental Protection Agency. "Waste Transportation." Available from http://www.epa.gov/ebtpages/wastwastetransportation.html.

Margrit von Braun and Deena Lilya

Waste Reduction

Waste reduction, also known as source reduction, is the practice of using less material and energy to minimize waste generation and preserve natural resources. Waste reduction is broader in scope than recycling and incorporates ways to prevent materials from ending up as waste before they reach the recycling stage. Waste reduction includes reusing products such as plastic and glass containers, purchasing more durable products, and using reusable products, such as dishrags instead of paper towels. Donating products, from office equipment to eyeglasses and clothing, reduces the amount of material manufactured overall. Purchasing products that replace hazardous materials with biodegradable ingredients reduces pollution as well as waste. In general, waste reduction offers several environmental benefits. Greater efficiency in the production and use of products means less energy consumption, resulting in less pollution. More natural resources are preserved. Products using less hazardous materials are used. Finally, less solid waste ends up in landfills.

Waste reduction also means economic savings. Fewer materials and less energy is used when waste-reduction practices are applied. Rather than using the traditional cradle-to-grave approach, a cradle-to-cradle system is adopted. In this cradle-to-cradle system, also called industrial ecology, products are not used for a finite length of time. Instead of disposing of materials, or the components of a product after a single use, products are passed on for further uses. This is considered a flow of materials. This can be applied within an organization, or between organizations that may be considered unrelated, on a cooperative basis. For example, a cotton manufacturer sends its unwanted scraps to an upholsterer, who uses the scraps as stuffing in chairs. When the life span of the chair is reached, the materials are returned to the manufacturer, who reuses the parts with endurance. The damaged upholstery, which was originally created using nonhazardous materials, is sold to a local farmer who uses it in composting. Money is also saved through reduced purchasing. Waste-disposal costs are decreased because fewer materials end up as waste.

Waste can be reduced by individuals, businesses, institutions such as hospitals or educational facilities, organizations, municipalities, or government agencies. There are several ways individuals can practice waste reduction: (1) Reusing products. This could mean reusing file folders rather than throwing them away after one use, or refilling water bottles; (2) Using products more efficiently. This could mean using both sides of paper in photocopying; and (3) Donating or exchanging products or materials that may seem useless, but that another party may find valuable. For example, the chair manufacturer mentioned above had no internal use for the scrap upholstery leftover after recycling the more durable parts of the used chairs. However, a cooperative agreement with a local farmer allowed the scraps to be used once again, benefiting the farmer by adding to his compost.

The EPA's WasteWise Program

The Environmental Protection Agency (EPA) lists waste reduction and reuse as top priorities in its solid waste management hierarchy, followed by recycling, composting, **waste-to-energy**, and **landfilling**. Many governments and businesses have adopted the practice of waste reduction.

The EPA offers a free, comprehensive waste-reduction program to businesses, organizations, and municipalities. The program, called WasteWise, offers educational and technical assistance in developing, executing, and measuring waste-reduction activities. Through WasteWise, groups can design and maintain a waste-reduction program that is flexible to their specific needs. The nationwide program was started in 1994, and it had over eleven hundred participating partners in 2002.

Large corporations, universities, and cities across the country have seen significant benefits, both economically and environmentally, by using WasteWise.

The National Recycling Coalition Recommendations

The National Recycling Coalition lists several steps that purchasing departments of organizations can use in their waste-reduction strategies:

1. Reduce product use. Adopt the practice of printing on both sides of office paper.

2. Rent or lease products or equipment. This includes leasing, rather than purchasing, equipment such as photocopiers, which can become obsolete, leaving the organization with old, unnecessary, and sometimes hazardous equipment to discard.

3. Purchase remanufactured or rebuilt products, or products that can be refurbished.

4. Purchase more durable products. Higher-quality products typically have a longer life cycle.

5. Purchase products that use nonhazardous materials. Nonhazardous materials are safer for individuals and landfills.

6. Purchase returnable, reusable, or refillable products. For instance, transport containers can be reused.

7. Purchase products in bulk.

8. Purchase products that reuse packaging or use less packaging.

ZERO EMISSIONS

In natural ecosystems, what is waste for one species is food for another. The concept of zero emissions, first elucidated in the early 1990s by Gunter Pauli, applies this principle to business endeavors and is being tested in Burlington, Vermont. At a 3,200 square meter eco-industrial complex enclosing a number of greenhouses. Waste heat from an existing power plant, fueled by discarded Christmas trees, will warm the greenhouses and fire up the brew kettle for a microbrewery. Pilot tests have shown that "wastes" from the brewing process can be efficiently transformed into nutritious growing medium for marketable mushrooms, salad greens and fish. What remains can be sold as cattle feed and soil amendment.

waste-to-energy to convert solid waste into a usable form of energy

landfills sanitary landfills are disposal sites for nonhazardous solid wastes spread in layers, compacted to the smallest practical volume, and covered by material applied at the end of each operating day; secure chemical landfills are disposal sites for hazardous waste, selected and designed to minimize the chance of release of hazardous substances into the environment

9. Share and reuse resources within the organization. Companies can implement an internal computer equipment and office supply exchange before purchasing new products.

The EPA reports that 232 million pounds of waste were generated in 2000. The amount of waste produced per person has grown over the last thirty-five years, from 2.7 to 4.6 pounds per day. In 1999, waste reduction saved over fifty million tons of municipal solid waste from being dumped into landfills. SEE ALSO ABATEMENT; COMPOSTING; GREEN; LIFESTYLE; RECYCLING; REUSE; TECHNOLOGY, POLLUTION PREVENTION.

Bibliography

National Recycling Coalition. (1999). *Purchasing Strategies to Prevent Waste and Save Money*. Alexandria, VA: Author.

Other Resources

U.S. Environmental Protection Agency. "WasteWise: Preserving Resources, Preventing Waste." Available from http://www.epa.gov/wastewise.

Terra Lenihan

Waste to Energy

Waste to energy (WTE) is the term used to describe the conversion of waste by-products into useful steam or steam-generated electricity. Typically, WTE is produced by converting municipal solid waste (MSW), which is defined as residential and commercial refuse, and makes up the largest source of waste in industrialized countries. This industry has been producing heat and power in the United States for a century, and there are currently more than one hundred WTE plants nationwide. Recently, however, the definition of waste has been expanded from MSW to include wastes such as wood, wood waste, peat, wood sludge, agricultural waste, straw, tires, landfill gases, fish oils, paper industry liquors, railroad ties, and utility poles. In 1999 these by-products produced approximately 3.2 quadrillion BTUs (i.e., 1×10^{15} British thermal units, which is also known as a quad) of energy out of approximately 97.0 quads of energy consumed in the United States.

Nearly thirty million tons of trash are processed each year in WTE facilities to generate steam and electricity. The benefits to society include the following: preventing the release of greenhouse gases such as methane into the atmosphere if the trash were landfilled; reducing the impact on landfills by reducing the volume of the waste 80 to 90 percent; providing an alternative to coal use, which prevents the release of emissions such as nitrogen oxides into the atmosphere; and saving the earth's natural resources by using less oil, coal, or natural gas for electricity generation.

The Process of Converting Waste to Energy

Generally, WTE facilities can be divided into two process types: mass burn and refuse-derived fuel (RDF). Mass burn facilities process raw waste that has not been shredded, sized, or separated before combustion, although large items such as appliances and hazardous waste materials and batteries are removed before combustion. In mass burn systems, untreated MSW is simply burned, with the heat produced converted into steam, which can then be passed through a steam **turbine** to generate electricity or used directly to supply heat to nearby industries or buildings.

RDF is a result of processing MSW to separate the combustible fraction from the noncombustibles, such as metals and glass. RDF is mainly composed of paper, plastic, wood, and kitchen or yard wastes, and has a higher energy content than untreated MSW. Like MSW, RDF is then burned to produce steam and/or electricity. A benefit of using RDF is that it can be shredded into uniformly sized particles or compressed into briquettes, both of which facilitate handling, transportation, and combustion. Another benefit of RDF rather than raw MSW is that fewer noncombustibles such as heavy metals are burned.

Energy Production from Waste in the United States and South America

South America, with its agrarian societies, surprisingly consumes very few wastes for the production of steam or electricity. Brazil is the largest country in South America and is also the largest energy consumer, consuming about 8.5 quads of energy each year as compared to 6.1 quads for Mexico, 12.5 quads for Canada, and 97.0 quads for the United States. Due to the large size of Brazil's agricultural sector, **biomass** is seen as the best future alternative energy source. Currently, Brazil produces about 4,000 gigawatt (1×10^9) hours annually (i.e., 0.1 quads equivalent) in the sugar industry to run its own refineries and distilleries. At the same time, Brazil produces up to 3.9 billion gallons of ethanol (i.e., 0.5 quads equivalent) for automobiles each year, although it is manufactured from sugar and not waste materials. No other South American countries produce significant quantities of energy from waste; however, Argentina's biomass energy use, like Brazil's, is expected to grow in the coming years.

In the United States, corn is the primary feedstock along with barley and wheat that is currently being used to produce ethanol, although neither corn or grains are considered wastes. Considerable ongoing research is exploring the use of true biomass wastes such as corn stover or wood chips and sawdust for ethanol production. One project at the U.S. Department of Energy involves the cofiring of sawdust and tires with coal in utility boilers. SEE ALSO RENEWABLE ENERGY.

Internet Resources

Energy Information Agency. "Energy in the Americas." Available from http://www.eia.doe.gov/emeu/cabs/theamericas.html.

Energy Information Agency. "Renewable Energy Annual 2000." Available from http://www.eia.doe.gov/cneaf/solar.renewables/page/rea_data/rea_sum.html.

Integrated Waste Services Association. "About Waste-to-Energy." Available from http://www.wte.org/waste.html.

Bruce G. Miller

Wastewater Treatment

Wastewater is simply water that has been used. It usually contains various pollutants, depending on what it was used for. It is classified into two major categories, by source:

turbine machine that uses a moving fluid (liquid or gas) to gas to turn a rotor, creating mechanical energy

biomass all of the living material in a given area; often refers to vegetation

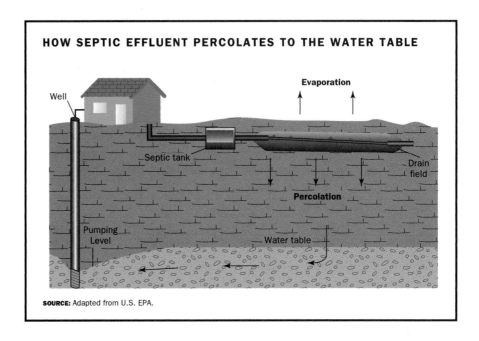

HOW SEPTIC EFFLUENT PERCOLATES TO THE WATER TABLE

SOURCE: Adapted from U.S. EPA.

1. Domestic or sanitary wastewater. This comes from residential sources including toilets, sinks, bathing, and laundry. It can contain body wastes containing intestinal disease organisms.

2. Industrial wastewater. This is discharged by manufacturing processes and commercial enterprises. Process wastewater can contain rinse waters including such things as residual acids, plating metals, and toxic chemicals.

Wastewater is treated to remove pollutants (contaminants). Wastewater treatment is a process to improve and purify the water, removing some or all of the contaminants, making it fit for reuse or discharge back to the environment. Discharge may be to surface water, such as rivers or the ocean, or to groundwater that lies beneath the land surface of the earth. Properly treating wastewater assures that acceptable overall water quality is maintained.

In many parts of the world, including in the United States, health problems and diseases have often been caused by discharging untreated or inadequately treated wastewater. Such discharges are called water pollution, and result in the spreading of disease, fish kills, and destruction of other forms of aquatic life. The pollution of water has a serious impact on all living creatures, and can negatively affect the use of water for drinking, household needs, recreation, fishing, transportation, and commerce.

Objectives and Evolution of Wastewater Treatment

We cannot allow wastewater to be disposed of in a manner dangerous to human health and lesser life forms or damaging to the natural environment. Our planet has the remarkable ability to heal itself, but there is a limit to what it can do, and we must make it our goal to always stay within safe bounds. That limit is not always clear to scientists, and we must always take the safe approach to avoid it.

Basic wastewater treatment facilities reduce organic and suspended solids to limit pollution to the environment. Advancement in needs and technology have necessitated the evolving of treatment processes that remove dissolved matter and toxic substances. Currently, the advancement of scientific knowledge and moral awareness has led to a reduction of discharges through pollution prevention and recycling, with the noble goal of zero discharge of pollutants.

Treatment technology includes physical, biological, and chemical methods. Residual substances removed or created by treatment processes must be dealt with and reused or disposed of in a safe way. The purified water is discharged to surface water or ground water. Residuals, called sludges or biosolids, may be reused by carefully controlled composting or land application. Sometimes they are incinerated.

Since early in history, people have dumped sewage into waterways, relying on natural purification by dilution and by natural bacterial breakdown. Population increases resulted in greater volume of domestic and industrial wastewater, requiring that we give nature a helping hand. Some so-called advancements in cities such as Boston involved collecting sewage in tanks and releasing it to the ocean only on the outgoing tide. Sludge was barged out to sea so as to not cause complaint.

Until the early 1970s, in the United States, treatment mostly consisted of removal of suspended and floating material, treatment of biodegradable organics, and elimination of pathogenic organisms by disinfection. Standards were not uniformly applied throughout the country.

In the early 1970s until about 1980, aesthetic and environmental concerns were considered. Treatment was at a higher level, and nutrients such as nitrogen and phosphorus were removed in many localities.

Since 1980, focus on health concerns related to toxics has driven the development of new treatment technology. Water-quality standards were established by states and the federal government and had to be met as treatment objectives. Not just direct human health but aquatic-life parameters were considered in developing the standards.

Wastewater Treatment Types

Rural unsewered areas, for the most part, use septic systems. In these, a large tank, known as the septic tank, settles out and stores solids, which are partially decomposed by naturally occurring anaerobic bacteria. The solids have to be pumped out and hauled by tank truck to be disposed of separately. They often go to municipal wastewater treatment plants, or are reused as fertilizer in closely regulated land-application programs. Liquid wastes are dispersed through perforated pipes into soil fields around the septic tank.

Most urban areas with sewers first used a process called primary treatment, which was later upgraded to secondary treatment. Some areas, where needed, employ advanced or tertiary treatment. Common treatment schemes are presented in the following paragraphs.

Primary Treatment. In primary treatment, floating and suspended solids are settled and removed from sewage.

Flow from the sewers enters a screen/bar rack to remove large, floating material such as rags and sticks.

The liquid and solid material removed from domestic septic tanks is called septage. Most septage is hauled to municipal sewage treatment facilities and most septage haulers must be licensed.

clarifier a tank in which solids settle to the bottom and are subsequently removed as sludge

It then flows through a grit chamber where heavier inorganics such as sand and small stones are removed.

Grit removal is usually followed by a sedimentation tank/**clarifiers** where inorganic and organic suspended solids are settled out.

To kill pathogenic bacteria, the final effluent from the treatment process is disinfected prior to discharge to a receiving water. Chlorine, in the form of a sodium hypochlorite solution, is normally used for disinfection. Since more chlorine is needed to provide adequate bacteria kills than would be safe for aquatic life in the stream, excess chlorine is removed by dechlorination. Alternate disinfection methods, such as ozone or ultraviolet light, are utilized by some treatment plants.

Sludge that settles to the bottom of the clarifier is pumped out and dewatered for use as fertilizer, disposed of in a landfill, or incinerated. Sludge that is free of heavy metals and other toxic contaminants is called Biosolids and can be safely and beneficially recycled as fertilizer, for example.

outfall the place where effluent is discharged into receiving waters

Secondary Treatment. Primary treatment provided a good start, but, with the exception of some ocean **outfalls**, it is inadequate to protect water quality as required by the Environmental Protection Agency (EPA).

With secondary treatment, the bacteria in sewage is used to further purify the sewage. Secondary treatment, a biological process, removes 85 percent or more of the organic matter in sewage compared with primary treatment, which removes about 50 percent.

The basic processes are variations of what is called the "activated sludge" process or "trickling filters," which provide a mechanism for bacteria, with air added for oxygen, to come in contact with the wastewater to purify it.

In the activated sludge process, flow from the sewer or primary clarifiers goes into an aeration tank, where compressed air is mixed with sludge that is recycled from secondary clarifiers which follow the aeration tanks. The recycled, or activated, sludge provides bacteria to consume the "food" provided by the new wastewater in the aeration tank, thus purifying it.

In a trickling filter the flow trickles over a bed of stones or synthetic media on which the purifying organisms grow and contact the wastewater, removing contaminants in the process. The flow, along with excess organisms that build up on the stones or media during the purification, then goes to a secondary clarifier. Air flows up through the media in the filters, to provide necessary oxygen for the bacteria organisms. Clarified effluent flows to the receiving water, typically a river or bog, after disinfection. Excess sludge is produced by the process and after collection from the bottom of the secondary clarifiers it is dewatered, sometimes after mixing with primary sludge, for use as fertilizer, disposed of in a landfill, or incinerated.

Advanced or Tertiary Treatment. As science advanced the knowledge of aquatic life mechanisms and human health effects, and the need for purer water was identified, technology developed to provide better treatment. Heavy metals, toxic chemicals and other pollutants can be removed from domestic and industrial wastewater to an increasing degree. Methods of advanced treatment include microfiltration, carbon adsorption, evaporation /distillation, and chemical precipitation.

Industrial Waste Treatment. Depending on the type of industry and the nature of its wastes, industries must utilize methods such as those used for advanced treatment of sewage to purify wastewater containing pollutants such as heavy metals and toxic chemicals before it can be discharged. Industries are permitted to discharge directly to receiving waters under the National Pollution Discharge Elimination System (NPDES) permit system or to municipal sewers under the Industrial Pretreatment Program. Pollution prevention programs are very effective in helping industries reduce discharged pollutants, by eliminating them at the source through recycling or through the substitution of safer materials. More and more industries are approaching or attaining zero discharge by cleaning and reusing their water over and over and over.

Combined Sewer Overflows

Combined sewer systems are sewers that are designed to collect rainwater runoff, domestic sewage, and industrial wastewater in the same pipe. Most of the time, combined sewer systems transport all of their wastewater to a sewage treatment plant, where it is treated and then discharged to a water body. During periods of heavy rainfall or snowmelt, however, the wastewater volume in a combined sewer system can exceed the capacity of the sewer system or treatment plant. For this reason, combined sewer systems are designed to overflow occasionally and discharge excess wastewater directly to nearby streams, rivers, or other water bodies. Some designs utilize an overflow at the treatment plant that diverts the excess flow to chlorination facilities for disinfection prior to discharge.

These overflows, called combined sewer overflows (CSOs), contain not only storm water but also untreated human and industrial waste, toxic materials, and debris. They are a major water pollution concern for the approximately 772 U.S. cities that have combined sewer systems.

CSO outfalls often result in violations of receiving stream-water quality standards and impairment to designated water uses. Violations can include aesthetics (including floatables, oil and grease, colors, and odor), solids, nutrients, harmful bacteria, metals, and reduced dissolved oxygen levels.

Historical and Regulatory Aspects

Environmental awareness and activism is not a present-day concept:

In the mid-1700s Benjamin Franklin and others petitioned the Pennsylvania Assembly to stop dumping waste and attempted to regulate waste disposal and water pollution. European countries were correlating sickness with lead and mercury in the late 1700s. In 1855, Chicago became the first U.S. city with a comprehensive sewer plan, and all U.S. towns with populations over 4,000 had city sewers by 1905.

In 1899 the Refuse Act prevented some obvious pollution of streams and placed the U.S. Army Corps of Engineers in charge of permits and regulation.

In 1914 U.S. government agencies began pollution surveys of streams and harbors. Reports filed by the early 1920s showed heavy damage from oil dumping, mine runoff, untreated sewage, and industrial wastes.

SETTLING POND

A settling pond, usually man-made, collects and slows water flow so that suspended solids (sediments) have time to precipitate or settle out of the water. Some applications of settling ponds include capturing runoff from farms (agricultural waste), construction projects (soil sediment) and mines (sediment and toxic waste). Settling ponds eventually fill and must be dredged to remain in operation. Polluted water from abandoned mines is diverted to settling ponds to remove solids such as iron oxide. When dredged, these sediments must be treated as contaminated waste. Pilot projects are underway to recapture iron oxide for use in paint pigments.

In 1924 the Oil Pollution Control Act prohibited discharge from any vessel within the three-mile limit, except by accident.

In 1948 the Federal Water Pollution Control Act and active House and Senate Public Works Committee in water pollution came about.

In 1956 Congress passed the Water Pollution Control Act, in 1961 the Clean Water Act, and in 1965 the Water Quality Act, setting standards for states.

In 1970 Congress and the president established the EPA.

In 1972 Congress passed the Federal Water Pollution Control Act (the "Clean Water Act").

In 1973 EPA issued the first NPDES permits.

In 1974 Congress passed the Safe Drinking Water Act.

The Clean Water Act of 1972.   Said to be one of the most significant pieces of environmental regulations ever enacted, the federal Clean Water Act of 1972 was prompted by growing national concern for the environment in the late 1960s, fueled by such concerns as the burning Cuyahoga River in Ohio, an unfishable, unswimmable Potomac River, and a nearly dead Lake Erie.

National goals and objectives were established "to restore and maintain the chemical, physical, and biological integrity of the Nation's waters." There were two major goals:

1. Eliminate the discharge of all pollutants into navigable waters of the United States; and

2. Achieve an interim level of water quality that provides for the protection of fish, shellfish, and wildlife and recreation (the "fishable, swimmable" goal).

To help do this, the following were established:

A state grant program to support the construction of sewage treatment plants; the NPDES program, whose goal was to eliminate discharges to U.S. waters; and technological standards or discharge limits that had to be met, based on water-quality standards set by the states.

A minimum required percent removal of pollutants was added in 1985.

Secondary treatment was required, and limits were set for three major effluent parameters: biological oxygen demand, suspended solids, and pH.

nonpoint source pollution pollution originating from a broad area, such as agricultural runoff or automobile emissions

The Water Quality Act of 1987 made several changes, addressing (1) excess toxic pollutants in some waters and (2) **nonpoint source pollution**. The construction grant program was phased out and replaced by financing projects with revolving fund, low-interest-rate loans. The amendments passed in 1987 also addressed storm-water controls and permits, regulation of toxics in sludge, and problems in estuaries. Penalties were added for permit violations. Also initiated were sludge-disposal regulations and funding for studies relative to nonpoint and toxic pollution sources.

The 1972 act has provided remarkable achievements, but there is still a long way to go. Forty percent of waters assessed by states still do not meet water-quality standards, mostly due to pollution from nonpoint sources. Other than from storm or combined storm sewer overflows, most of the remaining problem is not from pipes (point sources) but from sources such as farming and forestry runoff, construction sites, urban streets (storm water), automobiles, and atmospheric depositions, such as from power-plant air emissions (nonpoint sources). Current approaches to addressing nonpoint pollution include targeting and permitting by given watersheds and TMDL (total maximum daily load for a river stretch) assessments.

Many of the facilities funded by federal construction grants, which make up the wastewater collection and treatment infrastructure, are wearing out and are now undersized. Many, many dollars are needed to keep providing adequate treatment to maintain the status quo, let alone meet the needs of a growing populace.

Other Countries

Unfortunately, since the Industrial Revolution, most of Europe's rivers (not unlike in the United States) were utilized for transporting wastes to the sea, resulting in harm to human and aquatic health and causing coastal pollution. In earlier times, the rivers could handle the limited wastes discharged, through dilution and natural purification.

Significant progress has been made in treating the wastewater entering Europe's rivers, with measurable improvements in water quality. The agricultural sector (nonpoint pollution source) has not kept up, and nitrate levels are still high.

The fifteen-nation European Union's (EU) Urban Wastewater Treatment Directive has resulted in significant improvements in wastewater treatment capacity and methods. According to the European Environment Agency, increased treatment capacity has been realized in all EU countries except Sweden, Finland, and the Netherlands, where it is already efficient. The largest increase will be in southern Europe and Ireland. As a result, the EU's collection and treatment systems should be able to cope with all organic discharges from most member states by 2005. In Finland and Sweden most of the wastewater was being treated in tertiary plants in the 1980s. SEE ALSO ABATEMENT; BIOSOLIDS; NPDES; POLLUTION PREVENTION; WATER POLLUTION.

aerobic life or processes that require, or are not destroyed by, the presence of oxygen

anaerobic a life or process that occurs in, or is not destroyed by, the absence of oxygen

---

CONSTRUCTED WETLANDS

Constructed wetlands are wetlands that are specially built for the purpose of wastewater treatment and are utilized in place of naturally occurring wetlands. They provide a greater degree of wastewater treatment than natural wetlands, as their hydraulic loadings can be managed as required. Because these wetlands are constructed specifically for wastewater treatment, they should not be included in the jurisdictional group, which avoids the regulatory and environmental entanglement associated with natural wetlands. This is in accordance with Environmental Protection Agency regulations. The treatment process can be either **aerobic** or **anaerobic**, depending on whether the wetlands are constructed with an exposed water surface or one with subsurface flow. These wetlands can also be used to remove nitrogen, which is usually not removed during the standard wastewater treatment process. Nitrogen removal is accomplished by the growth of cattails and reeds, which utilize the highly nutrient wastewater and consequently remove nitrogen in the process. Sometimes the cattails and reeds must be harvested to complete the removal process.

Internet Resources

Environmental Protection Agency, Office of Water. (1993). "Constructed Wetlands for Wastewater Treatment and Wildlife Habitat." Available from http://www.epa.gov/owow/wetlands/construct.

Ohio State University Extension, Food, Agricultural, and Biological Engineering. "Wastewater Treatment Principles and Regulations." Available from http://ohio-line.osu.edu/aex-fact/0768.html.

Raymond Cushman and George Carlson

Water Pollution

Water covers more than 70 percent of Earth's surface. It is essential to all life. Organisms can survive longer without food than without water. It is one of our most valuable resources.

Pollute means to make impure or unclean. In that sense, water pollution has always occurred as a natural phenomenon. Forest fires, storms, volcanoes, or a heavy leaf fall can contaminate a water body. However, these organic materials are broken down or biodegraded naturally.

Pollution as we know it began when humans started discarding waste including sewage and toxic chemicals. By the middle of the twentieth century, the extent of water pollution became apparent when Ohio's Cuyahoga River caught fire as a result of widespread oil pollution.

The Clean Water Act of 1972 and its subsequent amendments reduced surface-water pollution by prohibiting the dumping of toxic chemicals and medical waste, and by establishing a permitting system to reduce the direct discharge of pollutants. The Safe Drinking Water Act of 1974, later amended in 1987, set maximum allowable contaminant levels for drinking water and called for the regular monitoring of groundwater. And the Ocean Dumping Ban prohibited the marine disposal of sewage and industrial waste after 1991.

Water pollution is described as point source or nonpoint source. Point source means one can pinpoint and reduce pollution at its source. Point source pollution may come from an industrial discharge pipe, a wastewater treatment plant, or a capsized oil tanker. Nonpoint source pollution occurs when substances such as fertilizer, pesticides, and soil from erosion enter water bodies through rain runoff. Other pollutants include heavy metals such as mercury, salt, acid rain, silt, hot water, petroleum products, excess nutrients such as nitrogen and phosphorus, sewage, and animal waste. Since polluted water is extremely difficult and costly to clean up, prevention is by far the best approach to this form of environmental pollution. SEE ALSO ACID RAIN; AGRICULTURE; CLEAN WATER ACT; CRYPTOSPORIDIOSIS; FISH KILLS; HYPOXIA; MERCURY; NONPOINT SOURCE POLLUTION; PCBs (POLYCHLORINATED BIPHENYLS); RIVERS AND HARBORS APPROPRIATIONS ACT; SEDIMENTATION; WATER POLLUTION; WATER TREATMENT; WASTEWATER TREATMENT.

Bibliography

EPA Clean Water Act. Available from http://www.epa.gov/r5water/cwa.htm.

U.S. Environmental Protection Agency. "Ocean Dumping Ban Act of 1988." Available from http://www.epa.gov/history.

Diana Strnisa

Liquid waste pouring from pipe into flowing river. (United States Environmental Protection Agency. Reproduced by permission.)

Water Pollution: Freshwater

Freshwater pollution is the contamination of inland water (not saline) with substances that make it unfit for its natural or intended use. Pollution may be caused by fecal waste, chemicals, pesticides, petroleum, sediment, or even heated discharges. Polluted rivers and lakes are unfit for swimming or fishing; polluted water is unsafe to drink.

Background

For centuries, fecal waste and other pollutants were dumped in rivers, with "dilution the solution" to pollution. In the mid-twentieth century, many

Steel mills in Indiana along the southern coast of Lake Michigan. (©Joel W. Rogers/ Corbis. Reproduced by permission.)

American rivers and streams were open sewers, choking on everything from human waste to highly toxic industrial discharges. New York City alone pumped a half billion gallons of raw sewage into its harbor every day. As pollution levels grew, so did the impacts. "No swimming" signs became the norm. Lake Erie was dying. The Hudson River's commercial striped bass fishery, once valued at $40 million a year, was closed and it became illegal to sell oysters from Oyster Bay, Long Island. And then, in June 1969, Ohio's Cuyahoga River caught fire.

The damning image of a river in flames is credited by many for passage of the Federal Water Pollution Control Act of 1972. The U.S. Environmental Protection Agency (EPA) set standards to regulate the discharge of industrial and municipal waste—so-called end-of-the-pipe pollution. With them came significant federal funding to help localities improve wastewater treat-

ment. Billions of dollars have been invested since 1972 building and upgrading sewage treatment facilities.

Improvements in municipal wastewater treatment have been matched by progress in the private sector. Nationally, more than thirty thousand major industrial dischargers pretreat their wastewater before it enters local sewers. By 2000, some 75 percent of toxic discharges, including heavy metals and PCBs, were being prevented.

Surface Water Pollution

Freshwater makes up less than three percent of earth's water, but is the source of virtually all drinking water. In 2002, each U.S. household used an average of 94,000 gallons of water per year. Some 55 percent of that water comes from reservoirs, rivers, and lakes, and a 2000 survey published in EPA's National Water Quality Inventory found almost 40 percent of U.S. rivers and 45 percent of lakes are polluted. These sources, called surface water, are vulnerable to pollution discharged out of pipes and precipitating out of the air but the primary source of their pollution today is runoff, pollutants washing off the land.

These nonpoint or scattered sources are not easily traceable. Pesticides and fertilizers used in agriculture and on golf courses and suburban lawns account for a major portion of nonpoint source pollution. Runoff from parking lots and roads flush spilled oil and gasoline and road salt into lakes and streams. Runoff containing manure from livestock and poultry producers has

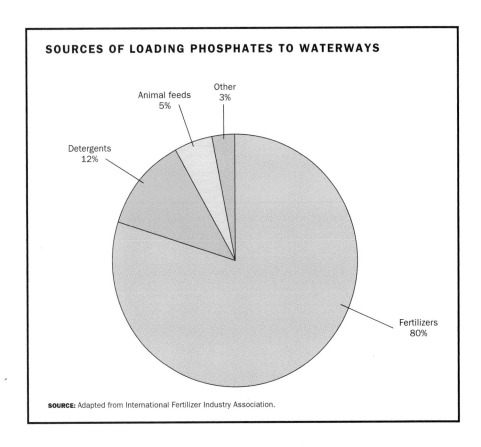

SOURCES OF LOADING PHOSPHATES TO WATERWAYS

Other
3%

Animal feeds
5%

Detergents
12%

Fertilizers
80%

SOURCE: Adapted from International Fertilizer Industry Association.

been a major source of surface water pollution. More than 150 pathogens found in livestock manure pose risks to humans. In 2003, concentrated animal feeding operation guidelines, or CAFO standards, were finalized requiring inspection of waste lagoons and outdoor manure tanks, as well as permits for applying manure on land.

Air pollutants such as dioxin and mercury along with sulfur and nitrogen oxides precipitate into lakes and rivers by rainfall in the form of acid rain. More than 95 percent of rainwater tested at four sites in Indiana between 2001 and 2002 contained unsafe levels of mercury according to a National Wildlife Federation report.

Point sources, such as chemical and municipal wastewater treatment plants, were the leading source of contamination for about ten percent of river and lake water according to the 2000 National Water Quality Inventory. Toxic chemicals, although now regulated, can still be discharged directly into surface water. AK Steel Corporation in Pennsylvania discharged the largest amount of any industrial pollutant, about 28 million pounds of nitrate compounds, to surface water between 1998 and 2000, according to the Toxic Release Inventory.

Other sources of surface water pollution include silt washed into streams and lakes that smothers organisms on the lake floor, upsetting or destroying aquatic ecosystems. Thermal pollution such as an influx of warm water from cooling towers for power plants also has a detrimental effect on aquatic ecosystems.

The recent discovery of surface-water contamination by minute amounts of pharmaceuticals and personal-care products, including synthetic hormones

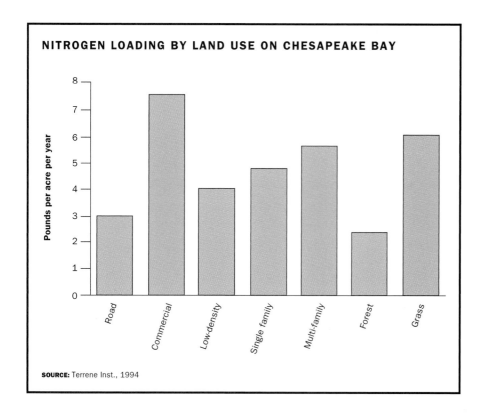

NITROGEN LOADING BY LAND USE ON CHESAPEAKE BAY

Pounds per acre per year

Road · Commercial · Low-density · Single family · Multi-family · Forest · Grass

SOURCE: Terrene Inst., 1994

from birth control pills, is being investigated to determine whether it poses a threat to humans, aquatic species, or wildlife. Water Quality Act amendments of 1987 established a $400-million program to help states to develop and implement nonpoint source management programs based on watershed protection.

Groundwater Pollution

Water contained in the pores of soil or in **aquifers** is called groundwater. About 40 percent of U.S. municipal water comes from groundwater and an additional forty million people, including most of the rural population, draw drinking water from domestic wells. Groundwater, while protected by the filtering action of soil, can be contaminated by leaking municipal landfills, sewage lagoons, and chemicals from industrial activity. Centers for Disease Control data shows that 318 waterborne disease outbreaks associated with groundwater systems occurred between 1971 and 1996. Leaking underground oil tanks and spills at gas stations account for oil and other chemicals such as benzene and methyl-tertiary-butyl ether (MBTE) found in groundwater. More than 400,000 leaking underground storage tanks were reported in the United States in 2001. Pesticides and agricultural fertilizers drain into groundwater polluting it with carcinogenic chemicals and nitrates.

The Safe Drinking Water Act of 1974 (SWDA) regulates groundwater. More than eighty possible contaminants are monitored, including carcinogens such as tetrachloroethylene, discharged from dry cleaners. Health effects of these contaminants range from increased cancer risk, intestinal lesions, kidney damage, and reproductive difficulties, to gastrointestinal distress. Municipal and private water suppliers are responsible for seeing that contaminants do not exceed the limits set by the EPA.

aquifer an underground geological formation, or group of formations, containing water; are sources of groundwater for wells and springs

Human and Environmental Health Effects

Fertilizer, animal manure, and waste-treatment plant effluent all contain nutrients that stimulate excessive plant and algal growth in freshwater bodies. When the plants die and decompose, dissolved oxygen is depleted, causing die-offs of fish and other species living in the water. Persistent organochlorine insecticides, such as DDT, deposited in lake sediments can bioaccumulate, harming the fish and birds that eat them. Pyrethroid insecticides, though derived from chrysanthemums, are extremely toxic to aquatic organisms. Estrogen-mimicking substances such as some pesticides and industrially produced chemicals have been shown to interfere with the reproductive system of fish.

Human and animal fecal waste contain disease-carrying organisms such as the bacterium *Escherichia coli (E. coli)* and pathogens that causes cholera, typhoid, and cryptosporidiosis. Cholera is rarely seen in the United States, but *E. coli* outbreaks are not rare, and in 1993, more than fifty people died, and an estimated 400,000 became ill from a massive outbreak of cryptosporidiosis in Milwaukee, Wisconsin. The outbreak was attributed to a failure in drinking water treatment, allowing the cyst form of the parasite, introduced by animal waste, to pass into tap water and be ingested. Ten outbreaks of cryptosporidiosis were reported in the United States between 1990 and 2000.

Mercury bioaccumulates in fish and can damage the nervous systems and brains of humans. It can interfere with normal behavior in birds, such as loons, causing them to spend less time looking for food or incubating eggs. About one-quarter of breeding adult loons have higher-than-normal (10 parts per million) levels of mercury. To protect people from eating contaminated fish, states and local governments post fish-consumption advisories when contaminant levels become unsafe. There were 2,800 advisories posted in the United States in 2002, alerting people to high levels of mercury, PCBs, chlordane, dioxins, and DDT in fish.

Prevention and Abatement

Once water is contaminated, it is difficult, expensive, and sometimes impossible to remove pollutants. Technologies to remove contaminants from groundwater are air stripping, granular activated carbon, and advanced oxidation. Air stripping involves pumping out the contaminated water, then heating it to evaporate the contaminant. The cleaned water is reinjected into the ground. Pumping out contaminated water and absorbing the pollutant on activated charcoal can remove less volatile compounds. Ninety percent of trichloroethylene was removed from NASA's launch complex thirty-four groundwater cleanup site on Cape Canaveral Air Force Station by thermal treatment. In this method an electric current heats soil and water, evaporating some water and the contaminant, which is carried out of the ground by the force of the steam and collected in recovery wells.

Preventing pollution is obviously important. Drinking water suppliers have discovered that **watershed** protection is cost-effective because it reduces pollution and cuts the cost of drinking water treatment. A watershed is the area that drains into surface or groundwater and keeping that area free from development and agricultural runoff are among the goals of watershed protection. The Barnes Aquifer in Massachusetts supplies water to sixty thousand residents and the aquifer's **recharge** area is under heavy development pressure from

watershed the land area that drains into a stream; the watershed for a major river may encompass a number of smaller watersheds

You can help prevent water pollution by simply not littering. Street trash that washes down storm drains is a major source of floatable debris. Properly dispose of used oil; oil poured down storm drains and sewers is a major source of petroleum pollution. Use nonphosphate detergents for dish and clothes washing. Don't overfertilize lawns and use integrated pest management practices to reduce pesticide use. Use hazardous waste collection programs to dispose of batteries, fluorescent lights that contain mercury, unused oil, paint remover, pesticides and old household chemicals.

large-scale residential subdivisions. Municipal wells have been contaminated with traces of ethylene dibromide and trichloroethylene. After learning about watershed protection, citizens voted against proposed changes to zoning that would have increased the number of new homes and increased the potential for groundwater pollution. And by investing $1 billion in watershed protection, New York City, with an enormous reservoir system, has avoided having to build water-filtration facilities, saving construction costs of some $8 billion.

Global

The United Nations (UN) theme for World Environment Day 2003 was "Water: Two Billion People are Dying for It!" It was not en exaggeration. The UN reports that one person in six lives without regular access to safe drinking water. Over twice that number—2.4 billion people—lack access to adequate sanitation. Water-related diseases kill a child every eight seconds, and are responsible for 80 percent of all illnesses and deaths in the developing world. Cholera outbreaks, due to water contaminated with raw sewage, occur regularly in India and Bangladesh and less frequently in many other countries. In Africa in 1997, 5,853 deaths due to cholera were reported to the World Health Organization. It is a situation, the UN said, "made all the more tragic by our long-standing knowledge that these diseases are easily preventable."

SEE ALSO: ACID RAIN; AGRICULTURE; CLEAN WATER ACT; CRYPTOSPORIDIOSIS; DDT (DICHLORODIPHENYL TRICHLOROETHANE); HEALTH, HUMAN; NONPOINT SOURCE POLLUTION; PCBS (POLYCHLORINATED BIPHENYLS); POINT SOURCE; SNOW, JOHN; WASTEWATER TREATMENT; WATER TREATMENT.

Bibliography

Pielou, E.C. (1998). *Fresh Water*. Chicago and London: The University of Chicago Press.

Internet Resources

Natural Resources Defense Council. "What's on Tap: Grading Water in 19 U.S. Cities." Available from http://www.nrdc.org/water/drinking/uscities/contents.asp.

U.S. Environmental Protection Agency. Browse EPA Topics. Available from http://www.epa.gov/ebtpages/alphabet.html.

U.S. Environmental Protection Agency. Clean Water Act. Available from http://www.epa.gov/r5water/cwa.htm.

U.S. Environmental Protection Agency. Concentrated Animal Feeding Operation Final Rule. Available from http://cfpub.epa.gov/npdes/afo/cafofinalrule.cfm.

U.S. Environmental Protection Agency. List of Drinking Water Contaminants and their MCLs. Available from http://www.epa.gov/safewater/mcl.html#mcls.

U.S. Environmental Protection Agency. Polluted Runoff (Nonpoint Source Pollution). Available from http://www.epa.gov/OWOW/NPS/facts/point1.htm.

U.S. Environmental Protection Agency. Proposed Groundwater Rule. Available from http://www.epa.gov/OGWDW/gwr.html.

U.S. Environmental Protection Agency. Safe Drinking Water Act. Available from http://www.epa.gov/safewater/sdwa/sdwa.html.

U.S. Environmental Protection Agency. 2000 National Water Quality Inventory. Available from http://www.epa.gov/305b/2000report.

U.S. Environmental Protection Agency's Water Science Great Lakes Initiative Topic. Available from http://www.epa.gov/ost/GLI/mixingzones/finalfact.html.

U.S. Environmental Protection Agency. Fish Advisories. Available from http://www.epa.gov/waterscience/fish.

U.S. Geological Survey National Water Quality Assessment Program. Available from http://water.usgs.gov/nawqa.

Patricia Hemminger

The Great Lakes Basin includes areas of the eight Great Lakes states: New York, Pennsylvania, Ohio, Minnesota, Indiana, Illinois, Wisconsin, and Michigan. In 1995, the U.S. Environmental Protection Agency (EPA) and the Great Lakes states agreed to a plan called the Great Lakes Initiative, aimed at reducing pollution and restoring the health of the Great Lakes. The plan included setting water quality standards for twenty-nine pollutants. In 2000, the EPA initiated a ten-year phase-out of the use of **mixing zones** for **bioaccumulative** chemicals in the Great Lakes. The EPA says this ruling will reduce discharges of toxic chemicals by 700,000 pounds a year.

recharge the process by which water is added to a zone of saturation, usually by percolation from the soil surface; e.g., the recharge of an aquifer

mixing zone an area of a lake or river where pollutants from a point source discharge are mixed, usually by natural means, with cleaner water

bioaccumulative relating to substances that increase in concentration in living organisms as they take in contaminated air, water, or food because the substances are very slowly metabolized or excreted

Water Pollution: Marine

Marine pollution is the release of by-products of human activity that cause harm to natural marine ecosystems. The pollutants may be sewage, farm waste, toxic chemicals, or inert materials that may smother, choke, or strangle living organisms.

Sewage, Animal Waste, and Fertilizers

Sewage, animal waste, and chemical fertilizers all have a high content of nitrogen and phosphorus. Artificially high levels of these substances in the water promote excessive growth of microscopic or **macroscopic** plants, in a process called **eutrophication**. When these plants accumulate, die, and decay, they cause low oxygen content in the water. Even if sewage is treated to remove solids, the liquid discharged contains high levels of nitrogen and phosphorus. Intensive cultivation of animals in feedlots, or application of more fertilizer than a crop can absorb, also cause runoff rich in nitrogen and phosphorus that find their way into rivers and estuaries. Vehicle exhausts and industrial chimneys are large sources of nitrogen compounds that are transported in the atmosphere and deposited in coastal waters.

On a global scale, agricultural runoff is the most important source of eutrophication, but atmospheric deposition is the fastest-growing source. It is the largest source of nitrogen off the coast of the northeastern United States, in the western Baltic Sea, and in the western Mediterranean Sea. International agencies consider that, worldwide, eutrophication is the most serious pollution problem in coastal waters. For example, in the Gulf of Mexico, off the mouth of the Mississippi River, water near the bottom has severely reduced oxygen content over a very large area, sixteen thousand square kilometers (6,200 square miles) by 1998. Mobile animals such as fish and shrimp leave the hypoxic area, but sedentary animals such as clams and worms are killed in large numbers.

A classic example of eutrophication and its treatment occurred in the estuary of the River Thames, near London, England. In the 1950s the water was severely hypoxic for thirty-five kilometers (twenty-two miles) below London Bridge. After several sewage treatment plants were built, the water returned to a well-oxygenated state and migratory fish such as salmon once again ascend the river. In the case of the Mississippi River, treatment of the eutrophication is more difficult because runoff from agricultural land is the major cause of the problem, and more than half of the agricultural land in the United States drains into the Mississippi basin. Cleaning up the pollution would involve changes in farming methods on a national scale.

Eutrophication has important indirect effects. The plants known as sea grasses, which grow in the shallow water of **estuaries**, provide food and shelter for a wide range of animals, including geese, turtles, manatees, and fish. In eutrophicated water, the dense microscopic plant life significantly reduces the penetration of light and smothers the sea grasses. In Chesapeake Bay, Maryland, eutrophication caused an area of sea grasses to decrease by two-thirds between 1960 and 1980, and there was a corresponding decrease in landings of fish and crabs. Similar effects have been observed in Australia.

Red tides, or harmful algal blooms, are associated with eutrophication. Single species of **phytoplankton** multiply at the expense of all other species

---

macroscopic large enough to be visible, in contrast to microscopic

eutrophication in nature, the slow aging process during which a lake, estuary, or bay evolves into a bog or marsh and eventually disappears; in pollution, excess algal growth or blooms due to introduction of a nutrient overload of nutrients, i.e., from un- or poorly treated sewage

estuary region of interaction between rivers and near-shore ocean waters, where tidal action and river flow mix fresh- and saltwater (i.e., bays, mouths of rivers, salt marshes, and lagoons); these ecosystems shelter and feed marine life, birds, and wildlife

phytoplankton that portion of the plankton community comprised of tiny plants; e.g. algae, diatoms

and become so abundant that the water is discolored. Many bloom species produce toxic substances. During the 1990s in estuaries located in the southeastern United States, there were numerous cases of blooms of *Pfiesteria piscida*, a **dinoflagellate** that produced a toxin which killed thousands of fish. The source of the nutrients support *Pfiesteria* is believed to be agricultural runoff or sewage discharge. Other types of blooms are ingested by shellfish, which become toxic for humans who consume them, causing partial paralysis, memory loss, or even death. Toxic blooms have been reported much more frequently in the 1990s than in the past, and the spread of eutrophication is believed to be a contributing factor.

dinoflagellate single-celled aquatic organism

Pollution and Coral Reefs

On coral reefs, eutrophication causes seaweed to grow and smother the corals. Several kinds of environmental problems interact with eutrophication to cause the deterioration of coral reefs. Overharvesting of the fish and invertebrates that eat seaweed accelerates the smothering. Careless development along coastlines and in river basins leads to soil erosion and the transport of heavy loads of silt and clay, which settle on the corals and smother them. Oil spills also take their toll. When corals are exposed to abnormally high water temperature, they respond by discharging the microscopic algae living within their tissues. Sometimes they recover, but often they die. These episodes, called coral bleaching, became much more frequent during the 1990s and are believed to be caused by global warming. The result of pollution and global warming is that at least half of the area of coral reefs in southeast Asia is in poor condition, and in parts of the Caribbean Sea only 5 percent of the reef area consists of living coral.

Metals and Organic Contaminants

effluent discharge, typically wastewater—treated or untreated—that flows out of a treatment plant, sewer, or industrial outfall; generally refers to wastes discharged into surface waters

Industrial **effluents** often contain metallic compounds. For example, Halifax, a small city in eastern Canada, discharged into its harbor during the 1990s about thirty-three tons of zinc and thirty-one tons of lead per year, with

Coral grows a new layer each year, much as a tree adds a new ring each year. Scientists analyzing layers of Bermudan coral have discovered an environmental record dating back to the mid-1800s. Marine pollution can be measured across the Industrial Revolution. Marine levels of lead have dropped dramatically since the phaseout of leaded gasoline but levels of lead in the Atlantic are still double their preindustrial concentrations.

organochlorine chemical containing carbon and chlorine

lesser amounts of copper and other metals. These metals are held in the sediment in a relatively inert form, but if stirred up into the water column, they become oxygenated and toxic. Tin is another common pollutant in harbors. It occurs as tributyltin (TBT), which is used as a component of antifouling paints on the undersides of ships. When taken up by shellfish, it accumulates in their tissues and has proved toxic to the shellfish and to organisms that consume them. The United States began to phase out TBT in 1988, and it will be banned internationally beginning in 2008.

Industry also produces organic compounds such as polychlorinated biphenyls (PCBs) and various pesticides. These accumulate in the fatty tissue of plants and animals low in the food chain, and as they pass through the food web to larger and long-lived animals, there is an increase in concentration of the substances in their fat, a process known as bioaccumulation. The St. Lawrence River, which drains the Great Lakes, has accumulated large amounts of **organochlorines**, which have amassed in the tissues of Beluga whales. During the 1990s, the level of this pollution was much reduced, and the whales have been protected from hunting, but their population fails to increase. Many animals have tumors and disease. There is mounting evidence that chronic exposure to contaminants causes suppression of the immune responses of marine mammals. Similar problems have occurred with seals in the Baltic Sea.

Oil Pollution

The most serious types of oil pollution occur when an oil tanker goes ashore or hits a reef and spills its contents. As the oil drifts ashore, great damage is done to beaches, rocky shores, salt marshes, or mangrove forests. Cleanup is often attempted using mechanical means, or the application of dispersants, with mixed results. Usually, a proportion of native organisms are killed, but given time, the lighter fractions of oil evaporate, while the heavier fractions are decomposed by photochemical processes and microorganisms. International law now requires that vessel owners be responsible for any loss of oil, damage to existing ecosystems, and the costs of recommended cleanup.

Chronic low levels of oil pollution, resulting from accidental spills when loading or unloading, or from washing out oil tanks, are widespread and of significant concern. For example, it has been determined that corals around an oil terminal in the Red Sea have experienced lower growth rates and poor reproduction as a result of chronic low-level oil pollution.

Oil pollution of the open ocean is also a major concern. When Thor Heyerdahl crossed the South Pacific on the raft *Kon-Tiki* in 1947 he reported pristine waters, but his *Ra* expedition across the Atlantic twenty-two years later encountered oil slicks on forty-three of fifty-seven days at sea. The International Convention for Prevention of Pollution from Ships was devised in 1973 and modified by the Protocol of 1978. Oceangoing vessels are subject to strict regulations concerning the discharge of oil, bilge water, and ballast water, and are forbidden to dump garbage and other solid waste. Accidental spills must be reported.

Marine Debris

Marine beaches serve as natural traps for marine debris. Globally, the most common materials are plastics, followed by glass and metal. The chief

When Thor Heyerdahl, a Norwegian biologist (1914–2002), sailed the balsa wood raft named *Kon-Tiki*, from Peru to Polynesia in 1947, he saw no pollution in the Pacific Ocean. Just over twenty years later, in 1970, when sailing a papyrus reed boat from Morocco to Barbados, Heyerdahl saw extensive marine pollution including oily wastes, plastic bottles and other trash floating in the water. He radioed the United Nations to report that floating lumps of solidified, asphalt-like oil polluted over one thousand miles of the Atlantic Ocean. After seeing the extent of the ocean's pollution first hand, Heyerdahl became actively involved in fighting marine pollution. In 1999, with the Norwegian Shipowners Organization, he initiated the Thor Heyerdahl International Maritime Environmental Award to be given for improvement of the global environment.

dangers to marine life result from the ingestion of these fragments, which may block the gut, and from entangling, which may cause suffocation or prevent **locomotion** and feeding. In a survey of U.S. beaches close to urban centers, cigarette butts were the most abundant debris, followed by packaging items (boxes, bags, caps, lids), medical waste, and sewage. A high proportion of this material reached the sea by way of sewers. Even street litter can be washed into surface drains and then to the sea. The dumping of sewage and waste by ships is another source. Public revulsion at the state of U.S. beaches was a key factor in the enactment of stronger environmental protection laws, like the Ocean Dumping Ban Act of 1988 that prohibited the dumping of sewage into the ocean. On sites more remote from cities, pieces of rope and netting are the most common types of marine debris.

locomotion self-powered movement

Reduction and Regulation of Marine Pollution

There is much that individuals can do to prevent marine pollution: avoid putting toxic substances into drains, avoid dropping litter, minimize the use of pesticides and fertilizers, reduce automobile emissions, and pressure your local government for sewage treatment in the community if it does not yet exist. Larger-scale problems require legislation and enforcement, ranging from the local laws of coastal states in the United States, through national laws such as the Clean Water Act and Clean Air Act, to international conventions such as the International Convention for the Prevention of Pollution from Ships. Such laws are effective only if they have the support of the people. SEE ALSO ACID RAIN; CLEAN WATER ACT; CRYPTOSPORIDIOSIS; FISH KILLS; HYPOXIA; MERCURY; OCEAN DUMPING; PCBs (POLYCHLORINATED BIPHENYLS); PETROLEUM; RIVERS AND HARBORS APPROPRIATIONS ACT; SNOW, JOHN; WATER TREATMENT; WASTEWATER TREATMENT.

Bibliography

Clark, R.B.; Frid, C; and Attrill, M. (2001). *Marine Pollution*, 5th edition. Oxford, UK: Oxford University Press.

Pelley, J. (1998). "Is Coastal Eutrophication out of Control?" *Environmental Science and Technology* 32:462A–466A.

Internet Resources

Global Investigation of Pollution in the Marine Environment (GIPME). "Marine Pollution Programme." Available from http://ioc.unesco.org/iocweb.

Ocean Conservancy. Available from http://www.oceanconservancy.org.

Kenneth H. Mann

Thousands of volunteers in every U.S. state and territory as well as in more than fifty other countries pick up tons of marine debris each fall in a one-day coastal cleanup. The Ocean Conservancy, which organizes the annual cleanup, collects data on the debris to determine sources of pollution. The most common item washed up on the shoreline? Cigarette butts and filters—a total of 1,640,614 were picked up in 2001. Volunteers also found 259 entangled animals, most snared in nylon fishing line.

Water Treatment

The goal of water treatment, usually from surface sources such as lakes, reservoirs, or rivers, is to remove contaminants and organisms through a combination of biological, chemical, and physical processes to make it safe for drinking. Some of these occur in the natural environment, whereas others occur in engineered and constructed water treatment plants. The engineered processes usually mimic or build on natural processes.

History of Water Treatment

Water-treatment concepts underlying those used today were developed in Europe during the 1700s. An outbreak of cholera in London was linked to a sewage-contaminated drinking water well in 1854. John Snow was credited with this finding. At the point in which the United States began using chlorine to disinfect drinking water (1908), Europe was also using chlorine but exploring the possibility of employing ozone to treat drinking water. The U.S. Public Health Service developed the first drinking-water regulations in the United States in 1914. The U.S. Environmental Protection Agency (EPA) later assumed responsibility for this task when it was established in 1970. The Safe Drinking Water Act (SDWA) became law in 1974, and was significantly revised in 1986 and 1996. The revisions reflected improvements in analytical methods to detect contaminants at lower levels and improvements in automated monitoring used to evaluate treatment plant performance. The revisions also started to address the need to balance immediate (acute) risks versus long-term (chronic) risks. The need to disinfect water to kill pathogens to protect against acute illnesses, versus the formation of disinfection by-products and their chronic health effects is an example of this risk balance.

The United States has continued to examine water treatment practices in Europe, particularly water-quality standards established by the World Health Organization (WHO). Although there are some philosophical differences between the United States and Europe relating to the treatment of the distribution system and its operations, the United States has benefited from the European experience. One such philosophical difference is that the European water treatment community does not see the maintenance of a disinfectant residual to the end of the distribution system as a necessary public health protection measure. The United States drinking water community sees this as an important step to protect customers and the water system from bacteriological regrowth or recontamination. As the United States entered the twenty-first century, researchers were collaborating with scientists around the world to continuously improve water quality and treatment, and openly share their research findings.

Water Quality Regulations in the United States

The EPA, under the requirements of the SDWA, regulates drinking water in the United States. The EPA additionally regulates wastewater, but under the requirements of the Clean Water Act (CWA). Storm water and discharges into surface water are also regulated under the CWA.

The SDWA sets maximum contaminant levels (MCLs) and treatment techniques (TTs) that drinking water must meet to be considered safe for

consumption. The list includes microorganisms, disinfectants and disinfection by-products, inorganic chemicals, organic chemicals, and radionuclides.

The Water Cycle

The requirements of the CWA and SDWA are different, but interrelated. Consider the water cycle and the water-use cycle. Water falls to the earth in the form of precipitation. It drains into rivers, lakes, and streams either naturally or via constructed storm-water-drainage systems. Industrial manufacturers and wastewater treatment plants discharge **effluent** from their processes into lakes and rivers. Under the CWA, these facilities have water-quality limits that their effluent must meet. These limits have been established to protect the water ecosystem and downstream users. Water suppliers withdraw water from lakes and rivers to be treated for human consumption and other uses. The water is treated and delivered to customers' taps through a system of pipes and storage facilities that make up the water distribution system. After the water is used, it is conveyed to a wastewater treatment plant and discharged back as effluent to a receiving water body. If the water is used outside, it either seeps into the ground or drains to a storm-water system, which may go to a treatment plant or directly to a river, lake, or another body of water. The cycle continues as the water flowing to the ocean evaporates,

Solid waste settling pond next to manure fiber piles at the Three Mile Canyon Farm near Boardman, Oregon. (AP/Wide World Photos. Reproduced by permission.)

effluent discharge, typically wastewater—treated or untreated—that flows out of a treatment plant, sewer, or industrial outfall; generally refers to wastes discharged into surface waters

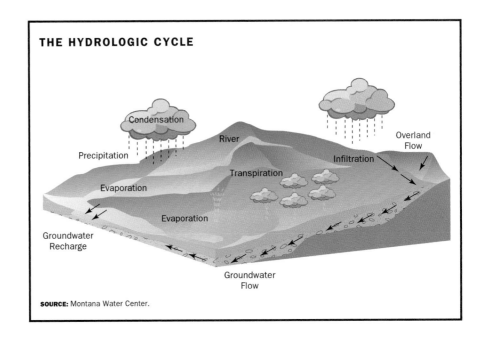

THE HYDROLOGIC CYCLE

SOURCE: Montana Water Center.

ultimately falling again as precipitation. See the illustration for a diagram of the water or hydrologic cycle.

Source Water Protection

Source water protection, often referred to as "watershed protection," is the reduction or prevention of water pollution at its source, represents a tradeoff between treatment plant construction and operation costs. This kind of protection is not always possible, but it has been very effectively implemented by several water systems. A water system that has access to a high-quality source may not need as extensive a treatment plant as a system with a poorer-quality source. This is especially true if a high-quality source, such as a reservoir in an isolated natural area, can be protected by limiting human activity close to that source. Water from such a source may not require the settling step, may involve fewer chemicals or smaller doses of them, or might be able to kill pathogens with strong disinfectants like ozone or ultraviolet light instead of providing filtration.

The Water-Treatment Process

Whether in the natural environment or a constructed water-treatment plant, there are several key processes that occur during water treatment: dilution, coagulation and flocculation, settling, filtration, disinfection, and other chemical treatments. The quality of the source water and the effectiveness of source-water protection and management have a direct bearing on the complexity of the treatment that is required. Source-water protection is the first step in water treatment, with the natural and engineered processes following. The processes in a water treatment plant are shown in the illustration.

Dilution. Prior to industrialization, the pollution of rivers and streams was not as significant a problem. Waste products were released into water bodies, but the quantity of such discharges was not as great as present-day levels. The receiving waters were large enough and the mixing or detention time was long enough that the contaminants were diluted to a level that reduced the

WATER–TREATMENT PLANT

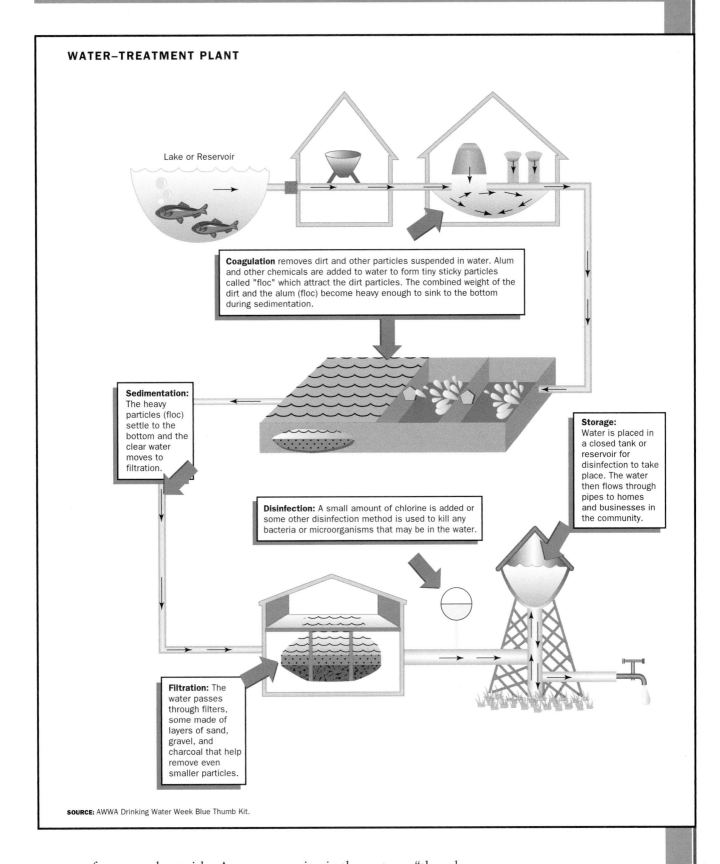

Lake or Reservoir

Coagulation removes dirt and other particles suspended in water. Alum and other chemicals are added to water to form tiny sticky particles called "floc" which attract the dirt particles. The combined weight of the dirt and the alum (floc) become heavy enough to sink to the bottom during sedimentation.

Sedimentation: The heavy particles (floc) settle to the bottom and the clear water moves to filtration.

Storage: Water is placed in a closed tank or reservoir for disinfection to take place. The water then flows through pipes to homes and businesses in the community.

Disinfection: A small amount of chlorine is added or some other disinfection method is used to kill any bacteria or microorganisms that may be in the water.

Filtration: The water passes through filters, some made of layers of sand, gravel, and charcoal that help remove even smaller particles.

SOURCE: AWWA Drinking Water Week Blue Thumb Kit.

amount of concern about risks. A common saying in the past was "the solution to pollution is dilution." This is not the most efficient treatment method because even small amounts of pollutants, such as some pesticides, can build up, or bioaccumulate, in body fat over time. It is also not the preferred approach because it sends the message that polluting the environment is an acceptable course of action.

Coagulation and Flocculation. Sometimes, the particles that need to be filtered out during water treatment are very small. This makes them less likely to settle out and less likely to be filtered out. Chemicals called coagulants and/or filter aids are added to the water and mixed in (flocculated) to make the fine particles stick together to form bigger particles that can better settle out or be filtered out more effectively. Depending on the microbial and chemical makeup of the water, different chemicals are used as coagulants. The purpose of these two steps is to improve the performance of the remaining treatment processes.

Settling. For facilities treating water that contains a lot of solids, settling or sedimentation is a common treatment step. The process slows the flow of the water in a pond or basin so heavier items can settle to the bottom. If the water is not sufficiently slowed down, these items are carried along to the next step in the process, which is not desirable. For plants treating very polluted **raw water**, settling may be used as the first step in the treatment plant (presedimentation) and again following the coagulation and flocculation steps.

raw water intake water prior to any treatment or use

Filtration. There are several methods of filtration used in water treatment. The selection of which type to use is generally a function of the raw water quality. As filtration implies, water flows through a material that removes particles, organisms, and/or contaminants. The material used is most often a granular medium such as sand, crushed anthracite coal, or activated carbon. Some facilities layer different types and sizes of media. Along with varying the size and type of filter media, facilities are also designed to operate at different flow rates through the filter media. Traditional filtration plants include slow sand filtration, high-rate filtration, and **diatomaceous earth** filtration.

diatomaceous earth a chalk-like material (fossilized diatoms) used to filter out solid waste in wastewater treatment plants; also used as an active ingredient in some powdered pesticides

Another type of filtration that was more widely used in the late 1990s and early 2000s is membrane filtration. It occurs by forcing water through a membrane barrier. A membrane is like a high-tech coffee filter. As water under pressure flows through the membrane, contaminants and organisms are captured on the membrane and not allowed to pass through. Membranes are not well suited to highly contaminated source waters because the solid materials clog up the membrane almost immediately. Membrane filtration is gaining use in the United States for special applications and in combination with other types of filtration.

Disinfection. Filtration and the steps prior to filtration focus on the physical removal of contaminants in the water. In addition to physical removal, it is still important to provide chemical disinfection. Disinfectants used include chlorine, chloramines (chlorine plus ammonia), ozone, ultraviolet light, and chlorine dioxide. Chlorine was first used in the United States in a water-treatment plant in 1908.

The advantage of chlorination is that it continues to kill bacteria as water moves through pipes to the tap. Its disadvantage is the possibility of disinfection by-products. Excess chlorine in water can combine with organic material in the water to form substances such as trihalomethanes, which can cause liver, kidney, or central nervous system problems, and are linked to an increased risk of cancer over a lifetime exposure.

Disinfection is needed to inactivate (kill) bacteria and viruses that make it through the physical removal (filtration) steps. Viruses and *giardia* are effectively killed by chlorine. Over time, scientists have found that some organisms such as *Cryptosporidium* are resistant to chlorine. *Cryptosporidium* rose to public attention in 1993 when it sickened over 400,000 people, killing a hundred, in Milwaukee, Wisconsin. Largely because of this scare, new or amended U.S. drinking-water regulations developed early in the twenty-first century that expanded water treatment requirements specifically to address *Cryptosporidium*. Although chlorine is not effective against *Cryptosporidium*, alternative disinfectants such as ozone and ultraviolet light do appear to be effective at killing it. In Europe, both of these disinfectants are often used without chlorination to kill bacteria in the water supply.

An amendment to the SDWA requires that all sources of potable water in the United States be filtered. In some locales throughout the nation, such as Boston and Seattle, reservoir water is essentially free of organic matter, and municipalities have been able to avoid filtration because they have extensive watershed protection and management programs in place.

Other Chemical Treatments. Chemicals are added to drinking water to adjust its hardness or softness, **pH**, and alkalinity. Water that is acidic is very corrosive to the pipes and materials with which it comes into contact. The addition of sodium hydroxide can reduce corrosivity and extend the service life of pipelines, storage tanks, and building plumbing systems. Pipes may also be coated with chemicals to prevent metals like copper from dissolving in the water. In addition, chemicals are used to reduce the leaching of lead from old lead pipes and lead-soldered copper supply pipes. Fluoride is frequently added to the water in many communities to improve the dental health of younger residents.

pH an expression of the intensity of the basic or acid condition of a liquid; may range from 0 to 14, where 0 is the most acid, 7 is neutral, and 14 is most base; natural waters usually have a pH between 6.5 and 8.

Groundwater Protection and Treatment

Wellhead protection is critical to preventing the contamination of groundwater supplies. Groundwater is pumped out of an aquifer, which is like a small underground lake surrounded by layers of rock and soil. Water from the surface flows through the rock and soil to get to the aquifer. The earth naturally provides filtration of microscopic pathogens. It does not always provide adequate protection against viruses or chemicals that are dumped on the ground. Groundwater typically contains higher concentrations of metals like iron and manganese because these metals occur naturally in the earth. Groundwater may also be much harder than surface water. Processes similar to those outlined above are also used to treat groundwater, except that the filtration steps are often focused on removing chemicals or metals rather than pathogens. Some groundwater supplies are not treated at all, while others may be filtered and disinfected. As with surface waters, the quality of the source dictates what treatment steps are required.

Regulatory Reporting and Public Education

Water systems in the United States submit reports each month to state or federal regulatory agencies, summarizing treatment-plant performance and sampling results. The majority of medium and large water systems in the United States have staff working twenty-four hours a day. If something were to go wrong at the plant, the plant operators have procedures that they would follow to shut down the plant, switch to alternate equipment, adjust chemical dosages, or collect additional samples. State and federal regulations

One of the problems in protecting drinking water is that by the time results of tests for *E. coli* or *Cryptosporidium* or even anthrax are known, an urban population can already be at risk. Inventors Gregory Quist and Hanno Ix are out to change that. They use laser beams to scan a flow of water; particles in the water scatter the light beam and each scatter pattern is different. A computer analyzes the pattern and provides continuous real-time identification of microorganisms. The system is being tested at a Los Angeles, California, water facility.

specify when the water plant operator must notify the state or federal agency, and these requirements are built into the plant's procedures. The regulations also specify when the public must be notified. Orders to boil the water are usually jointly issued by the state health agency and the drinking-water system quickly after a problem has been discovered (most likely via telephone and radio). Public notices about problems with routine monitoring results or the failure to collect required samples would generally be distributed in the newspaper or via the water utility's annual water quality report (also called a consumer confidence report). The requirement that all water systems compile and distribute a user-friendly report began in 1998. This report provides an overview of the water-system activities and compliance with regulations for the year, as well as identifying ways that customers can get involved or acquire more information. SEE ALSO AGRICULTURE; CRYPTOSPORIDIOSIS; GROUNDWATER; HEALTH, HUMAN; NONPOINT SOURCE POLLUTION; SNOW, JOHN; WASTEWATER TREATMENT; WATER POLLUTION.

Bibliography

American Water Works Association. (1999). *Water Quality and Treatment, A Handbook of Community Water Supplies,* 5th edition. San Francisco: McGraw-Hill.

American Water Works Association and American Society of Civil Engineers. (1998). *Water Treatment Plant Design,* 3rd edition. San Francisco: McGraw-Hill.

Peavy, Howard S.; Rowe, Donald R.; and Tchobanoglous, George. (1985). *Environmental Engineering.* McGraw-Hill Series in Water Resources and Environmental Engineering. San Francisco: McGraw-Hill.

Symons, James M. (1992). *Plain Talk about Drinking Water: Answers to 101 Important Questions about the Water You Drink.* Boulder, CO: American Water Works Association.

Internet Resource

U.S. Environmental Protection Agency, Office of Water Web site. Available from http://www.epa.gov/ow.

Julie Hutchins Cairn

Whistleblowing

Employees are the eyes and ears of environmental protection. They bury waste, operate incinerators, and witness the discharge of pollutants into the environment. However, employees who "blow the whistle" and report environmental wrongdoing are often subject to harassment, dismissal, and blacklisting.

In 1972 as part of the Federal Water Pollution Control Act, Congress recognized the critical role workers play in ensuring the enforcement of environmental laws and enacted the first environmental whistleblower law. Retaliating against environmental whistleblowers was made illegal and the victims of such misconduct finally benefitted from a government remedy at the federal level. By 1980 Congress passed six other environmental whistleblowers laws, protecting employees who blow the whistle on violations of the Toxic Substances, Safe Drinking Water, Solid Waste Disposal, Clean Air, Atomic Energy, and Comprehensive Environmental Response (Superfund) Acts.

Nearly every American worker is protected under these laws. The types of employees who have successfully filed suit include high-level managers who exposed the dumping of raw sewage into rivers, quality-assurance inspectors who reported nuclear safety violations, federal EPA scientists who published papers documenting flaws in risk assessments, and state inspectors who blew the whistle on a school built on a toxic waste dump. In each case the public was able to learn about and prevent significant threats to the environment and public health.

The federal environmental protection laws offer a remedy to the victims of retaliation. The employee initiates the process by filing a complaint with the U.S. Department of Labor (DOL) within thirty days of learning of the discriminatory action. The Occupational Safety and Health Administration (OSHA) investigates the complaint, and the whistleblower is entitled to a full evidentiary hearing before the DOL. If the whistleblower wins, that individual is entitled to reinstatement, back pay, attorney fees, damages for loss of reputation, and emotional distress. Under the Toxic Substances and Safe Drinking Water laws, whistleblowers may also be awarded punitive damages.

Although hundreds of employees have obtained relief under these laws (including some multimillion-dollar judgments), whistleblower cases are hard fought, and many environmental whistleblowers with valid cases lose in court. Some cannot afford attorneys experienced in this special area of the law, whereas others miss the thirty-day statute of limitations for filing complaints. Many whistleblowers are overwhelmed by the personal crises they must face after losing their jobs.

Environmental whistleblowers have also received significant support outside the legal system. For example, press coverage of Karen Silkwood's whistleblowing in the early 1970s called attention to the hazards of nuclear power plants. Additionally, public interest groups, such as the National Whistleblowers Center and Public Employees for Professional Responsibility, provide resources and assistance to environmental whistleblowers. SEE ALSO ACTIVISM; LAWS AND REGULATIONS, INTERNATIONAL; LAWS AND REGULATIONS, UNITED STATES

Bibliography

Kohn, Stephen. (2001). "Environmental and Nuclear Whistleblowing." In *Concepts and Procedures in Whistleblower Law*. Westport, CT: Greenwood Publishing Group.

Internet Resources

National Whistleblower Center. "Environmental Issues and Nuclear Safety." Available from www.whistleblowers.org.

U.S. Department of Labor Office of Administrative Law Judges. "Whistleblower." Available from www.oalj.dol.gov.

Stephen M. Kohn

Wise-Use Movement

The *wise-use movement* is a general term relating to an approach to the management of federal lands in the United States that encompasses many themes, but emphasizes a preference for extractive (e.g., mining, oil drilling) or utilitarian (e.g., grazing) uses over ecological, scenic, wildlife, or aesthetic values. The movement was founded in 1988 by Ron Arnold and Alan Gottlieb, who run the Center for the Defense of Free Enterprise based in Seattle, Washington. The movement is a loose coalition of individuals and organizations that initially advocated increased access to and development of federal lands and resources. Although the movement has enlisted some support nationwide, its appeal has existed primarily in the West, where the percentage of land owned by the federal government is the highest. The federal government owns approximately one-third of U.S. lands, but the percentage is much higher in many western states, a fact that has engendered considerable resentment among corporations and individuals who want to use or develop the resources on those lands. The movement had its ideological origins in the Sagebrush Rebellion of the late 1970s and 1980s that focused on eliminating federal ownership of many lands in the West. However, the wise-use movement focuses less on ownership issues and more on changing public and corporate access to and uses of federal lands, and encompasses other issues as well.

"Wise use" was a phrase originally used by Gifford Pinchot, an early conservationist and the first head of the Forest Service in the early 1900s, who advocated the use of federally owned natural resources for the greatest good of the greatest number. However, the phrase is used by the wise-use movement to encompass a wide range of issues, from eliminating environmental controls, to defense of private property rights with compensation for all environmental regulation, to local control of federal lands in order to permit unrestricted logging, grazing, drilling, and mineral development—even in national parks and wilderness areas. The movement is largely sustained by corporate funding and contributions from other organizations. The movement deliberately adopted the grassroots techniques and terminology of the environmental movement to create a proworker and community image for policies that actually furthered corporate and industrial goals (i.e., mining).

Many of the positions advocated by the wise-use movement continue to be influential. Anti–big-government policies in general, greater nonfederal control of federal lands, self-audits by corporations to determine environmental compliance, increased emphasis on commodity development, and the weakening of environmental laws are but a few examples. Some of the laws the movement seeks to reverse or eliminate include the Clean Air and Clean Water Acts, and the Endangered Species Act. Many wise-use movement organizations have adopted names that camouflage the organization's prodevelopment, antienvironmentalist stance, such as the National Wetlands Coalition, the Public Lands Council, Citizens for the Environment, Environmental Conservation Organization, and Defenders of Property Rights. Some aspects of movement positions also reflect the policies of other organizations. For example, the American Enterprise Institute and Political Economy Research Center advocate the privatization of natural resources through "free market environmentalism"—policies that overlap with some of those of the wise-use movement. On the other end of the spectrum, the movement has ties to more extreme organizations, such as militia groups. Its writings range from constitutional interpretations supporting its viewpoint to vitriolic attacks on "pagan" and "communist" environmentalists whose alleged goal is a "totalitarian one-world government."

Bibliography

Arnold, Ron, and Gottlieb, Alan M. (1998). *Trashing the Economy: How Runaway Environmentalism Is Wrecking America*. Bellevue, WA: Free Enterprise Press.

Helvarg, David. (1994). *The War Against the Greens: The Wise Use Movement, the New Right, and Anti-Environmental Violence*. San Francisco: Sierra Club Books.

Pendley, Perry. (1995). *War on the West: Government Tyranny on America's Great Frontier*. Washington, D.C.: Regnery.

Internet Resources

Arnold, Ron. "Overcoming Ideology." Available from Center for the Defense of Free Enterprise Web site, http://www.cdfe.org/wiseuse.htm.

Environmental Working Group Clearinghouse on Environmental Advocacy and Research (CLEAR). "The Wise Use Movement: Strategic Analysis and Fifty-State Review." Available from http://www.ewg.org/pub.

Pamela Baldwin

Workers Health Bureau

The Workers Health Bureau of America (WHB), active from 1921 to 1928, was a grassroots organization run by Grace Burnham, Harriet Silverman, and Charlotte Todes. Primarily an advocacy organization, WHB is known for focusing public attention on occupational health and safety issues for the first time. The bureau endorsed local trade labor unions' efforts to improve workers' health conditions. WHB conducted investigations, wrote informational reports, and organized union movements. During its eight years, WHB enjoyed the membership of approximately 180 local trade unions and garnered support from leading public health experts.

WHB contended that workers' health problems resulted from a combination of industrial employment and urban living. The bureau had little confidence in government agencies' abilities to improve working conditions, although it did advocate for changes in national labor laws. Considering workers' problems a class issue, WHB solicited memberships among workers and unions in exchange for help in improving work conditions at the local level. WHB advised employees and labor unions to solve problems at their source. By advocating that unions add health and safety clauses into their employment contracts, WHB hoped that employers would proactively improve conditions in their plants.

The bureau concentrated on the most common occupational health problems of the time. Some of WHB's major campaigns addressed workplace exposures to benzol, carbon dioxide, coal and silica dust, lead, and mercury. The bureau used scientific studies and terminology to strengthen their arguments in highly politicized debates. Ironically, WHB ended its work in 1928 because it was too successful. The Affiliated Federation of Labor (AFL) pressured local unions to withdraw from WHB, perhaps to rein in their influence over unions. In the end, WHB is best remembered for bringing labor health issues to national attention, beginning the movement that eventually led to the creation of the Occupational Safety and Health Bureau (OSHA) in 1970. SEE ALSO ACTIVISM; INDUSTRY; OCCUPATIONAL SAFETY AND HEALTH ADMINISTRATION (OSHA); PUBLIC POLICY DECISION MAKING.

Bibliography

Rosner, David, and Markowitz, Gerald. (1987). "Safety and Health as a Class Issue: The Workers Health Bureau of America during the 1920s." In *Dying for Work: Workers' Safety and Health in Twentieth-Century America*, edited by David Rosner and Gerald Markowitz. Bloomington: Indiana University Press.

Internet Resource

Robert F. Wagner Labor Archives and Tamiment Archives. Available from http://www.nyu.edu/library.

Mary Elliott Rollé

World Trade Organization

The General Agreement on Tariffs and Trade (GATT) was an international organization created in 1947 to reduce trade barriers through multilateral negotiations. The World Trade Organization (WTO) was organized in January 1995 to replace GATT and improve international trade. Its membership in 2002 totaled more than 140 nations.

Whereas GATT focused on tariff reduction, the WTO works to eliminate so-called nontariff barriers, which can include environmental, health, and other public-interest regulations that are considered impediments to international trade. Any member country has the right to challenge other members' laws under the WTO dispute-settlement process. When this occurs, the WTO forms a three-person tribunal to hold hearings on the case, which take place in secret in Geneva, Switzerland. If the tribunal finds that the law is illegal within the context of WTO policy, it has the power to order the country to change the law or face trade sanctions.

The WTO's first ruling involved a successful challenge to the U.S. Clean Air Act. Brazil and Venezuela had complained that a part of the act that required all foreign sources of U.S. gas imports to meet a certain cleanliness standard was discriminatory. The U.S. government was ordered to amend its regulation or face retaliatory trade sanctions of approximately $150 million per year. It opted to modify the law.

This ruling unleashed a flood of other challenges against environmental laws, such as U.S. dolphin and sea turtle protections, Japan's ban on fruit imports carrying invasive species, and the European Union's ban on U.S. beef injected with growth hormones.

The WTO does allow some exceptions for laws that are "necessary to protect human, animal or plant life and health." However, this exception has proved virtually useless, since WTO panels have interpreted the language to mean that laws must represent the "least trade-restrictive" way to achieve the environmental goal.

Although WTO rulings have most often targeted environmental protections, the organization has also drawn strong criticism from labor unions. Among other complaints, they argue that the WTO should adopt rules in support of internationally recognized labor rights as a way to prevent corporations and governments from gaining an unfair trade advantage by abusing workers.

In December 1999 tens of thousands of environmentalists joined with trade unionists and other activists to protest the WTO Ministerial Meeting in Seattle, Washington. The "Teamsters and Turtles" united in the streets, combined with disputes among some member countries, forced the organization to abandon plans to launch a new round of negotiations. The "Battle in Seattle" also thrust the WTO into the public limelight for the first time. The next WTO meeting was sited far from angry crowds and international media attention in isolated Doha, Qatar. Although it concluded with the announcement of plans for a new round of discussion, the meeting was fraught with tensions and the WTO's future appears anything but smooth.

SEE ALSO ECONOMICS; ENVIRONMENTAL CRIME; ENVIRONMENTAL JUSTICE; LAWS AND REGULATIONS, INTERNATIONAL; TREATIES AND CONFERENCES.

Bibliography

Shrybman, Steven. (1999). *The World Trade Organization: A Citizen's Guide.* Toronto, Ontario: Canadian Centre for Policy Alternatives.

Internet Resource

World Trade Organization Web site. Available from http://www.wto.org.

Sarah Anderson

Writers

Although writers have explored the relationship between humans and the natural world for centuries, they primarily viewed the environment as subordinate to the needs of civilization and human progress. However, by the middle of the nineteenth century, writers such as Henry David Thoreau and Ralph Waldo Emerson began to reinterpret the significance of nature and our relationship to it.

Although not writing of pollution specifically, these writers laid the groundwork for an evolution in environmental thought and ethics in which the environment was seen as more than just a natural resource. For example, in *Walden* and other writings, Thoreau pointed out that our natural environment had far more to offer than material resources to be exploited. Rather, Thoreau noted that nature and the environment were sources of spiritual truth and support.

Although these writers and others, like John Muir and Aldo Leopold, helped educate the public about nature and the environment, one of the first "environmental" books published in the United States to include a discussion of pollution was *Man and Nature.* In his 1865 book, author George Perkins Marsh presented a comprehensive discussion of ecological problems brought on by the impact of human civilization, including the growing problem of water pollution.

By the beginning of the twentieth century, most writings about pollution focused on how industrial pollution was affecting those in the workplace. For example, Alice Hamilton, a University of Michigan Medical School graduate of 1893, conducted studies of occupational diseases, including those brought on by industrial pollution in the lead, rubber, and munitions industry. Her books included the 1925 work called *Industrial Poisons in the United States*, an early and compelling scientific look into pollution in the workplace and its effects on workers.

Upton Sinclair's novel *The Jungle* exposed the horrendous working and living conditions of slaughterhouse workers in the meat packing industry of Chicago. His exposure of the unsanitary slaughterhouse conditions led to the first U.S. meat protection laws and raised public awareness of corporate greed and the plight of poorly paid workers and their families in dense, polluted sections of large urban areas.

Leopold's conversational essays compiled in *A Sand County Almanac* (1949) argued persuasively that nature is not a machine of interchangeable parts but an interdependent community and that humans are part of this community, not detached from it. Because we are part of it and have the power to impact it so profoundly, we have an ethical obligation to act in ways

> "I went to the woods because I wished to live deliberately, to front only the essential facts of life, and see if I could not learn what it had to teach, and not, when I came to die, discover that I had not lived."
>
> —Henry David Thoreau, *Walden* (1854)

> "We abuse the land because we regard it as a commodity belonging to us. When we see land as a community to which we belong, we may begin to use it with love and respect."
>
> —Aldo Leopold

> "As crude a weapon as the cave man's club, the chemical barrage has been hurled against the fabric of life."
>
> —Rachel Carson, *Silent Spring*

that preserve the integrity of the whole community. Leopold called this integration of science, aesthetics, and ethics "the land ethic" and it laid the practical foundation for systems thinking and the ecological perspective.

Few would argue that Rachel Carson's book *Silent Spring* was a seminal work that launched both a growing public concern about pollution and the modern environmental movement. First published in serialized form in the *New Yorker* magazine, *Silent Spring* was published in 1962 and exposed the dangers posed by numerous pesticides and fertilizers, including DDT. The book was a catalyst for a new view of industry and pollution that would overturn the long-held belief that scientific progress was always for the good.

Silent Spring's publication set off a storm of controversy. The pesticide industry tried to suppress the book's publication and challenged its findings. Carson's book eventually led to a presidential commission to study the effects of pesticides. The commission verified Carson's findings, which eventually led to the banning of DDT in 1972. More importantly, the book led to a wave of public concern over the use of chemicals and pollution and how they impact the environment and life. When the Modern Library published its list of the 100 Best Nonfiction Books of the Century, *Silent Spring* was listed as the fifth-most-important book of the twentieth century.

As public concern and interest in the environment and pollution grew, more writers began to explore the possible catastrophic impact of ever-increasing pollution. Many writers pointed out that it was important to look at more than the impact of specific chemicals and pollutants on the environment. For example, in his 1968 book *The Population Bomb*, Paul Ehrlich took the writings of Thomas Malthus (1766–1834) about overpopulation and expanded them. Ehrlich stated that overpopulation not only would lead to widespread starvation in the world but also affected the environment by creating more garbage and other pollutants.

In his 1971 book *The Closing Circle*, social commentator and one-time presidential candidate Barry Commoner also placed the environmental message about pollution into a broader context as he connected the growth of technology to environmental degradation. Not only did Commoner discuss the environmental crises in terms of population and "affluence" but also provided in-depth looks at how growing population and advancing technology were the culprits behind such specific environmental problems as the hazardous air pollution in Los Angeles and the polluted waters of Lake Erie. Commoner argued three principles that became rallying cries of early environmentalists: Everything has to go somewhere, nature knows best, and there is no such thing as a free lunch.

Arne Naess, the Norwegian philosopher and founder of the deep ecology movement, is another notable philosopher and writer who created a philosophy of how to deal with issues such as pollution. In a 1973 article in *Inquiry* magazine, Naess laid out her philosophy of shallow ecology versus deep ecology. Naess argues that a shallow ecology philosophy fights against pollution but continues to inadvertently support those who cause much of the world's pollution, namely those in healthy and affluent nations. Naess notes that deep ecology, on the other hand, focuses on changing the relationship between civilization and the natural world, not only by fighting pollution but also by establishing a philosophy of human respect for all species and nature.

> "To err is human, but to really foul things up you need a computer."
>
> —Paul Ehrlich

George Sessions played an important role in popularizing deep ecology in North America and its emphasis on biocentrism (nature-centered, not human-centered philosophy) and social justice from exploitation of poorer countries by affluent ones. Deep ecology argues that nature has inherent value apart from human use. It traces the destruction of nature to industrial society. Gary Snyder's poetry reflects deep ecology through sensual images.

In 1989, Bill McKibben, prolific nature writer and environmental commentator and historian, followed in the tradition of *Silent Spring* when he changed the public's deepest perceptions of the world with his book *The End of Nature*. McKibben brought to the forefront a public and policy discussion about the latest scientific evidence concerning pollutants such as acid rain and their impact on the greenhouse effect, the depletion of the ozone layer, and global warming. In the book, McKibben points out that industrial society with all of its pollutants has altered the chemistry of the atmosphere and changed the most elemental process of life everywhere. In the end, McKibben states that the only hope in stopping pollution and saving the environment is that people will come to fully understand the dangers caused by pollution and other environmental problems and make a conscious decision to live with less, thus creating less pollution.

About two years after McKibben's book was published, *Earth in the Balance: Ecology and the Human Spirit* appeared on bookshelves. The book's author Al Gore, at the time a United States senator, states that a fundamental change in how we view the world and interact with it is necessary if we are to save the earth's ecology for future generations. Gore discusses the deteriorating quality of air, water, and soil due to a variety of pollutants, including those that cause a rise in carbon dioxide levels, which is leading to a deteriorating ozone layer. Like McKibben, Gore also points out that pollution problems are no longer local or regional but global. Because of Gore's high-profile career and place of power on the American scene, the book received worldwide attention from the public as well as political circles.

Like deep ecologists, ecofeminist authors, both men and women, are varied in their understandings of the philosophy. Nevertheless, all ecofeminists begin with the premise that the exploitive and abusive treatment of nature is linked to the patriarchal (male-dominated) exploitation and violence towards women. For example, see Susan Griffen's *Woman and Nature: The Roaring Inside Her* for a discussion of the metaphors of "the rape of nature" and "virgin land," and "Mother Earth/Nature." Another theme of many ecofeminist writings is the rediscovery and celebration of the goddess that was once the center of earlier cultures and Native American spiritual teachings.

Sandra Steingrabber's popular book *Living Downstream* highlights the cumulative risks faced by river human and nonhuman communities living downstream from pollution. She argues for the responsibility of upstream communities to act environmentally responsible for their downstream neighbors.

Nature and environmental writing has exploded since the first Earth Day in 1970. This article presents just a few representative writers of fiction, nonfiction, and poetry.

Because of writers like Carson, McKibben, and many others, few people would argue that pollution is not a threat to the environment and the health of people and other species. More books and articles than ever are being

> "The invention of nuclear weapons may actually have marked the beginning of the end of nature: we possessed, finally, the capacity to overmaster nature, to leave an indelible imprint everywhere all at once."
>
> —Bill McKibben, *The End of Nature*

> "We can believe in that future and work to achieve it and preserve it, or we can whirl blindly on, behaving as if one day there will be no children to inherit our legacy. The choice is ours; the earth is in the balance."
>
> —Al Gore, *Earth in the Balance*

"When considering hormones such as estradiol, the most potent estrogen, forget parts per million or parts per billion. The concentrations are typically parts per trillion, one thousand times lower than parts per billion. One can begin to imagine a quantity so infinitesimally small by thinking of a drop of gin in a train of tank cars full of tonic. One drop in 660 tank cars would be one part in a trillion; such a train would be six miles long."

—Theo Colborn, *Our Stolen Future*

written about pollution issues, and many writers are carrying on the legacy of Carson. For example, in 1996's *Our Stolen Future*, the authors discuss the various ways in which chemicals are disrupting human reproductive patterns and causing such problems as birth defects, sexual abnormalities, and reproductive failure. In 2002, Devra Lee Davis was nominated for the National Book Award in nonfiction for her book *When Smoke Ran Like Water: Tales of Environmental Deception and the Battle against Pollution*. In the book, Davis discusses how industry and government have conspired to conceal the true effects of pollution on public health.

Man and Nature (1865) by George Perkins Marsh; Early scientific look at the environment and how humans influence it, including the effects of human pollution on water

Industrial Poisons in the United States (1925) by Alice Hamilton; Scientific study of industrial pollutants and how they affect workers

Silent Spring (1962) by Rachel Carson; Groundbreaking look at pesticides that helped create the modern environmental movement and led to government banning of DDT

The Closing Circle (1971) by Barry Commoner; Connected growth of technology to pollution and environmental degradation

The End of Nature (1990) by Bill McKibben; Helped raise worldwide concern over pollution, the greenhouse effect, and the depletion of the ozone layer

When Smoke Ran Like Water (2002) by Devra Lee David; National Book Award finalist that tells of government and industry coverups concerning the effects of pollution on the populace

Bibliography

Carson, Rachel. (1962). *Silent Spring*. Boston: Houghton Mifflin.

Caulfield, Henry. (1989). "The Conservation and Environmental Movements: An Historical Analysis." In *Environmental Politics and Policy: Theories and Evidence*, ed. James Lester. Durham, NC: Duke University Press.

Davis, Devra Lee. (2002). *When Smoke Ran Like Water: Tales of Environmental Deception and the Battle against Pollution*. New York: Basic Books.

Ehrlich, Paul. (1968). *The Population Bomb*. New York: Ballantine.

Gore, Al. (1992). *Earth in the Balance: Ecology and the Human Spirit*. Boston: Houghton Mifflin.

McKibben, Bill. (1989). *The End of Nature*. New York: Random House.

Naess, Arne. (1973), "The Shallow and the Deep: The Long-Range Ecology Movement," *Inquiry* 16:95-100.

Netzley, Patricia D., compiler. (1999). *Environmental Literature: An Encyclopedia of Works, Authors, and Themes*. Santa Barbara, CA: ABC-CLIO.

Thoreau, Henry David. (1995). *Walden, or, Life in the Woods*. New York: Dover Publications.

David Petechuk

Yucca Mountain

The United States has accumulated more than forty thousand tons of spent nuclear fuel and high-level radioactive wastes from commercial, research, and defense activities with an estimated two thousand tons added every year. The materials are currently stored in thirty-nine states at 131 temporary above-ground facilities, requiring constant monitoring and maintenance. Worldwide, scientific consensus holds that deep geologic disposal, with robust engineered barriers, can best contain and isolate these materials from the accessible environment. The Nuclear Waste Policy Act of 1982 established this approach as U.S. policy. If ultimately licensed by the Nuclear Regulatory Commission, Yucca Mountain, in southern Nevada, could become the first U.S. geologic repository for such materials. The Department of Energy (DOE) plans to open the proposed repository by 2010 if a license is granted.

Between about fifteen and twelve million years ago, large volcanic eruptions deposited hot ash that solidified into the rock composing Yucca Mountain. The proposed repository would be built about one thousand feet underground and, on average, about one thousand feet above the water table in rock that has remained undisturbed for millions of years. For about two thousand feet under the mountain's surface the rock is very dry, or unsaturated, meaning its pore spaces are not completely filled with water.

Waste Forms and Other Engineered Barriers

All materials sent to a repository would be in solid form. Spent nuclear fuel comprises hard ceramic pellets in sealed corrosion-resistant metal tubes. Liquid wastes from defense-related activities would be solidified into glass logs, inside sealed metal containers, before shipment.

At the repository, the materials would be sealed inside double-walled containers, called waste packages, made of stainless steel and a corrosion-resistant alloy. Once underground, each waste package would be placed on its own individual pallet, in one of dozens of miles of tunnels carved deep within the rock. In addition, corrosion-resistant titanium drip shields would be placed above the sealed containers as an added barrier to water. (See illustration.)

Potential Problems at Site

Groundwater contamination. Yucca Mountain's climate is very dry, with annual precipitation averaging about 7.5 inches (190 millimeters or mm). About 95 percent either runs off, evaporates, or is taken up by vegetation. Overall, very little water infiltrates the mountain and reaches the repository level. The bulk of any water moves very slowly through the unsaturated rock. Some data, however, suggest that water may reach the repository level in a few decades by moving through fractures that are large enough to permit this. Therefore, the sophisticated computer calculations used to estimate the repository's likely performance assume the presence of such fractures and their impact. After water has infiltrated the repository level, it must move down through approximately one thousand more feet of unsaturated rock to reach the saturated zone. Only from this zone can water be pumped to the surface.

Earthquake activity. Southern Nevada has low to moderate seismic activity. Experts have analyzed potentially active faults within sixty miles of Yucca Mountain. Although scientists expect earthquakes to occur at or near the mountain, those working on the design of the Yucca Mountain repository think that with modern techniques, repository facilities can be designed and constructed to withstand the effects of earthquakes and other natural phenomena. Contributing to underground safety is the fact that seismic

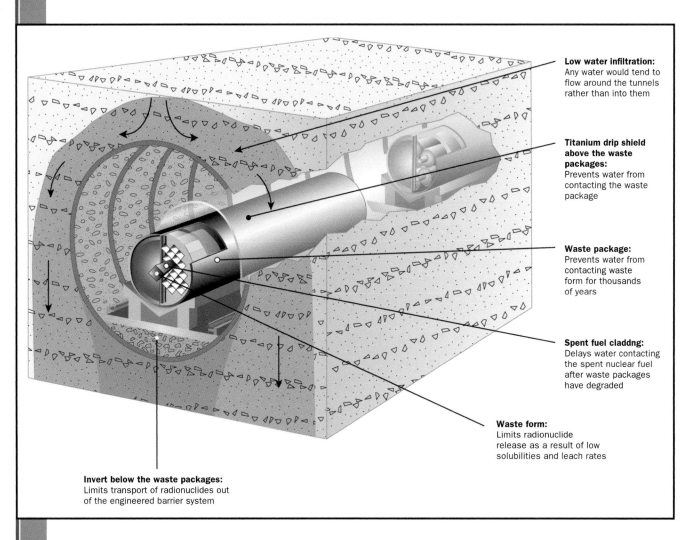

Low water infiltration: Any water would tend to flow around the tunnels rather than into them

Titanium drip shield above the waste packages: Prevents water from contacting the waste package

Waste package: Prevents water from contacting waste form for thousands of years

Spent fuel claddng: Delays water contacting the spent nuclear fuel after waste packages have degraded

Waste form: Limits radionuclide release as a result of low solubilities and leach rates

Invert below the waste packages: Limits transport of radionuclides out of the engineered barrier system

Cutaway illustrating natural and engineered barriers working together in an emplacement tunnel. Capillary action would cause most available water to flow around, rather than into, the tunnels. Federal law limits the proposed repository to seventy thousand metric tons of heavy metal "until such time as a second repository is in operation," unless the law is changed. (From Office of Public Affairs, U.S. Department of Energy. (2002). *Why Yucca Mountain? Frequently Asked Questions.* Washington, D.C., p. 10.)

ground motion diminishes with depth, so earthquakes have less impact deep underground than they do on or near the surface.

Transportation. Some people fear that vehicles moving nuclear waste across the country could be subject to accidents or become a target for terrorists. Federal regulations require that transportation cask designs be certified to withstand a series of severe impacts and extreme conditions without leaking radioactive materials. The regulations also require that shipments be monitored and tracked by satellite twenty-hour hours a day and accompanied by trained escorts, who must report in regularly. Armed escorts would be required through heavily populated metropolitan areas.

Other Nations' Approaches. Some nations using nuclear power do not have economical sources of fresh uranium to make nuclear fuel. France and the United Kingdom, for example, reprocess their own spent nuclear fuel for a second usage; they also do reprocessing for other countries, such as Japan and Switzerland. Current techniques for reprocessing involve complex chemical and physical procedures and actually produce additional radioactive waste. Most nations with nuclear power intend to build their own geologic repositories.

Health and safety. For more than twenty years, scientists and engineers have gathered technical data about the rock in Yucca Mountain, water movement through it, expected earthquakes, and the potential for volcanic disturbance of the proposed repository. Applying advanced software and high-powered computers to these data, scientists have estimated radiation doses due to the repository for tens of thousands of years. The radiation protection standards set by the Environmental Protection Agency (EPA) require that the calculations estimate the likely level of radiation that the most exposed member of the public would receive from the repository for ten thousand years after its closure. The standards require that this hypothetical person be assumed to live about fifteen miles from the repository, to eat some foods grown with local groundwater, and to drink two liters of water per day drawn from the most concentrated plume of repository-caused contamination in the aquifer. The estimates indicate that, for at least ten thousand years, the level of repository-yielded radioactivity this hypothetical person would likely receive, through all potential exposure pathways, would be far below fifteen millirem per year, the radiation protection standard for public health and safety. SEE ALSO CANCER; HEALTH, HUMAN; RADIOACTIVE WASTE; WASTE, TRANSPORTATION OF.

Bibliography

Board on Radioactive Waste Management, National Research Council, National Academy of Sciences. (2001). *Disposition of High-Level Waste and Spent Nuclear Fuel: The Continuing Societal and Technical Challenges.* Washington, D.C.: National Academy Press. Also available from http://www.books.nap.edu/books.

International Atomic Energy Agency. (2002). *Institutional Framework for Long Term Management of High Level Waste and/or Spent Nuclear Fuel.* (IAEA-TECDOC-1323) Vienna, Austria: IAEA Press. Also available from http://www.pub.iaea.org/mtcd.

Office of Public Affairs, U.S. Department of Energy. (2002). *Why Yucca Mountain? Frequently Asked Questions.* Washington, D.C. Also available from http://www.ocrwm.doe.gov/ymp.

Wheelwright, Jeff. (2002). "Welcome to Yucca Mountain." *Discover* 23(9):66–75.

Internet Resource

Herne Data Systems Web site. "WasteLink: Guide to Radioactive Waste Resources on the Internet." Available from http://www.radwaste.org.

Donald J. Hanley

Zero Population Growth

Malthus's *Essay on Population,* published in 1798, still plays a role in environmental policymaking. The discrepancy between rates of human population growth and agricultural productivity lies at the heart of Malthusianism. One dynamic leads to ever-increasing population; the other to diminishing food and ecological degradation.

To avoid human suffering, Malthusians pursue worldwide zero population growth (ZPG). Because worldwide mortality levels are low, a society can attain ZPG through replacement fertility. If the average number of children born to women in a particular society equals two, then it has reached replacement fertility or ZPG. Governments today annually spend a total of

$900 million to promote ZPG, mostly in less developed countries. ZPG is also the name of a nongovernmental organization that advocates for population awareness. Its origin in 1968 was inspired by Paul Ehrlich's *Population Bomb*. SEE ALSO EHRLICH, PAUL; MALTHUS, THOMAS ROBERT; POPULATION; SMART GROWTH.

Bibliography

Bongaarts, John. (1998). "Demographic Consequences of Declining Fertility." *Science* 282:419–420.

Humphrey, Craig R.; Lewis, Tammy L.; and Buttel, Frederick H. (2002). *Environment, Energy, and Society: A New Synthesis.* Belmont, CA: Wadsworth.

United Nations Fund for Population Action. (2000). *Financial Resource Flows for Population Activities in 1999.* New York.

Craig R. Humphrey

Glossary

24-hour standard: in regulations: the allowable average concentration over 24 hours

absorption spectrum: "fingerprint" of a compound generated when it absorbs characteristic light frequencies

absorption: the uptake of water, other fluids, or dissolved chemicals by a cell or an organism (as tree roots absorb dissolved nutrients in soil)

acetylcholine: a chemical that transmits nerve signals to muscles and other nerves

acute: in medicine, short-term or happening quickly

adherence: substances: sticking to; regulation: abiding by

adjudicative: involving the court system

adsorption: removal of a pollutant from air or water by collecting the pollutant on the surface of a solid material; e.g., an advanced method of treating waste in which activated carbon removes organic matter from wastewater

advise and consent: the formal responsibility of a government body to provide counsel and approval for the actions of another body, especially the Senate to the president

aerate: process of injecting air into water

aerobic: life or processes that require, or are not destroyed by, the presence of oxygen

affinity: physical attraction

afforestation: conversion of open land to forest

air scrubbers: pollution-control devices that remove pollutants from waste gases before release to the atmosphere

air stripping: a treatment system that removes volatile organic compounds (VOCs) from contaminated groundwater or surface water by forcing an airstream through the water and causing the compounds to evaporate

allergen: a substance that causes an allergic reaction in individuals sensitive to it

alloy: mixture of two or more metals

alluvial: relating to sediment deposited by flowing water

alpha radiation: fast-moving particle composed of two protons and two neutrons (a helium nucleus), emitted by radioactive decay

ambient: surrounding or unconfined; air: usually but not always referring to outdoor air

anaerobic: a life or process that occurs in, or is not destroyed by, the absence of oxygen

antagonistic: working against

anthropogenic: human-made; related to or produced by the influence of humans on nature

antimicrobial: an agent that kills microbes

aquaculture: practice of growing marine plants and raising marine animals for food

aquifer: an underground geological formation, or group of formations, containing water; are sources of groundwater for wells and springs

archetype: original or ideal example or model

arithmetic: increase by addition, e.g., 2, 4, 6, 8 . . . as opposed to geometric, in which increase is by multiplication, e.g., 2, 4, 8, 16 . . .

arthropod: insects, spiders, and other organisms with jointed appendages and hard outer coverings

asbestosis: a disease associated with inhalation of asbestos fibers; the disease makes breathing progressively more difficult and can be fatal

asymmetrical warfare: conflict between two forces of greatly different sizes; e.g., terrorists versus superpower

autoimmune: reaction of the body's immune system to the body's own tissues

baghouse: large fabric bag, usually made of glass fibers, used to eliminate intermediate and large particles

ballast: material in a ship used for weight and balance

bed load transport: movement of sediments that remain at the bottom of a moving water body

beta radiation: high-energy electron, emitted by radioactive decay

bilge: deepest part of a ship's hold

bioaccumulation: buildup of a chemical within a food chain when a predator consumes prey containing that chemical

bioaccumulative: relating to substances that increase in concentration in living organisms as they take in contaminated air, water, or food because the substances are very slowly metabolized or excreted

bioaerosol: very fine airborne particles produced by living organisms

bioassay: a test to determine the relative strength of a substance by comparing its effect on a test organism with that of a standard preparation

bioavailability: degree of ability to be absorbed and ready to interact in organism metabolism

bioconcentrate: chemical buildup in an organism, i.e., fish tissue, to levels higher than in the surrounding environment

biodegradation: decomposition due to the action of bacteria and other organisms

biodegrade: to decompose under natural conditions

biodiversity: refers to the variety and variability among living organisms and the ecological complexes in which they occur; for biological diversity, these items are organized at many levels, ranging from complete ecosystems to the biochemical structures that are the molecular basis of heredity; thus, the term encompasses different ecosystems, species, and genes

biogeochemical interaction: interactions between living and nonliving components of the biosphere

biological capital: oceans, forests, and other ecosystems that provide resources or other values

biological effects: effects on living organisms

bioluminescence: release of light by an organism, usually a bacterium

biomass: all of the living material in a given area; often refers to vegetation

biomonitoring: the use of living organisms to test the suitability of effluents for discharge into receiving waters and to test the quality of such waters downstream from the discharge; analysis of blood, urine, tissues, etc. to measure chemical exposure in humans

bioremediation: use of living organisms to clean up oil spills or remove other pollutants from soil, water, or wastewater; use of organisms such as non-harmful insects to remove agricultural pests or counteract diseases of trees, plants, and garden soil

biosolid: solid or semisolid waste remaining from the treatment of sewage

bituminous: soft coal, versus the harder anthracite coal

boom: a floating device used to contain oil on a body of water; or, a piece of equipment used to apply pesticides from a tractor or truck

boreal: northern, subarctic

botanical: derived from or relating to plants

breakdown product: part of a whole resulting from a chemical transformation

breakdown: degradation into component parts

brine: salty water

British thermal unit (BTU): unit of heat energy equal to the amount of heat required to raise the temperature of one pound of water by one degree Fahrenheit at sea level

bush-fallow: practice of alternating between cultivating a piece of land and leaving it unplanted

cabinet: in government: collective name for the heads of federal departments that report directly to the president

carbamate: class of chemicals widely used as pesticides

carcinogen: any substance that can cause or aggravate cancer

carcinogenic: causing or aggravating cancer

cascade: waterfall; a system that serves to increase the surface area of the water to speed cooling

casing: the exterior lining of the well

catalyst: a substance that changes the speed or yield of a chemical reaction without being consumed or chemically changed by the chemical reaction

catalytic: of a substance that promotes reaction without being consumed

cesspool: holding compound for sewage in which bacterial action breaks down fecal material

chelating agents: chemicals that trap metal ions (*chele* = claw)

chemically active: able to react with other chemicals

chloramination: use of chlorine and ammonia to disinfect water

chromatography: means of resolving a chemical mixture into its components by passing it through a system that retards each component to a varying degree

chronic: in medicine, long-term or happening over time

claim: legal statement of intent

clarifier: a tank in which solids settle to the bottom and are subsequently removed as sludge

codify: put into law

coke: carbon fuel, typically derived from bituminous coal, used in blast furnaces for the conversion of iron ore into iron

combustion: burning, or rapid oxidation, accompanied by release of energy in the form of heat and light

complex emergency: a humanitarian crisis in which there is a breakdown of political authority

compliance: in law: meeting the terms of a law or regulation

computer model: a program that simulates a real event or situation

concordance: state of agreement

condenser: apparatus used to condense vapors

congener: a member of a class of chemicals having a of similar structure

consensus-building: negotiation to create agreement

consent order: a legal agreement requiring specific actions to remedy a violation of law

conservation easement: legal agreement restricting a landowner's development rights to preserve long-term conservation and environmental values

conservationist: a person who works to conserve natural resources

containment: prevention of movement of material beyond the immediate area

contaminant: any physical, chemical, biological, or radiological substance or matter that has an adverse effect on air, water, or soil

control rod: a rod containing substance that absorbs neutrons inserted into a nuclear reactor to control the rate of the reaction

conversion: chemical modification to another form

counterculture: a culture with social ideas that stand in opposition to the mainstream culture

criteria pollutant: a pollutant for which acceptable levels can be defined and for which an air quality standard has been set

crop rotation: alternation of crop species on a field to maintain soil health

cultivar: a plant variety that exists only under cultivation

DDT: the first chlorinated hydrocarbon insecticide chemical name: Dichloro-Diphenyl-Trichloroethane); it has a half-life of fifteen years and can collect in fatty tissues of certain animals; for virtually all but emergency uses, DDT was banned in the U.S. in 1972

defoliant: an herbicide that removes leaves from trees and growing plants

defoliation: loss of vegetation

deicer: chemical used to melt ice

denitrification: the biological reduction of nitrate or nitrite to nitrogen gas, typically by bacteria in soil

deposit: concentration of a substance, i.e., mineral ore

desertification: transition of arable land to desert

desiccant: a chemical agent that absorbs moisture; some desiccants are capable of drying out plants or insects, causing death

despoliation: deprivation of possessions by force

deuterium: a hydrogen atom with an extra neutron, making it unstable and radioactive

diatomaceous earth: a chalk-like material (fossilized diatoms) used to filter out solid waste in wastewater treatment plants; also used as an active ingredient in some powdered pesticides

diffuser: something that spreads out or dissipates another substance over a wide area

dinoflagellate: single-celled aquatic organism

dioxin: any of a family of compounds known chemically as dibenzo-p-dioxins; concern about them arises from their potential toxicity as contaminants in commercial products; tests on laboratory animals indicate that it is one of the more toxic anthropogenic (man-made) compounds

disaster cycle: phases in the public response to a disaster: preparedness, disaster, response, recovery, and mitigation of effects

dissolution into the oceans: dispersion in ocean water

dissolved oxygen: the oxygen freely available in water, vital to fish and other aquatic life and for the prevention of odors; DO levels are considered a most important indicator of a water body's ability to support desirable aquatic life; secondary and advanced waste treatment are generally designed to ensure adequate DO in waste-receiving waters

distillation: the act of purifying liquids through boiling, so that the steam or gaseous vapors condense to a pure liquid; pollutants and contaminants may remain in a concentrated residue

double containment: use of two independent protection systems around a potential pollutant

drier: a compound that increases the drying rate

drilling waste: material (soil, ground rock, etc.) removed during drilling

drinking water: water used or with the potential to be used for human consumption

ecosystem: the interacting system of a biological community and its nonliving environmental surroundings

effluent: discharge, typically wastewater—treated or untreated—that flows out of a treatment plant, sewer, or industrial outfall; generally refers to wastes discharged into surface waters

efflux pump inhibitors: a drug that prevents a cell from expelling another drug; used with antibiotics to increase their effectiveness

electoral consensus: the will of the voters

electrode: conductor used to establish electrical contact with a substance by delivering electric current to it or receiving electric current from it

electromagnetic spectrum: the range of wavelengths of light energy, including visible light, infrared, ultraviolet, and radio waves

emissions: substances, often polluting, discharged into the atmosphere

endocrine: the system of glands, hormones, and receptors that help control animal function

endocrine disruption: disruption of hormone control systems in the body

environmental stewardship: human commitment to care for the environment

epidemic: rapid spread of disease throught a population, or a disease that spreads in this manner

epidemiological: epidemiology: study of the incidence and spread of disease in a population

epidemiology: study of the incidence and spread of disease in a population

epilepsy: seizure disorder

estrogenic: related to estrogens, hormones that control female sexual development

estuary: region of interaction between rivers and near-shore ocean waters, where tidal action and river flow mix fresh- and saltwater (i.e., bays, mouths of rivers, salt marshes, and lagoons). These ecosystems shelter and feed marine life, birds, and wildlife

eutrophication: in nature, the slow aging process during which a lake, estuary, or bay evolves into a bog or marsh and eventually disappears; in pollution, excess algal growth or blooms due to introduction of a nutrient overload of nutrients, i.e., from un- or poorly treated sewage

evaporative: relating to transition from liquid to gas

excavate: dig out

excess death: deaths over the expected number

exothermic: releasing heat

fatalistic: of a person who believes that nothing one does can improve a situation

fecal matter: animal or human excrement

fetus: unborn young of vertabrate animals; human: developing child in the womb from eighth week to birth

filtration: process for removing particulate matter from water by means of porous media such as sand or synthetic filter

flammable: any material that ignites easily and will burn rapidly

flux: 1. a flowing or flow; 2. a substance used to help metals fuse together

French drain: buried plastic tubing with numerous holes, to collect or disperse water

friable: capable of being crumbled, pulverized, or reduced to powder by hand pressure

fungicide: pesticide used to control, deter, or destroy fungi

gamma radiation: very high-energy light with a wavelength shorter than x rays

gelling agent: chemical used to thicken a substance, i.e., oil, to prevent it from spreading out

genetic diversity: the broad pool of genes that insures variety within a species

Geneva Conventions: humanitarian rules governing treatment of soldiers and civilians during war

geometric: by multiplication, e.g., 2, 4, 8, 16 . . ., as opposed to arithmetic, in which increase is by addition, e.g., 2, 4, 6, 8 . . .

global warming: an increase in the near-surface temperature of the Earth; the term is most often used to refer to the warming believed to be occuring as a result of increased emissions of greenhouse gases

grassroots: individual people and small groups, in contrast to government

green choice: a product that is not harmful for the environment

greenhouse gas: a gas, such as carbon dioxide or methane, which contributes to potential climate change

groundwater: the supply of freshwater found beneath the Earth's surface includes; aquifers, which supply wells and springs

guano: solid or semisolid waste from birds and bats, rich in nutrients

Hague Conventions: international agreements governing legal disputes between private parties

half-life: the time required for a pollutant to lose one-half of its original concentration; for example, the biochemical half-life of DDT in the environment is fifteen years

halogenated organic compounds: organic (carbon-containing) compounds containing fluorine, chlorine, bromine, iodine, or astatine

HAZMAT team: hazardous materials response group

heavy metals: metallic elements with high atomic weights; (e.g. mercury, chromium, cadmium, arsenic, and lead); can damage living things at low concentrations and tend to accumulate in the food chain

hemoglobin: oxygen-carrying protein complex in red blood cells

herbicide: a chemical pesticide designed to control or destroy plants, weeds, or grasses

heterotrophic phytoplankton: floating microorganisms that consume other organisms for food

hexavalent: an oxidation state characterized by the ability to make six bonds; symbolized by (VI)

hormone receptors: cell proteins that respond to hormones to influence cell behavior

hormone: a molecule released by one cell to regulate development of another

host: in genetics, the organism, typically a bacterium, into which a gene from another organism is transplanted; in medicine, it is an animal infected or parasitized by another organism

humus: rich soil component derived from plant breakdown and bacterial action

hybridization: formation of a new individual from parents of different species or varieties

hydraulic: related to fluid flow

hydrocarbon: compounds of hydrogen and carbon

hydrodynamic condition: related to flow of water

hydrology: the science dealing with the properties, distribution, and circulation of water

hydromodification: any process that alters the hydrologic characteristics of a body of water

immobile: not moving

immunocompromised: having a weakened immune system

impact: a change to the environment resulting from a human activity or product

impermeable: not easily penetrated; the property of a material or soil that does not allow, or allows only with great difficulty, the movement or passage of water

in situ: in its original place; unmoved or unexcavated; remaining at the site or in the subsurface

incident solar: sun energy that hits a particular spot

industrial metabolism: flow of resources and energy in an industrial system

inertness: inability to react chemically

infrastructure: the basic facilities, services, and installations needed for the functioning of a system, i.e., the various components of a water supply system

ingest: take in through the mouth

inhalation: drawing into the lungs by breathing

injection well: a well into which fluids are pumped for purposes such as underground waste disposal, improving the recovery of crude oil, or solution mining

inorganic: compounds not containing carbon

integrative commons governance: a governing system which recognizes and protects publicly shared resources, usually under local control

integrity: wholeness and stability

interest groups: corporate or citizen groups with a stake in influencing legislation

intergenerational sustainability: ability of a system to remain stable and productive over several generations

ion: an electrically charged atom or group of atoms

isotope: a variation of an element that has the same atomic number of protons but a different weight because of the number of neutrons; various isotopes of the same element may have different radioactive behaviors, some are highly unstable

labor market: the area or pool of workers from which an employer draws employees

lake acre: an acre of lake surface

land subsidence: sinking or settling of land

landfills: sanitary landfills are disposal sites for nonhazardous solid wastes spread in layers, compacted to the smallest practical volume, and covered by material applied at the end of each operating day; secure chemical landfills are disposal sites for hazardous waste, selected and designed to minimize the chance of release of hazardous substances into the environment

late-onset: occurring in adulthood or old age

leach pad: in mining: a specially prepared area where mineral ore (especially gold) is heaped for metal extraction

leach solution: in mining: chemical solution sprayed on ore to extract metal

leach: dissolve out

leachate: water that collects contaminants as it trickles through wastes, pesticides, or fertilizers; leaching may occur in farming areas, feedlots, and landfills, and may result in hazardous substances entering surface water, ground water, or soil

leguminous: members of the pea family, or legumes

lipophilicity: solubility in or attraction to waxy, fatty, or oily substances

locomotion: self-powered movement

loess: soil deposited by wind

low tillage: reduced level of plowing

maceral: organic remains visible in coal

macroscopic: large enough to be visible, in contrast to microscopic

Magna Carta: English charter giving landowners rights under the king's authority

malleable: able to be shaped and bent

Malthusian hypothesis: idea that populations always grow faster than their food supply, from Thomas Malthus

maximum contaminant level: in water: the maximum permissible level of a contaminant in water delivered to any user of a public system; MCLs are enforceable standards

media: specific environments—air, water, soil—which are the subject of regulatory concern and activities

mediation: dispute resolution in which a neutral third party helps negotiate a settlement

megawatt: one million watts

mesothelioma: malignant tumor of the mesothelium, a cell layer within the lungs and other body cavities

metabolism: physical and chemical reactions within a cell or organism necessary for maintaining life

metabolite: any substance produced by biological processes, such as those from pesticides

metabolize: chemically transform within an organism

methanogenesis: creation of methane gas by microbes

microorganism: bacteria, archaea, and many protists; single-celled organisms too small to see with the naked eye

mine workings: the parts of a quarry or mine that is being excavated

mineralize: convert to a mineral substance

mitigation: measures taken to reduce adverse impacts

mixing zone: an area of a lake or river where pollutants from a point source discharge are mixed, usually by natural means, with cleaner water

mole: a chemical quantity, 6×10^{23} molecules. For oxygen, this amounts to 32 grams

molecule: the smallest division of a compound that still retains or exhibits all the properties of the substance

molluscicide: chemical that kills mollusks

monoculture: large-scale planting of a single crop species

multilateral treaty: treaty between more than two governments

multisite: several sites

mutagenic: capable of causing permanent, abnormal genetic change

natural attenuation: reduction in a pollutant through combined action of natural factors

nematocide: a chemical agent which is destructive to nematodes

nematode: worm-like organisms common in soil

neo-Malthusians: modern adherents to the ideas of Thomas Malthus

neonate: newborn

neural: related to nerve cells or the nervous system

neurodegeneration: loss of function and death of brain cells

neurology: medical science relating to the nervous system

neurotoxic: harmful to nerve cells

neurotoxicant: chemical that is toxic to neurons, or brain cells

nitrate catch crop: crop planted to harvest soil nitrates

nitrification: the process whereby ammonia, typically in wastewater, is oxidized to nitrite and then to nitrate by bacterial or chemical reactions

nonpoint source pollution: pollution originating from a broad area, such as agricultural runoff or automobile emissions

nucleotide: building block of DNA and RNA in a cell

off-gas control: control of gases released into the air

open path monitor: detection device that employs a beam of light passing through an open space

organic: referring to or derived from living organisms; in chemistry, any compound containing carbon

organochlorine: chemical containing carbon and chlorine

organophosphate: pesticide that contains phosphorus; short-lived, but some can be toxic when first applied

outfall: the place where effluent is discharged into receiving waters

overburden: rock and soil cleared away before mining

ovoid: shaped like an oval or egg

oxidize: react with oxygen

oxygenate: increase the concentration of oxygen within an area

ozonation: application of ozone to water for disinfection or for taste and odor control

PAHs: polyaromatic hydrocarbons; compounds of hydrogen and carbon containing multiple ring structures

particulate: fine liquid or solid particles such as dust, smoke, mist, fumes, or smog, found in air or emissions; they can also be very small solids suspended in water, gathered together by coagulation and flocculation

patent: legal document guaranteeing the right to profit from an invention or discovery

pathogenic: causing illness

pathway: the physical course a chemical or pollutant takes from its source to the exposed organism

PCBs: polychlorinated biphenyls; two-ringed compounds of hydrogen, carbon, and chlorine

per capita: per individual person in the population

percolating: moving of water downward and radially through subsurface soil layers, usually continuing downward to groundwater; can also involve upward movement of water

persistent bioaccumulative toxics: a group of substances that are not easily degraded, accumulate in organisms, and exhibit an acute or chronic toxicity

pH: an expression of the intensity of the basic or acid condition of a liquid; may range from 0 to 14, where 0 is the most acid, 7 is neutral, and 14 is most base; natural waters usually have a pH between 6.5 and 8.5

photochemical: light-induced chemical effects

phthalate: particular class of complex carbon compounds

physical removal: digging up and carting away

phytoplankton: that portion of the plankton community comprised of tiny plants; e.g. algae, diatoms

planktonic: that portion of the plankton community comprised of tiny plants; e.g. algae, diatoms

plume: a visible or measurable discharge of a contaminant from a given point of origin; can be visible, invisible, or thermal in water, or visible in the air as, for example, a plume of smoke

PM-10: airborne particles under 10 micrometers in diameter

polymer: a natural or synthetic chemical structure where two or more like molecules are joined to form a more complex molecular structure (e.g., polyethylene)

polyvinyl chloride (PVC): class of complex carbon compounds containing chlorine

pore waters: water present in the pores or cavities in sediments, soil, and rock

porosity: degree to which soil, gravel, sediment, or rock is permeated with pores or cavities through which water or air can move

priority pollutant: a designated set of common water pollutants

protein: complex nitrogenous organic compound of high molecular weight made of amino acids; essential for growth and repair of animal tissue; many, but not all, proteins are enzymes

protocol: in government: agreement establishing rules or code of conduct; science: a series of formal steps for conducting a test

pyrethroid: chemicals derived from chrysanthemums and related plants

radionuclide: radioactive particle, man-made or natural, with a distinct atomic weight number; can have a long life as soil or water pollutant

ratification: formal approval

raw water: intake water prior to any treatment or use

reactive chemicals: chemicals likely to undergo chemical reaction

recharge: the process by which water is added to a zone of saturation, usually by percolation from the soil surface (e.g., the recharge of an aquifer)

reclamation: in recycling: restoration of materials found in the waste stream to a beneficial use which may be for purposes other than the original use

reevaporate: return to the gaseous state

refractory: resistant (to heat: difficult to melt; also to authority)

refrigerant: liquid or gas used as a coolant in refrigeration

regenerative: able to be regenerated or created anew

remediate: reduce harmful effects; restore contaminated site

remediation: cleanup or other methods used to remove or contain a toxic spill or hazardous materials from a Superfund site or for the Asbestos Hazard Emergency Response program

residue: the dry solids remaining after evaporation

respiratory: having to do with breathing

river mile: one mile, as measured along a river's centerline

royalty: money paid by a user to an owner

scrubber: an air pollution control device that uses a spray of water or reactant or a dry process to trap pollutants in emissions

sedative: substance that reduces consciousness or anxiety

sediment impoverishment: loss of sediment

sedimentary: related to or formed by deposition of many small particles to form a solid layer

seep: movement of substance (often a pollutant) from a source into surrounding areas

septic tank: an underground holding tank for wastes from homes not connected to a sewer line

sick building syndrome: shared health and/or comfort effects apparently related to occupation of a particular building

sink: hole or depression where a compound or material collects; thermodynamics: part of a system used to collect or remove heat

smelting: the process in which a facility melts or fuses ore, often with an accompanying chemical change, to separate its metal content; emissions cause pollution

solubility: the amount of mass of a compound that will dissolve in a unit volume of solution; aqueous solubility is the maximum concentration of a chemical that will dissolve in pure water at a reference temperature

soluble: able to be dissolved in

solvent: substance, usually liquid, that can dissolve other substances

sorbent: a substance that absorbs (within) or adsorbs (on the surface) another substance

source reduction: reducing the amount of materials entering the waste stream from a specific source by redesigning products or patterns of production or consumption (e.g., using returnable beverage containers); synonymous with waste reduction

spatial: related to arrangement in space

spent radioactive fuel: radioactive fuel rods after they has been used for power generation

spray dryers: dryer used to remove heavy metals and other pollutants from incineration gases

standing: the legal right to pursue a claim in court

stenothermic: living or growing within a narrow temperature range

stewardship: care for a living system

stratosphere: the portion of the atmosphere ten to twenty-five miles above the earth's surface

subset: a smaller group within a larger one

subsidence: sinking of earth surface due to underground collapse

substrate: surface on which an organism, i.e. mold, grows

Superfund: the fund established to pay for the cleanup of contaminated sites whose owners are bankrupt or cannot be identified

supersonic: faster than the speed of sound

suppression: reduction in or prevention of an effect

surface water: all water naturally open to the atmosphere (rivers, lakes, reservoirs, ponds, streams, seas, estuaries, etc.)

sustainable development: economic development that does not rely on degrading the environment

sustainable: able to be practiced for many generations without loss of productivity or degradation of the environment

synergistic: combination of effects greater than the sum of the parts

systemic: throughout the body

tailings: residue of raw material or waste separated out during the processing of mineral ores

Takings impacts analysis: analysis of the impacts due to government restriction on land use

teach-in: educational forum springing from a protest movement (derived from sit-in protests)

temperature inversion: temporary trapping of lower warm air by higher cold air

teratogen: something that causes birth defects, may be radiation, a chemical or a virus

teratogenic: causing birth defects

thermal infrared imaging: photographs in which contrast depends on differences in temperature

thermal shock: rapid temperature change beyond an organism's ability to adapt

thermodynamic limitations: tendency of chemical reactions to reverse when products remain in the reaction mixture

thermotolerance: ability to withstand temperature change

titleholder: the person or entity holding the legal title or deed to a property

toluene: carbon-containing chemical used in fuel and as a solvent

topography: the physical features of a surface area including relative elevations and the position of natural and man-made (anthropogenic) features

transient: present for a short time

transuranic waste: waste containing one or more radioactive elements heavier than uranium, created in nuclear power plants or processing facilities

tribunal: committee or board appointed to hear and settle an issue

trophic: related to feeding

turbid: containing suspended particles

turbine: machine that uses a moving fluid (liquid or gas) to gas to turn a rotor, creating mechanical energy

ultraviolet radiation: high-energy, short-wavelength light beyond human vision

unitary system: a centralized system or government

unreactivity: lack of chemical reactivity

unsaturated: capable of dissolving more solute, i.e., water

variable vale control: a system for automatically adjusting engine valve timing for better fuel efficiency

vector: an organism, often an insect or rodent, that carries disease; plasmids, viruses, or bacteria used to transport genes into a host cell: a gene is placed in the vector; the vector then "infects" the bacterium

volatility: relating to any substance that evaporates readily

volatilize: vaporize; become gaseous

Warsaw Pact: nations allied with the former Soviet Union

waste-to-energy: to convert solid waste into a usable form of energy

water table: the level of water in the soil

watershed: the land area that drains into a stream; the watershed for a major river may encompass a number of smaller watersheds

wetland: an area that is saturated by surface or ground water with vegetation adapted for life under those soil conditions, as swamps, bogs, fens, marshes, and estuaries

Index

Deregulation, in electric power
    generation, 1:169–170
Derris, 2:97
DES (Diethylstilbestrol), 1:176
Desalinization, 2:139
Desertification, 1:24, 1:132
Desiccants, defined, 1:278
Design modifications, for
    abatement, 1:2
Despoliation, defined, 1:159
Detectors, 1:74–75, 2:194–197
Detergents, pollution from, 2:209,
    2:310
Deterrent enforcement, 1:192–193
Detjen, Jim, 2:35
Deuterium, defined, 2:278
Developed countries
    agriculture in, 1:27
    vs. developing countries,
        2:141–143
    Kyoto Protocol and, 1:229
    natural disasters in, 1:132–133
    See also European Union;
        specific countries
Developing countries
    agriculture in, 1:27–28
    compliance with international
        standards by, 2:7
    vs. developed countries,
        2:141–143
    Earth Summit and, 1:152
    eco-apartheid in, 2:25
    energy sources in, 2:175
    green revolution and,
        1:240–241
    injection wells in, 1:292
    natural disasters in, 1:132
    nonpoint source pollution in,
        2:76
    P2 technologies in, 2:234
    poverty and environment in,
        2:140–142
    resource consumption in, 2:23,
        2:24
    trade disparities and, 2:24–25
    use of DDT in, 1:120,
        1:211–212
    waste trade in, 1:263,
        2:291–292, 2:293
    water-contaminated diseases in,
        2:311
    See also Africa; specific countries
Development, of mines, 2:47
Diablo Canyon nuclear power
    plant, 1:40, 1:61

Diapers, 2:212
Diarrheal diseases, 1:251
Diatomaceous earth, defined, 2:320
Diazinon, 2:96
Dichloropropene, 2:97
Dichlorodiphenyl dichloroethylene
    (DDE), 1:252
Dichlorodiphenyl trichloroethane.
    See DDT
Dieldrin, 2:94, 2:96
Diesel, 1:120–121, 2:275
Diesel, Rudolf Christian Karl,
    1:120
Diethylstilbestrol (DES), 1:176
Diffusers, defined, 2:240
Dilution, 1:121, 2:318–320
Dimethoate, 2:96
Dinoflagellates, defined, 2:313
Dinosaur National Monument,
    1:61
Dintrophenols, 2:98
Dioxins, 1:121–124, 1:123, 2:94
    bioaccumlation of, 1:51
    in chemical accidents, 1:127
    chemical structure of, 1:122
    cleanup of, 1:95
    defined, 2:216
    FDA on, 2:265–266
    gas chromatography for, 2:194
    from open trash burning, 1:65
    as PBT chemicals, 2:93
    as priority pollutant, 2:117
    reporting requirements, 2:248
    at Times Beach, Missouri,
        2:243–244
    from World Trade Center
        terrorist attack, 2:235
Diquat, 2:194
Direct current, 1:165
Direct methanol fuel cells, 1:217
Direct push technologies, 1:98
Direct-action groups, 1:9, 1:204,
    1:206, 2:71
    See also names of specific groups
Disaster cycle, defined, 1:132
Disasters
    chemical accidents and spills,
        1:124–129, 1:125, 1:127,
        1:173, 1:206
    environmental engineers and,
        1:79
    environmental mining
        accidents, 1:129–130
    natural, 1:130–134, 1:131,
        1:133

nuclear accidents, 1:40,
    1:134–138, 1:137, 1:205–206,
    1:264–265, 2:160–161
    See also Industrial accidents; Oil
        spills; specific accidents
Discrimination, environmental,
    1:208–209
Disease clusters, 1:69–71, 1:89–90,
    1:109, 2:187
Disinfection, in water treatment,
    2:320–321
Disinfection by-products (DBPs),
    2:270, 2:316, 2:320
Disposal. See Waste disposal
Dispute resolution, by consensus
    building, 1:110
Dissolution into the oceans,
    defined, 1:72
Dissolved oxygen (DO), defined,
    2:116
Distillation
    column reboilers, 1:2
    in solvent recovery, 1:146
Distributive justice, 1:210, 1:212
Disulfoton, 2:96
DNA
    cancer cells and, 1:66
    of molds, 2:54
    UV radiation and, 2:266
DNAPLs (Dense nonaqueous
    phase chemicals), 1:97, 1:99,
    2:69
DOE. See U.S. Department of
    Energy
Doe Run Smelting, 2:15, 2:15–16
DOJ. See U.S. Department of
    Justice
DOL. See U.S. Department of
    Labor
Domestic wastewater. See Sanitary
    wastewater
Donora, Pennsylvania, 1:30, 1:142,
    1:143, 1:201–202, 1:251, 1:261,
    2:207
Dose, of radiation, 2:161
Dose-response relationship,
    2:186–187, 2:188, 2:250–251,
    2:251
DOT. See U.S. Department of
    Transportation
Double containment, defined,
    1:292
The Doubly Green Revolution, 1:241
Douglas, Michael, 2:133
Doyle, Jack, 1:284

hypoxia and, 1:270
See also Eutrophication
Oxygen demand, biochemical, **2:84**
Oxygenate, defined, 2:269
Ozonation, defined, 1:255
Ozone hole, 1:15, 1:245, 1:261, 2:71–72, 2:*85*, 2:*86–87*
See also Montréal Protocol
Ozone (O₃), 2:**84–88**, 2:*85*, 2:*86*
air quality standards on, 2:118, 2:195
carbon monoxide and, 1:74
CFCs and, 1:87–88
control, 1:37
as criteria pollutant, 1:33, 1:36
ground-level, 2:119
nitrogen oxides and, 2:64
petroleum and, 2:106
smog from, 2:207–208
UV spectra and, 2:196, 2:266
from vehicle emissions, 2:272
from VOCs, 2:281
in water treatment, 2:320–321

P

P2. *See* Pollution prevention
PACCE (People Against a Chemically Contaminated Environment), 2:61
Packaging, plastic, 2:110–114, 2:*113*
PAHs. *See* Polycyclic aromatic hydrocarbons
Paine, Thomas, 1:239
Paints
lead, 1:276, 2:14, 2:249
recycling, 2:172
reuse, 2:183
Palmerton, Pennsylvania, 1:59
Palo Alto Hardware, 2:*176*
Paper products, 2:212, 2:213
as energy source, 2:176
recycling, 2:171, 2:215
Paracelsus, 2:250
Paradise in the Sea of Sorrow, 1:294
Paraquat, 2:98
Parathion, 2:96
Park, Marion Edward, 1:*246*
Parkinson's disease, 1:255
Parks, national, 2:59, 2:278–279
Parliamentary governments, 1:232–233
Particulate matter (PM). *See* Particulates

Particulates, 2:**88–91**, 2:*89*
air quality standards on, 2:118, 2:195
from coal, 1:103
control of, 1:37
as criteria pollutant, 1:33, 1:35
defined, 2:47
ground level, 2:119
from incineration, 1:274
from petroleum, 2:106
scrubbers for, 2:199
smog from, 2:207–208
sulfates, 1:36
from vehicle emissions, 2:272–273
Passive cleanup. *See* Natural attenuation
Patents, defined, 2:51
Pathogenic, defined, 2:97
Pathways (Chemical), defined, 2:16
Patterson, Clair, 1:281
PBT chemicals. *See* Persistent bioaccumulative and toxic (PBT) chemicals
PCBs (Polychlorinated biphenyls), 2:**91–93**, 2:*92*, 2:*93*, 2:*94*
bioaccumulation of, 1:50–51, 1:*51*
chemical structure of, 2:*91*
defined, 2:209
detection of, 2:194
disposal of, 1:220, 2:287–288
PCDF in, 1:122
as priority pollutant, 2:117
reporting requirements, 2:248
sidescan sonars for, 2:198
soil pollution from, 2:209
thyroid hormones and, 1:177
Toxic Substances Control Act and, 2:249
water pollution from, 2:314
PCC (Primary combustion chamber), 1:272
PCDD. *See* Polychlorinated dibenzo[1,4]dioxins
PCDF. *See* Polychlorinated dibenzofurans
PCNs (Polychlorinated naphthalenes), 2:93
PCP. *See* Pentachlorophenol
PCSD (President's Council on Sustainable Development), 2:147, 2:158
Peace Corps, 1:163
Peanut industry, 1:84–85

Pearl Harbor National Monument, 2:287
Peat, coal from, 1:100–101
Peccei, Aurelio, 1:10
PEM (Proton exchange membrane) fuel cells, 1:216, 1:*217*
Pennsylvania Alliance for Aquatic Resource Monitoring, 1:290
Pennsylvania Department of Health, 1:135–136
Pentachlorophenol (PCP), 1:122, 2:98, 2:227
People Against a Chemically Contaminated Environment (PACCE), 2:61
Per capita, defined, 2:33
PERC (Perchloroethylene), 1:145–146
Perception, public. *See* Public perception
Perchloroethylene (PERC), 1:145–146
Percolating, defined, 1:243
Pereira, Fernando, 1:243
Performance Partnership Grants Program, 2:128
Permethrin, 2:97
Permits, 2:9, 2:124, 2:127, 2:191–192
Persian Gulf, oil from, 2:102
Persian Gulf War (1991), 2:284
Kuwaiti oil fields, 1:185, 2:239, 2:*283*, 2:284
oil spill, 1:139
Persistent bioaccumulative and toxic (PBT) chemicals, 2:**93–94**, 2:126, 2:247–248
Persistent organic pollutants (POPs), 1:51, 1:122, 2:6, 2:93, 2:**94**
Personal care products. *See* Pharmaceuticals and personal care products
Personal responsibility, 2:26–27
Personnel protective equipment (PPE), 1:95, 1:*95*
Pervious concrete, 2:77
Pesticide Education Center, 1:2
Pesticides, 2:*95*, 2:**95–101**, 2:*96*
in agriculture, 1:*25*, 1:26–27
Chávez, César and, 1:88–89, 2:1–2
dioxin in, 1:122
electron capture for, 2:195